MODERN ASPECTS OF ELECTROCHEMISTRY

No. 18

LIST OF CONTRIBUTORS

L. I. BOGUSLAVSKY
A. N. Frumkin Institute of Electrochemistry
Academy of Sciences of the USSR
Moscow, USSR

LAURENCE D. BURKE
Chemistry Department
University College Cork
Cork, Ireland

F. HINE
Nagoya Institute of Technology
Nagoya, Japan

HIDEAKI KITA
Department of Chemistry
Faculty of Science
Hokkaido University
Sapporo, Japan

MICHAEL E. G. LYONS
Chemistry Department
Trinity College
Dublin, Ireland

JOHN NEWMAN
Materials and Molecular Research Division
Lawrence Berkeley Laboratory
and Department of Chemical Engineering
University of California
Berkeley, California

MARK E. ORAZEM
Department of Chemical Engineering
University of Virginia
Charlottesville, Virginia

PAUL J. SIDES
Department of Chemical Engineering
Carnegie-Mellon University
Pittsburgh, Pennsylvania

B. V. TILAK
Occidental Chemical Corporation
Research Center
Grand Island, New York

KOHEI UOSAKI
Department of Chemistry
Faculty of Science
Hokkaido University
Sapporo, Japan

K. VISWANATHAN
Occidental Chemical Corporation
Research Center
Grand Island, New York

A Continuation Order Plan is available for this series. A continuation order will bring delivery of each new volume immediately upon publication. Volumes are billed only upon actual shipment. For further information please contact the publisher.

MODERN ASPECTS OF ELECTROCHEMISTRY

No. 18

Edited by

RALPH E. WHITE
Department of Chemical Engineering
Texas A&M University
College Station, Texas

J. O'M. BOCKRIS
Department of Chemistry
Texas A&M University
College Station, Texas

and

B. E. CONWAY
Department of Chemistry
University of Ottawa
Ottawa, Ontario, Canada

PLENUM PRESS • NEW YORK AND LONDON

The Library of Congress cataloged the first volume of this title as follows:

Modern aspects of electrochemistry. no. [1]
    Washington, Butterworths, 1954–
    v. illus., 23 cm.
    No. 1–2 issued as Modern aspects series of chemistry.
    Editors: no. 1–   J. Bockris (with B. E. Conway, No. 3–   )
    Imprint varies: no. 1, New York, Academic Press.—No. 2, London, Butterworths.

    1. Electrochemistry—Collected works. I. Bockris, John O'M. ed. II. Conway, B. E. ed.   (Series: Modern aspects series of chemistry)
QD552.M6   54-12732 rev

ISBN 0-306-42312-X

© 1986 Plenum Press, New York
A Division of Plenum Publishing Corporation
233 Spring Street, New York, N.Y. 10013

All rights reserved

No part of this book may be reproduced, stored in a retrieval system, or transmitted in any form or by any means, electronic, mechanical, photocopying, microfilming, recording, or otherwise, without written permission from the Publisher

Printed in the United States of America

Preface

This volume of *Modern Aspects of Electrochemistry* contains six chapters. The first four chapters are about phenomena of interest at the microscopic level and the last two are on phenomena at the macroscopic level.

In the first chapter, Uosaki and Kita review various theoretical models that have been presented to describe the phenomena that occur at an electrolyte/semiconductor interface under illumination. In the second chapter, Orazem and Newman discuss the same phenomena from a different point of view. In Chapter 3, Boguslavsky presents state-of-the-art considerations of transmembrane potentials and other aspects of active transport in biological systems. Next, Burke and Lyons present a survey of both the theoretical and the experimental work that has been done on hydrous oxide films on several metals.

The last two chapters cover the topics of the production of chlorine and caustic and the phenomena of electrolytic gas evolution. In Chapter 5, Hine *et al.* describe the engineering aspects of the three processes used in the chlor-alkali industry, and in Chapter 6, Sides reviews the macroscopic phenomena of nucleation, growth, and detachment of bubbles, and the effect of bubbles on the conductivity of and mass transfer in electrolytes.

Texas A&M University	R. E. White
Texas A&M University	J. O'M. Bockris
University of Ottawa	B. E. Conway

Contents

Chapter 1

THEORETICAL ASPECTS OF SEMICONDUCTOR ELECTROCHEMISTRY

Kohei Uosaki and Hideaki Kita

I. Introduction	1
II. Electronic Energy Levels of Semiconductor and Electrolyte	2
1. Junctions between Two Electronic Conductors	2
2. Semiconductor/Electrolyte Interface	4
III. Potential Distribution at Semiconductor/Electrolyte Interface	13
1. Schottky Barrier	14
2. Effect of the Redox Potential on the Potential Drop in the Semiconductor at the Semiconductor/Electrolyte Interfaces	15
3. Distribution of Externally Applied Potential at the Semiconductor/Electrolyte Interfaces	20
IV. Distribution of Energy States for Ions of Redox System in Solution	23
1. Importance of the Energy Levels of Redox Couples	23
2. Gurney's Model	24
3. Gerischer's Model	29
4. Continuum Solvent Polarization Fluctuation Model	32
5. Validity of the Models	35

V. Rate Expressions for Electron Transfer at Illuminated
   Semiconductor/Electrolyte Interfaces 37
   1. Phenomenological Description 37
   2. The Model of Bockris and Uosaki 38
   3. Butler's Model—Semiconductor/Electrolyte
      Interface as a Schottky Barrier 42
   4. Competition between Surface Recombination and
      Charge Transfer 45
   5. Effects of Recombination in Space Charge Region 51
   6. Effects of Grain Boundary Recombination 53
VI. Concluding Remarks 54
References 55

Chapter 2

PHOTOELECTROCHEMICAL DEVICES FOR SOLAR ENERGY CONVERSION

Mark E. Orazem and John Newman

I. Semiconductor Electrodes 62
   1. Physical Description 63
   2. The Mechanism of Cell Operation 66
II. Mathematical Description 69
   1. Semiconductor 70
   2. Electrolyte 76
   3. Semiconductor-Electrolyte Interface 78
   4. Boundary Conditions 82
   5. Counterelectrode 83
III. Photoelectrochemical Cell Design 84
   1. Choice of Materials 85
   2. Solution of the Governing Equations 87
   3. The Influence of Cell Design 91
IV. Conclusions 98
Notation 99
References 101

Chapter 3

ELECTRON TRANSFER EFFECTS AND THE MECHANISM OF THE MEMBRANE POTENTIAL

L. I. Boguslavsky

I. Introduction 113
II. Modeling Nonenzymatic Systems of Electron Transfer in the Initial Part of the Respiratory Chain of Mitochondria 115
   1. Respiratory Chain 115
   2. Potentials of the Respiratory Chain Elements 116
   3. Ubiquinones in the Respiratory Chain 118
   4. Participation of Membrane Lipids in the Functioning of the Respiratory Chain 119
   5. Transmembrane Potentials in the Chain NADH–Coenzyme Q–O_2 122
   6. Participation of FMN in the Oxidation of Membrane Lipids 127
   7. Participation of FMN in the Transmembrane Transport of Protons 130
   8. Participation of FMN in the Transmembrane Transport of Electrons 134
   9. Interaction of FMN with Other Chain Components ... 135
III. Potential Generation on Bilayer Membranes Containing Chlorophyll 137
   1. Chlorophyll at the Membrane/Electrolyte Interface ... 137
   2. Redox Potentials of Chlorophyll 138
   3. Reactions of Chlorophyll Inserted in the Membrane with the Redox Components in an Aqueous Solution under Illumination 138
   4. Transmembrane Potentials in the Chain OX–CHL–RED 141
IV. Possible Mechanisms of the Motion of Electrons and Protons in the Membrane 144
   1. Hypotheses on the Mechanism of Electron Motion in Biological Membranes 145

2. Ion Permeability of Bilayer Membranes in the Iodine/Iodide System Controlled by Redox Reactions at the Interface 148
3. Electron and Proton Transport in Bilayers Containing Chlorophyll and Quinones under Illumination 151
4. Possible Conductance Mechanisms in Bilayers Containing Ubiquinone 152
5. Hypothesis on the Mechanism of Proton Transport in Biological Membranes 153
6. Potentials of Coupling Membranes 155
V. Conclusions 161
References 163

Chapter 4

ELECTROCHEMISTRY OF HYDROUS OXIDE FILMS

Laurence D. Burke and Michael E. G. Lyons

I. Introduction 169
II. Formation of Hydrous Oxides 171
III. Acid-Base Properties of Oxides 173
IV. Structural Aspects of Hydrous Oxides 179
V. Transport Processes in Hydrous Oxide Films 182
VI. Theoretical Models of the Oxide-Solution Interphase Region 188
    1. Classical Models 189
    2. Nonclassical Models 189
VII. Platinum 191
    1. Monolayer Oxidation 191
    2. Hydrous Oxide Growth on Platinum 198
VIII. Palladium 205
IX. Gold 208
    1. Monolayer Behavior 208
    2. Hydrous Oxide Growth 210

X.	Iridium	213
	1. Monolayer Growth	213
	2. Hydrous Oxide Films	214
XI.	Rhodium	224
	1. Hydrous Oxide Growth	224
	2. Behavior of Rh/Pt Alloys	226
XII.	Ruthenium	227
XIII.	Some Nonnoble Metals	230
	1. Iron and Cobalt	230
	2. Nickel and Manganese	233
	3. Tungsten	238
XIV.	Conclusion	239
	Addendum	241
	References	243

Chapter 5

CHEMISTRY AND CHEMICAL ENGINEERING IN THE CHLOR-ALKALI INDUSTRY

F. Hine, B. V. Tilak, and K. Viswanathan

I.	Introduction	249
II.	Chemical and Electrochemical Principles Involved in Chlor-Alkali Production	250
III.	Manufacturing Processes	253
	1. Importance of Brine Purification	253
	2. Diaphragm Cell Process	256
	3. Membrane Cell Process	259
	4. Mercury Cell Process	260
IV.	Electrode Materials and Electrode Processes	263
	1. Anodes	263
	2. Cathodes	268
V.	Engineering Aspects in Chlor-Alkali Operations	270
	1. Chemical Engineering Aspects of Amalgam Decomposition	270
	2. Chemical Engineering Aspects of Porous Diaphragms	279

VI. Ion-Exchange Membranes and Membrane Technology … 287
   1. General Requirements of Membranes for Chlor-Alkali Production … 287
   2. Properties of Membranes and Their Performance Characteristics … 288
   3. Engineering Design Aspects of Ion-Exchange Membrane Cell Technology … 292
   4. Advantages Afforded by Membrane Technology … 295
   5. State of the Art of Membrane Cell Technology … 296
Notation … 297
References … 298

Chapter 6

PHENOMENA AND EFFECTS OF ELECTROLYTIC GAS EVOLUTION

Paul J. Sides

I. Introduction … 303
II. Nucleation, Growth, and Detachment of Bubbles … 304
   1. Nucleation … 304
   2. Growth … 306
   3. Detachment … 312
   4. Effect of Additives and Operating Parameters … 315
III. Electrical Effects of Gas Evolution … 318
   1. Conductivity of Bulk Dispersions … 318
   2. Electrical Effects of Bubbles on Electrodes … 330
IV. Mass Transfer at Gas-Evolving Electrodes … 342
   1. Penetration Theory … 342
   2. The Hydrodynamic Model … 344
   3. The Microconvection Model … 347
   4. Microscopic Investigation … 348
V. Summary … 348
Notation … 349
References … 352

Index … 355

1

Theoretical Aspects of Semiconductor Electrochemistry

Kohei Uosaki and Hideaki Kita

Department of Chemistry, Faculty of Science, Hokkaido University, Sapporo 060, Japan

I. INTRODUCTION

The history of semiconductor photoelectrochemistry started over 100 years ago when Becquerel found that an electric current flowed if one of the electrodes immersed in a dilute acid was illuminated by light.[1] Although the concept of a semiconductor did not exist at that time, it is now clear that the electrodes which Becquerel used had semiconductor properties. In 1955, Brattain and Garrett used germanium as the first semiconductor electrode.[2] Since then the knowledge of semiconductor electrochemistry has grown steadily and several reviews and books on this subject have appeared.[3-10] The research activity in this field, however, has grown quite significantly in the last decade or so, following the suggestion by Fujishima and Honda that solar energy may be directly converted to a chemical energy, hydrogen, by using semiconductor/aqueous electrolyte solution/metal cells.[11] Just after their paper was published, a worldwide search for new alternative energy sources began, and this process, *photoelectrochemical energy conversion*, attracted many research groups as one of the possible means to convert solar energy to electrical or chemical energy. After 10 years of intensive research, not only has the solar conversion efficiency of photoelectrochemical devices exceeded 10%[12] and the application of this process widened,[13] but also the progress of the theoretical understanding of the process is quite significant.[14,15] Because of its multidisciplinary nature, research workers in this field have a very wide variety of backgrounds such as solid state physics, inorganic chemistry, photochemistry, catalytic chemistry, and, of course, electrochemistry. This diverse background is an advantage on the one hand but makes communication between the research workers with different backgrounds difficult.[16] Many theories and models developed for semiconductor/metal and semiconductor/semiconductor interfaces have been applied to semiconductor/electrolyte interfaces without considering the unique nature of this system. Some of the confusion has been resolved already, but there are still many points to be made clear. Thus, the present authors thought it would be useful to write an article on theoretical aspects of semiconductor electrochemistry which can, hopefully, clarify many theoretical ambiguities in the semiconductor electrochemistry and bridge the gap between solid state physics oriented research workers and electrochemistry oriented ones.

The stress will be given to the similarities and the differences between the semiconductor/metal and semiconductor/semiconductor interfaces where charge carriers are electrons and holes in both phases and the semiconductor/electrolyte interfaces where charge carriers are electrons and holes in a semiconductor phase but cations and anions in an electrolyte.

II. ELECTRONIC ENERGY LEVELS OF SEMICONDUCTOR AND ELECTROLYTE

One of the most important theoretical problems in semiconductor electrochemistry is the understanding of the nature of the semiconductor/electrolyte interfaces. Knowledge of the electronic energy levels of the semiconductor and the electrolyte is essential to describe the interface.

1. Junctions between Two Electronic Conductors

When metal/metal, semiconductor/semiconductor, or semiconductor/metal junctions are under consideration, the relative positions of the Fermi levels before contact or the work functions of two phases determine the nature of contact. Let us first consider the metal/metal contact.[17] Figure 1a shows the electronic energy levels of two metals before contact. Φ^{M_1} and Φ^{M_2} represent the work function of M_1 and M_2, respectively. When two metals are in contact, the Fermi levels of two metals must be the same. In other words,

$$\bar{\mu}_e^{M_1} = \bar{\mu}_e^{M_2} \tag{1}$$

where $\bar{\mu}_e^M$ is the electrochemical potential of an electron in a metal, M. To achieve this, electrons should flow from M_1 to M_2 because originally $E_{F,M_1} > E_{F,M_2}$. At equilibrium, M_2 is charged negatively while M_1 is positively charged and a potential difference builds up at the interface. The electrochemical potential of the electron is given by[18]

$$\bar{\mu}_e = -\Phi - e\Psi = \alpha_e - e\Psi \tag{2}$$

where α_e is the reverse of the electron work function which is the minimum work required to extract electrons from an uncharged

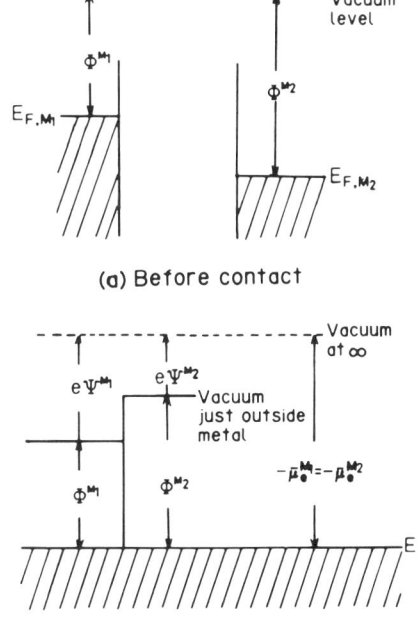

Figure 1. Energy diagram of metal/metal junction (a) before and (b) after contact. See text for notations.

metal[19] and is called the real potential and Ψ is the Volta potential governed by the charge on the metal surface. From (1) and (2),

$$\alpha_e^{M_1} - e\Psi^{M_1} = \alpha_e^{M_2} - e\Psi^{M_2}$$

$$e(\Psi^{M_1} - \Psi^{M_2}) = \alpha_e^{M_1} - \alpha_e^{M_2} = \Phi^{M_2} - \Phi^{M_1} \qquad (3)$$

The situation after contact is schematically shown in Fig. 1b. ($\Psi^{M_1} - \Psi^{M_2}$) is known as a contact potential and is an absolute, measurable quantity. Here one must note, however, that Ψ^{M_1} and Ψ^{M_2} separately depend on the size of the phases M_1 and M_2, respectively, and are not absolute quantities. Thus, $\bar{\mu}_e^M$, which contains Ψ^M, is not an absolute quantity either.

The above arguments are easily applicable to semiconductor/metal and semiconductor/semiconductor junctions where the electronic energy level of both phases can also be represented by the Fermi level. The major difference between these two junctions and the metal/metal contact is that while in the case of the latter system, $\Delta\Psi(=\Psi^{M_2}-\Psi^{M_1})$ develops only within several angstroms of the interface, $\Delta\Psi$ extends to very deep into the bulk of semiconductor in the case of the junctions which involve a semiconductor phase because the semiconductor has fewer frees carriers than a metal. The situation is shown schematically in Figs. 2a and 2b.[20]

2. Semiconductor/Electrolyte Interface

Now let us turn to the semiconductor/electrolyte interface, which is our main concern. In the electrolyte, there are no electrons nor holes but oxidized and/or reduced forms of a redox couple. Thus, the Fermi statistics developed for the solid state phase are not applicable and rather careful consideration of the electronic energy level of the electrolyte is required.

(i) The Electronic Energy Level of the Electrolyte and the Redox Potential

In electrochemistry the energy level of a solution is represented by the redox potential. The measurement of the redox potential is carried out in an arrangement schematically shown in Fig. 3.[21] The electrode M is in electrochemical equilibrium with a redox couple

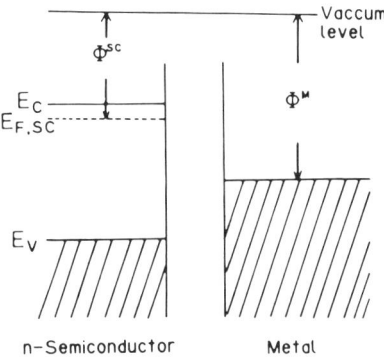

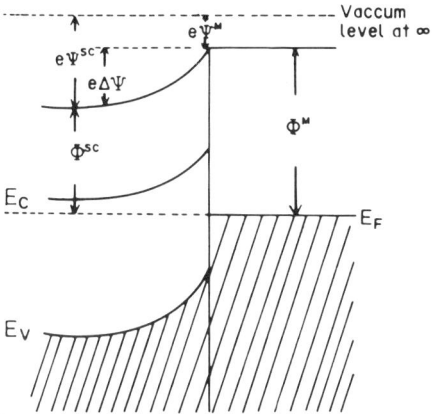

Figure 2. Energy diagram of semiconductor/metal junction (a) before and (b) after contact. See text for notations.

in the solution and the electrode M_1 is a reference electrode such as the hydrogen electrode, the saturated calomal electrode (SCE) or Ag/AgCl electrode. M and M' differ by the electrical state only and there is an electronic equilibrium between M_1 and M'. In this arrangement, the measured electric potential difference, E, between the two terminals of the cell, M and M', is given by[22]

$$eE = e(\phi^{M'} - \phi^{M}) = -\bar{\mu}_e^{M'} + \bar{\mu}_e^{M} \qquad (4)$$

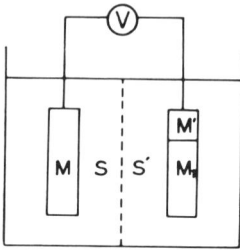

Figure 3. Schematic representation for the measurement of the redox potential in solution S. Liquid junction potential is assumed to be removed.[21]

where $\phi^{M'}$ and ϕ^{M} are the Galvani potentials of M' and M, respectively. Since

$$\bar{\mu}_e^{M_1} = \bar{\mu}_e^{M'} \quad (5)$$

we have

$$eE = -\bar{\mu}_e^{M_1} + \bar{\mu}_e^{M} \quad (6)$$

Thus, E is a measure of the energy level of an electron in metal M with respect to that in M_1. In other words, E represents the energy level of the redox couple with respect to a reference electrode since $\bar{\mu}_e^{M} = \bar{\mu}_e^{S}$, where $\bar{\mu}_e^{S}$ is the electrochemical potential of electron in a solution S, and is called a redox potential. For the convention, E with respect to the hydrogen electrode is most often used. Once a value with respect to the hydrogen electrode or any other reference electrode is known, a value with respect to some other reference electrode is easily calculated by using

$$V_{\text{redox}}(\text{ref. scale}) = V_{\text{redox}}(\text{H scale}) - V_{\text{ref}}(\text{H scale}) \quad (7)$$

where $V_{\text{redox}}(\text{ref. scale})$ and $V_{\text{redox}}(\text{H scale})$ are the redox potential of the redox couple with respect to the reference electrode and the hydrogen electrode, respectively, and $V_{\text{ref}}(\text{H scale})$ is the equilibrium potential of the reference electrode with respect to the hydrogen electrode. There is no need whatsoever to determine the absolute energy level of the redox couple to describe the semiconductor/electrolyte interface as far as both the flat band potential, V_{FB}, of the semiconductor electrode at which the bands within the semiconductor are flat throughout and the redox potential of the couple are known with respect to a given reference electrode. As shown in Fig. 4, if the V_{FB} which represents the electrochemical potential or the Fermi level of the semiconductor before contact is more negative

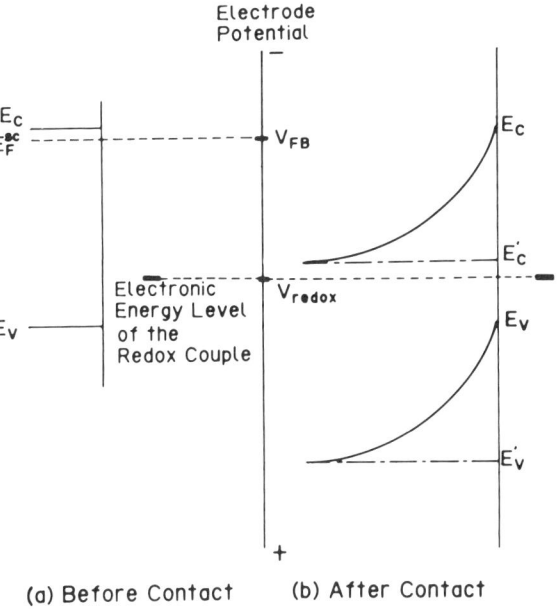

Figure 4. Energy diagram of semiconductor/electrolyte interface (a) before and (b) after contact. E_c and E_v are the surface energy levels of the conduction band and the valence band, respectively, when all the potential drop occurs within the semiconductor, and E'_c and E'_v are those when all the potential drop occurs within the solution.

than the potential of the redox couple, electrons flow from the semiconductor to the redox couple when the semiconductor and the solution are in contact and the potential difference builds up at the interface to attain the equilibrium.† At the equilibrium, the potential of the semiconductor electrode is the same as that of the redox couple. At a metal/metal contact, the excess charge builds up just within a few angstroms of the interface but at a metal/semiconductor interface, the charge extends very deep into the bulk of semiconductor and the potential drop occurs only in semiconductor if no surface state exists at the semiconductor surface. At an electrolyte/semiconductor interface, the potential distribution is more complicated because the electrolyte has a low density of mobile

† Note that the equilibrium at a semiconductor/electrolyte or even a metal/electrolyte interface is not necessarily attained just by contact.

charge carriers compared with the metal. Thus, the potential drop of this junction may occur entirely within the semiconductor as at a metal/semiconductor interface or within the electric double layer in solution (Helmholtz layer) as at a metal/electrolyte interface. But in the actual situation, the potential drop must occur some in the semiconductor and some in the double layer in solution depending upon the carrier density, dielectric constant, surface states concentration, etc. of the semiconductor and the dielectric constant and the concentration of the electrolyte. The details of this will be discussed in Section III.

(ii) Fermi Level of Solution and Absolute Electrode Potential

For a solid state physicist, it is rather uncomfortable to use the redox potential with respect to a reference electrode, and "to link the language of solid state physicists with that of electrochemists,"[23] the concept of the Fermi level of solution was introduced by Gerischer.[24-26] Unfortunately this word has been used by many research workers without considering its physical significance and Bockris questioned the use of this concept recently.[27-31]

According to Gerischer, the Fermi level of the redox system is correlated to the standard potential, V_{redox}, of the redox couple by

$$E_F = -eV_{redox} + \text{const.} \qquad (8)$$

Since $E_F \equiv \bar{\mu}_e^M$, the redox potential on the vacuum scale, V_{redox}(vac. scale), should give the Fermi level of the electrolyte with respect to vacuum. It is obvious that

$$V_{redox}(\text{H scale}) = V_{redox}(\text{vac. scale}) - V_{NHE}(\text{vac. scale}) \qquad (9)$$

where V_{NHE}(vac. scale) is the normal hydrogen electrode potential (NHE) on the vacuum scale. Thus, if one knows the value of V_{NHE}(vac. scale), the Fermi level of the solution can be known. V_{NHE}(vac. scale) = 4.5 V and $E_F = -e[V_{redox}(\text{H scale}) + 4.5 \text{ V}]$ are often used.[32]

In these discussions the physical meaning of E_F and V_{redox}(vac. scale) is not well considered. What does the redox potential with respect to vacuum level really mean? This problem has been discussed by many authors,[33-40] but the concept became clearer only quite recently. In 1984, the recommendation for "The Absolute Electrode Potential" was published by IUPAC.[41]

The reversible work to take an electron from the Fermi level of the metal to vacuum at infinity is the reverse of the electrochemical potential, $\bar{\mu}_e^M$ and the reversible work to extract an electron from the Fermi level of uncharged metal is the work function Φ^M. Thus,[18,42]

$$-\bar{\mu}_e^M = -\mu_e^M + e\phi^M$$
$$= -\mu_e^M + e(\Psi^M + \chi^M) \quad (10)$$

and

$$\Phi^M = -\mu_e^M + e\chi^M \quad (11)$$

where χ^M is the surface potential of the metal. The work function represents the energy difference between the Fermi level and the vacuum level just outside the metal. When the metal is in solution, the reversible work required to take an electron from the Fermi level of the metal to the vacuum level just outside the solution through metal/solution interface is given by $-\mu_e^M + e\chi^M + e(\Psi^M - \Psi^S)$, where Ψ^S is the Volta potential of the solution.[22,41] This value actually represents the Fermi level of the metal and, if the electrochemical equilibrium between the metal and the redox couple is attained, the electronic energy level of the redox couple with respect to the vacuum level just outside the solution. Thus, this is the redox potential on the vacuum scale which Trasatti called the absolute electrode potential, E_k[22] or $E(M)/abs$.[41]

$$eE_k = -\mu_e^M + e\chi^M + e(\Psi^M - \Psi^S)$$
$$= e\Delta_S^M \phi - \mu_e^M + e\chi^S = -\bar{\mu}_e^M - e\Psi^S$$
$$= eV_{redox}(\text{vax. scale}) \quad (12)$$

where χ^S is the surface potential of the solution and $\Delta_S^M \phi$ is the Galvani potential difference between the metal and the solution. By considering the potential difference across the interface of a real electrochemical cell, Bockris and Khan obtain the following relation (cf. Trasatti[43]):

$$eV_{redox}(\text{vac. scale}) = e\Delta_S^M \phi - \mu_e^M \quad (13)$$

which was actually called absolute potential by Trasatti in his earlier publication[43] but is now called reduced absolute potential, E_T,[22] or $E(M)/r$,[41] which is

$$eE_T = e\Delta_S^M \phi - \mu_e^M = eE_k - e\chi^S = -\bar{\mu}_e^M - e\phi^S \qquad (14)$$

and differs from $eE_k = eV_{\text{redox}}(\text{vac. scale}) = e\Delta_S^M \phi - \mu_e^M + e\chi^S$. As was the case at metal/metal contact, the situation at metal/electrolyte contact is shown schematically in Fig. 5. Since

$$\bar{\mu}_e^M = \alpha_e^M - e\Psi^M = -\Phi^M - e\Psi^M \qquad (15)$$

$$\bar{\mu}_e^S = \alpha_e^S - e\Psi^S = -eE_k - e\Psi^S \qquad (16)$$

and

$$\bar{\mu}_e^M = \bar{\mu}_e^S \qquad (17)$$

we have

$$-\Phi^M - e\Psi^M = -eE_k - e\Psi^S \qquad (18)$$

Thus,

$$e(\Psi^M - \Psi^S) = eE_k - \Phi^M = eV_{\text{redox}}(\text{vac. scale}) - \Phi^M \qquad (19)$$

This is very similar to Eq. (3) and gives

$$eV_{\text{redox}}(\text{vac. scale}) = e\Delta_S^M \Psi + \Phi^M \qquad (20)$$

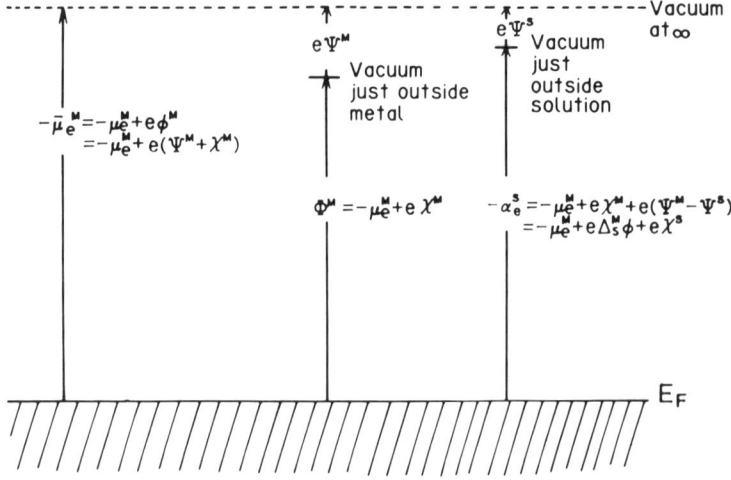

Figure 5. Energy diagram of metal/electrolyte interface after contact.

where $\Delta_S^M\Psi$ is the Volta potential difference between the metal and the solution. Since $\Delta_S^M\Psi$ and Φ^M are measurable quantities $V_{\text{redox}}(\text{vac. scale})$ is also a measurable quantity. As shown before, if $V_{\text{NHE}}(\text{vac. scale})$ is known, $V_{\text{redox}}(\text{vac. scale})$ can be obtained by using Eq. (9).

(iii) Absolute Potential of the Normal Hydrogen Electrode

Several attempts were made to determine $V_{\text{NHE}}(\text{vac. scale})$. The knowledge of the work function of the electrode metal is essential to obtain $V_{\text{NHE}}(\text{vac. scale})$ by using Eq. (20). Physical quantities are best known for the perfectly polarizable Hg electrode and it is possible to write for the potential of zero charge of this metal

$$V_{\sigma=0}^{\text{Hg}}(\text{H scale}) = V_{\sigma=0}^{\text{Hg}}(\text{vac. scale}) - V_{\text{NHE}}(\text{vac. scale}) \quad (21)$$

and

$$V_{\sigma=0}^{\text{Hg}}(\text{vac. scale}) = \Phi^{\text{Hg}}/e + \Delta_S^{\text{Hg}}\Psi_{\sigma=0}^0 \quad (22)$$

where $V_{\sigma=0}^{\text{Hg}}(\text{H scale})$ and $V_{\sigma=0}^{\text{Hg}}(\text{vac. scale})$ are the potentials of zero charge of Hg in NHE scale and vacuum scale, respectively, Φ^{Hg} is the work function of Hg and $\Delta_S^{\text{Hg}}\Psi_{\sigma=0}^0$ is the contact potential difference at Hg electrode at the potential of zero charge. From the above two equations, one obtains[41,44]

$$V_{\text{NHE}}(\text{vac. scale}) = \Phi^{\text{Hg}}/e + \Delta_S^{\text{Hg}}\Psi_{\sigma=0}^0 - V_{\sigma=0}^{\text{Hg}}(\text{H scale}) \quad (23)$$

Frumkin and Damaskin determined $V_{\text{NHE}}(\text{vac. scale})$ as 4.44 V[44] by using this equation with the assumption $\Phi^{\text{Hg}} = 4.51$ eV and the values of $\Delta_S^{\text{Hg}}\Psi_{\sigma=0}^0 = -0.26$ V[45] and $V_{\sigma=0}^{\text{Hg}}(\text{H scale}) = -0.19$ V. The same equation was used by Trasatti[41] with the values of $\Phi^{\text{Hg}} = 4.50 \pm 0.02$ eV,[43] $V_{\sigma=0}^{\text{Hg}}(\text{H scale}) = -0.192 \pm 0.01$ V,[46] and $\Delta_{\text{H}_2\text{O}}^{\text{Hg}}\Psi^0 = -0.248 \pm 0.001$ V.† $V_{\text{NHE}}(\text{vac. scale})$ thus obtained is 4.44 ± 0.02 V. He also showed that[41,43]

$$V_{\text{NHE}}(\text{vac. scale}) = (\Delta G_{\text{at}}^0 + \Delta G_{\text{ion}}^0 + \alpha_{\text{H}^+}^{0,\text{H}_2\text{O}})/F \quad (24)$$

† This value was obtained by measuring the standard potential difference of the cell: Hg|air|H$^+$(aq)|(H$_2$)Pt, which is given by $E = V_{\text{NHE}}(\text{vac. scale}) - \Phi^{\text{Hg}}/e = \Delta_{\text{Hg}}\Psi^0 - V_{\sigma=0}^{\text{Hg}}(\text{H scale})$. E was determined by Farrell and McTigue as -0.0559 ± 0.0002 V.[47]

where ΔG_{at}^0 and ΔG_{ion}^0 are the atomization ($\frac{1}{2}H_2 \to H$) and ionization ($H \to H^+ + e^-$) free energy of hydrogen, respectively, and $\alpha_{H^+}^{0,H_2O}$ is the real solvation free energy of H^+ in water. By using the values of $\Delta G_{at}^0 = 230.30$ kJ mol^{-1},[48] $\Delta G_{ion}^0 = 1313.82$ kJ mol^{-1},[43,48] $\alpha_{H^+}^0 = -1088 \pm 2$ kJ mol^{-1},[47] V_{NHE}(vac. scale) $= 4.44 \pm 02$ V was obtained.†

Trasatti critically examined other values very often referred to in the literature.[41] The value most often used in semiconductor electrochemistry as V_{NHE}(vac. scale) is 4.5 V,‡ which was calculated by Lohmann.[39] This became popular because it was quoted by Gerischer, whose reviews are read by most semiconductor electrochemists. His calculation is based on the application of Eq. (24) to Ag with final conversion to the NHE using V_{Ag/Ag^+}(H scale) $= 0.800$ V and is conceptually correct. According to Trasatti, his value is less accurate for two reasons: (1) He used ΔH^0 instead of ΔG^0 for Ag ionization. (2) His value of $\alpha_{Ag^+}^0$ differs by about 2 kJ mol^{-1} from that obtained by subtracting the $\alpha_{H^+}^0$ value recommended by Trasatti from the relative value of the free energy of hydration of Ag^+.[49]

Reiss suggested[50] using 4.8 V, which is based on the experiment by Gomer and Tryson.[51] They measured the electrode/solution contact potential difference in an electrochemical cell with a static liquid surface. Since no specific purification procedure was adopted for the solution and no particular precautions were used to protect the surface from impurities, their experimental arrangement does not ensure the necessary conditions of cleanliness of the solution phase and, therefore, their value of 4.78 V is not recommended by Trasatti.[43]

Hansen and Kolb obtained 4.7 V as V_{NHE}(vac. scale) by measuring the work function of an immersed electrode with a NHE by the Kelvin method.[52] This value may be affected by contamination of the electrode[41] since more recent results show that the surface emerging from the solution is contaminated.[53]

Although Trasatti recommended 4.44 V for V_{NHE}(vac. scale), it contains some uncertainties which mainly come from the value

---

† This value agrees with the value obtained by using Eq. (22), but it must be stressed that calculations based upon Eqs. (22) and (23) are independent only apparently since the quantities involved are derived from the same set of experimental data.[41]

‡ Lohman actually gave the value of 4.48 V.

of the solvation energy of a single ion and χ^S,† and, therefore, the absolute value of V_{redox}(vac. scale) should be used with caution.

(iv) The Real Meaning of the Fermi Level of Solution

Although V_{redox}(vac. scale) is determined as a measure for the Fermi level of a metal which is in equilibrium with a redox couple, it has a unique value for the redox couple, and, therefore, it can be considered as a measure of the electronic energy level of the redox couple on a vacuum scale. Thus, as at a metal/metal or a metal/semiconductor interface, $\Delta_S^M \Psi$ can be determined at the solid phase/electrolyte interface as a difference between Φ^M and eV_{redox}(vac. scale), which can be considered as a reverse of the real potential or the effective work function of the redox couple. At equilibrium, the Fermi level of the solid phase and the electronic energy level of the redox couple is the same ($\bar{\mu}_e^M = \bar{\mu}_e^S$) and sometimes the energy level of the redox couple is called the Fermi level of the redox couple in analogy to that of the solid phase.[5,23-26,32,54,55] As already mentioned, Fermi statistics is not applicable to the redox couple and, therefore, there is no Fermi level in an electrolyte, but one may accept this terminology with the understanding that the Fermi level of the redox couple actually means the Fermi level of the solid phase in equilibrium with the redox couple.

III. POTENTIAL DISTRIBUTION AT SEMICONDUCTOR/ELECTROLYTE INTERFACE

From the discussion in the last section, it is possible to predict the potential difference between a semiconductor and an electrolyte phase. The next problem is how the potential is distributed at the semiconductor/electrolyte interface. Let us again first deal with the semiconductor/metal interface (Schottky barrier), which is less complicated and is well described.

---

† The presence of electrolytes which may give rise to preferential penetration of anions or cations may lead to some variation in χ^{H_2O}.[18]

1. Schottky Barrier

From Eq. (3), the Galvani potential difference between the semiconductor and the metal, $\Delta_M^{sc}\Psi$, is given by

$$\Delta_M^{sc}\Psi = (\Phi^M - \Phi^{sc})/e \qquad (25)$$

where Φ^{sc} is the work function of the semiconductor. This potential difference can be also written as

$$\Delta_M^{sc}\Psi = \Delta V_{sc} + \Delta V_I$$
$$= \phi_B/e + \Delta V_I \qquad (26)$$

where ΔV_{sc} is the potential difference developed within the semiconductor which is equivalent to the Schottky barrier height, ϕ_B,[56] and ΔV_I is the potential difference developed at the interface due to the surface states. The effect of the metal work function on the barrier height or ΔV_{sc} is easily understood from Eqs. (25) and (26) as

$$\frac{d(e\Delta_M^{sc}\Psi)}{d\Phi^M} = \frac{d(e\Delta V_{sc})}{d\Phi^M} + \frac{d(e\Delta V_I)}{d\Phi^M} = \frac{d}{d\Phi^M}(\Phi^M - \Phi^{sc}) = 1 \qquad (27)$$

When the surface state concentration is low, the variation of work function of the contact metal, Φ^M, does not affect the interfacial potential drop, ΔV_I, i.e., $d(e\Delta V_I)/d\Phi^M = 0$. In this case, $d(e\Delta V_{sc})/d\Phi^M = 1$. This means that the change of the work function affects only the potential drop within the semiconductor, and the energy of the band edge at the surface before and after the contact is constant with respect to the Fermi level of the contact metal (band edge pinning). On the other hand, if the amount of the surface state is large and $d(e\Delta V_I)/d\Phi^M \neq 0$, this simple model does not apply. At the extreme situation, Φ^M does not affect ΔV_{sc} at all, i.e., $d(e\Delta V_{sc})/d\Phi^M = 0$. This is the situation where the change of the work function affects only the potential drop at the interface and the potential drop within the semiconductor is independent of the contact metal. Thus, the presence of a high density of surface states on the semiconductor usually "pins" the Fermi level at the surface,† resulting in the formation of a Schottky barrier even before the contact of the semiconductor with a metal.

† This situation is sometimes called a "Fermi level pinning."

When the semiconductor/metal contact is formed, any charge needed to achieve the equilibrium of the Fermi levels can be supplied mainly from the surface states without shifting the pinning of the Fermi level at the interface with respect to the band edges.[57] Figure 6 shows the barrier height, ϕ_B, i.e., $e\Delta V_{sc}$, at semiconductor/metal junctions as a function of electronegativity (work function) of the contact metal.[58] In the case of ZnS, the barrier height correlates linearly with the electronegativity of the metal but almost constant barrier height is shown for GaAs/metal interfaces. This is known to be due to the high concentration of surface states at a GaAs surface,[58,59] which is mainly associated with an As deficit.[60]

2. Effect of the Redox Potential on the Potential Drop in the Semiconductor at the Semiconductor/Electrolyte Interfaces

The Galvani potential difference between the semiconductor bulk and the electrolyte bulk, $\Delta_S^{sc}\Psi$, is given by Eq. (28) in analogy to Eq. (25) and to Eq. (26)[50]:

$$\Delta_S^{sc}\Psi = V_{redox}(\text{vac. scale}) - \Phi^{sc}/e$$
$$= \Delta V_{sc} + \Delta V_I + \Delta V_H + \Delta V_G \qquad (28)$$

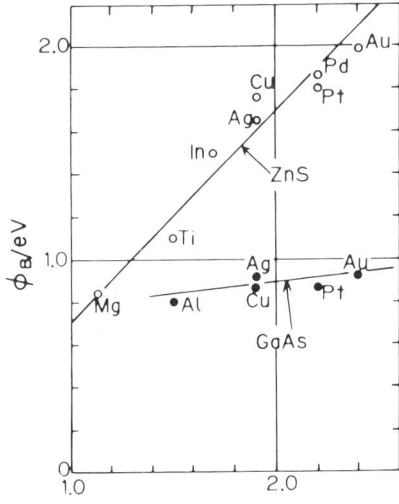

Figure 6. Barrier height of ZnS (electronegativity controlled) and GaAs (surface state controlled) as a function of electronegativity of contact metals.[57]

where ΔV_H and ΔV_G are the potential drop in the Helmholtz layer and that in the Gouy layer, respectively [cf. Eq. (26)]. ΔV_G can be minimized by using sufficiently concentrated solutions where most of the solution charge is squeezed into the Helmholtz plane and little charge is scattered diffusely into the solution.[21,61] In most experiments, a very high concentration of electrolyte is added, and, therefore, ΔV_G is neglected in the following discussions.[50] Then Eqs. (27) and (28) become

$$\Delta_S^{sc}\Psi = V_{redox}(\text{vac. scale}) - \Phi^{sc}/e$$
$$= \Delta V_{sc} + \Delta V_I + \Delta V_H \quad (29)$$

As one can easily recognize by comparing Eqs. (26) and (29), the one big difference between the semiconductor/electrolyte and the semiconductor/metal interfaces is the existence of the potential drop in the Helmholtz layer, ΔV_H, even if ΔV_G can be neglected. When one considers the effect of the redox potential on ΔV_{sc}, it is better to use $V_{redox}(\text{ref. scale})$ rather than $V_{redox}(\text{vac. scale})$ since only the former values can be obtained directly by experiments. From Eqs. (7) and (9)

$$V_{redox}(\text{ref. scale}) = V_{redox}(\text{vac. scale}) + \text{const.} \quad (30)$$

From Eqs. (29) and (30),

$$\frac{d\Delta_S^{sc}\Psi}{dV_{redox}(\text{ref. scale})} = \frac{d}{dV_{redox}(\text{ref. scale})}(\Delta V_{sc} + \Delta V_I + \Delta V_H)$$
$$= \frac{d}{dV_{redox}(\text{ref. scale})}\left[V_{redox}(\text{vac. scale}) - \frac{\Phi_{sc}}{e}\right]$$
$$= 1 \quad (31)$$

(i) Ideal Situation

As is the case at a metal/semiconductor interface, when the surface state density is very low and the carrier density of the semiconductor is not too high,

$$\frac{d}{dV_{redox}(\text{ref. scale})}(\Delta V_I + \Delta V_H) = 0$$

Theoretical Aspects of Semiconductor Electrochemistry

and

$$\frac{d\Delta V_{sc}}{dV_{redox}(\text{ref. scale})} = 1$$

(Ref. 62).

In this case, the semiconductor/electrolyte interface behaves as the Schottky barrier. Thus, ΔV_{sc} gives the barrier height in the semiconductor and should be equal to the photovoltage, V_{ph}. Therefore, the photovoltage should vary with the redox potential following

$$\frac{dV_{ph}}{dV_{redox}(\text{ref. scale})} = 1$$

One experimental result proving this relation is shown in Fig. 7.[63]

(ii) Fermi Level Pinning

With few exceptions, most of the papers published before 1980 treated the semiconductor/electrolyte interface as an ideal Schottky barrier and relations similar to Fig. 7 were often presented and used to predict the photovoltage of photoelectrochemical cells. Recently, however, many experimental results which do not follow the above prediction were reported[64-70] and it became clear that the other terms in Eq. (31), i.e.,

$$\frac{d\Delta V_I}{dV_{redox}(\text{ref. scale})} \quad \text{and} \quad \frac{d\Delta V_H}{dV_{redox}(\text{ref. scale})}$$

should be taken into account, although the importance of these terms was stressed before.[3] Bard et al.[71] considered the importance of the surface states on the potential distribution at the semiconductor/electrolyte interface and applied the concept of Fermi level pinning developed for a semiconductor/metal interface to this system. Qualitatively Fermi level pinning is important when the charge in the surface states, q_{ss}, is appreciably larger than that in the space charge region, q_{sc}.[71] A q_{sc} value can be calculated as a function of carrier level, N, and ΔV_{sc} from the equation[7,72]

$$q_{sc} = (2kTn_i\varepsilon\varepsilon_0)^{1/2} F(\lambda, y) \tag{32}$$

$$F(\lambda, y) = [\lambda(e^{-y} - 1) + \lambda^{-1}(e^y - 1) + (\lambda - \lambda^{-1})y]^{1/2} \tag{33}$$

$$\lambda = n_i/N, \qquad y = e\Delta V_{sc}/kT \tag{34}$$

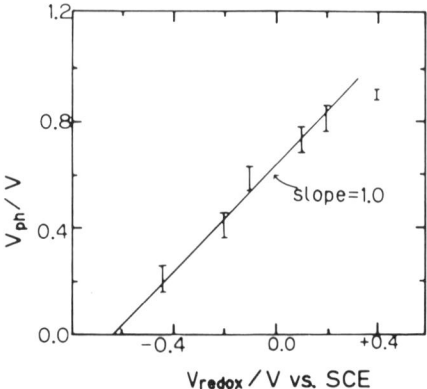

Figure 7. Open-circuit photovoltage for the a-Si:H/CH$_3$OH interface as a function of solution redox potential.[63] The redox couples employed are N,N'-dimethyl-4,4'-bipyridinium$^{+/0}$ (0.95 V); N,N'-dimethyl-4,4'-bipyridinium$^{2+/+}$ (−0.45 V); [η^5-C$_5$(CH$_3$)$_5$]$_2$Fe$^{+/0}$ (−0.21 V); N,N,N',N'-tetramethylphenylenediamine$^{+/0}$ (−0.11 V); 1,1'-dimethylferrocene$^{+/0}$ (+0.10 V); ferrocene$^{+/0}$ (+0.20 V); acetylferrocene$^{+/0}$ (+0.39 V). All determinations were in 1.0 M LiClO$_4$ except for the N,N'-dimethyl-4,4'-bipyridinium system, which required 1.0 M KCl for solubility. Measurements are the result of several independent determinations for a number of anodes.

where k is the Boltzmann constant, n_i is the intrinsic carrier density, ε is the dielectric constant of the semiconductor, and ε_0 is the permittivity of free space. Also, q_{ss} can be correlated with ΔV_{sc} but the form of the equation depends on the nature of the distribution of the surface state energies, i.e., whether the surface states are uniformly distributed in energy or are localized at a single energy level. In the case of uniformly distributed acceptor surface states with centering energy of E_0, q_{ss} is given by

$$q_{ss} \simeq eN_{ss}\frac{(E_0 - e\Delta V_{sc} - E_F)}{E_0} \qquad (35)$$

On the other hand, in the case of acceptor surface states at a single energy level E_{ss}, q_{ss} is given by

$$q_{ss} \simeq \frac{eN_{ss}}{1 + g_{ss}^{-1}\exp[(E_{ss} - e\Delta V_{sc} - E_F)/kT]} \qquad (36)$$

where g_{ss} is the degeneracy of the energy level. The total charge in the electrolyte, q_{el}, is equal in magnitude to $q_{ss} + q_{sc}$. By assuming

q_{el} is arranged at a distance d from the electrode surface, ΔV_H is given by

$$\Delta V_H = \frac{q_{el}}{\varepsilon_H \varepsilon_0} d \tag{37}$$

where ε_H is the dielectric constant in the Helmholtz layer. When the concentration of surface states is very high so that q_{ss} becomes larger than q_{sc} at a given ΔV_{sc}, Fermi level pinning by surface states occurs. In this case, even before the contact with an electrolyte solution, band bending within the semiconductor occurs ($q_{el} = 0$, $q_{sc} = -q_{ss}$)[71] and the charge required to attain an equilibrium with a redox couple upon contact is provided by the surface states. This situation is quite similar to the metal/semiconductor junctions with a high density of surface states discussed before. Bard et al.[71] calculated q_{sc} and q_{ss} to demonstrate the conditions required for this situation with values of $n_i = 1.3 \times 10^6 \text{ cm}^{-3}$, $\varepsilon = 12$, $kT = 0.0257$ eV and with the assumption of half occupancy of N_{ss} (Fig. 8). One must note that when N_{ss}[71] becomes larger than ca. 10^{12} cm^{-2}, which represents only ~1% surface coverage, q_{ss} is greater than q_{sc} for any moderately doped semiconductors, leading to Fermi level pinning. Experimental results at several semiconductor elec-

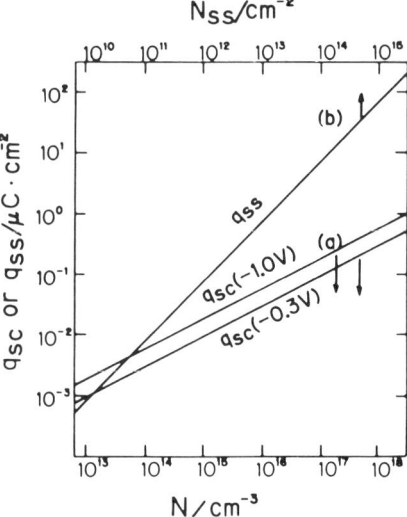

Figure 8. Surface charge in an n-type semiconductor: (a) space charge, q_{sc}, at various doping levels (N) at ΔV_{sc} of -0.3 and -1.0 V; (b) surface state charge (q_{ss}) as a function of surface state density, N_{ss}, assuming half-occupancy. Potential drop across Helmholtz layer due to semiconductor charge can be calculated by Eq. (37).

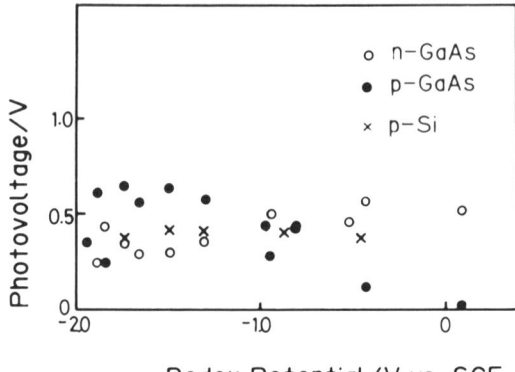

Figure 9. Open-circuit photovoltage for the n-GaAs,[64] p-GaAs,[64] and p-Si[67] in $CH_3CN/[n-Bu_4N]ClO_4$ solutions as a function of solution redox potential.

trodes have shown that this is a rather common situation.[64-70] Some typical results of Fermi level pinning are shown in Fig. 9.

Gerischer[73] and Nozik[74] have stated that the formation of an inversion layer can also cause a shift of band edges. For small bandgap semiconductors, this effect may be more important. The formation of an inversion layer has been shown experimentally at several semiconductor/electrolyte interfaces.[75-78]

3. Distribution of Externally Applied Potential at the Semiconductor/Electrolyte Interfaces

(i) Surface States of Free Semiconductors

So far only the effect of the surface states is considered, but if the carrier density is high, even when the surface state density is low, the potential drop in the Helmholtz layer may be more significant than that in the semiconductor.

Uosaki and Kita calculated $\Delta V_H/(\Delta V_{sc} + \Delta V_H)$ as a function of total potential change, $\Delta V_{sc} + \Delta V_H$, by using Eqs. (32)–(34) and (37) and demonstrated that the contribution of ΔV_H is surprisingly high[79] (Fig. 10).†

† Gerischer demonstrated the effect of the dielectric constant of the semiconductor on $[\Delta V_H/(\Delta V_{sc} + \Delta V_H)]$.[80]

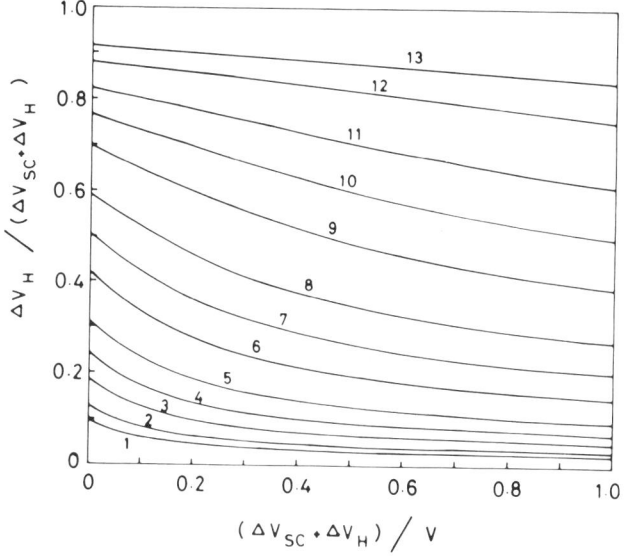

Figure 10. The ratio of the potential change in the Helmholtz layer to the total potential change as a function of the total potential change.[79] $\varepsilon = 173$. $C_H = 10 \,\mu\text{F cm}^{-2}$. Carrier densities are 1, 10^{16} cm^{-3}; 2, 2×10^{16} cm^{-3}; 3, 5×10^{16} cm^{-3}; 4, 10^{17} cm^{-3}; 5, 2×10^{17} cm^{-3}; 6, 5×10^{17} cm^{-3}; 7, 10^{18} cm^{-3}; 8, 2×10^{18} cm^{-3}; 9, 5×10^{18} cm^{-3}; 10, 10^{19} cm^{-3}; 11, 2×10^{19} cm^{-3}; 12, 5×10^{19} cm^{-3}; and 13, 10^{20} cm^{-3}.

The space charge, q_{sc}, is often expressed[81] as

$$q_{sc} = (2\varepsilon\varepsilon_0 kTN)^{1/2}\left(\Delta V_{sc} - \frac{kT}{e}\right)^{1/2} \tag{38}$$

by using the Mott-Schottky approximation. The comparison is made between $\Delta V_H/(\Delta V_{sc} + \Delta V_H)$ calculated by using Eqs. (37) and (38) and that obtained by using Eqs. (32)-(34) and (37) in Fig. 11. When the potential change, $(\Delta V_{sc} + \Delta V_H)$, is relatively large, the two calculations give similar results, but the difference is significant when $(\Delta V_{sc} + \Delta V_H) < 0.15$ V.[79]

The linearity of the Mott-Schottky plot is often considered to be the evidence of the band-edge pinning.[82] The measured capacitance, C, can be written as

$$C^{-1} = C_{sc}^{-1} + C_H^{-1} \tag{39}$$

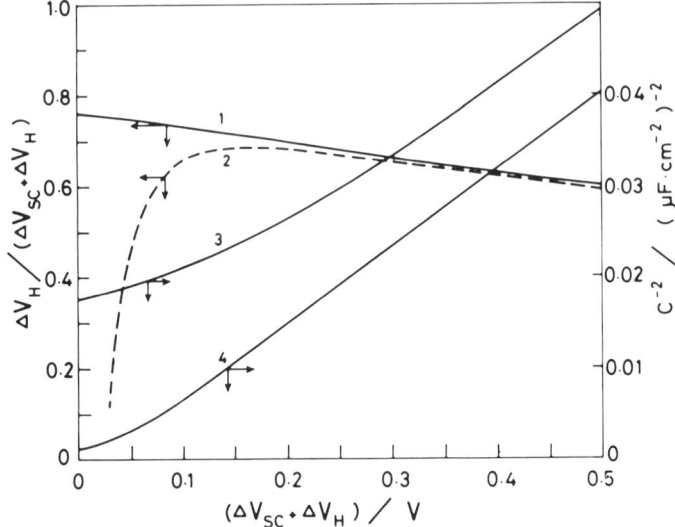

Figure 11. Dependence of the ratio of the potential change in the Helmholtz layer to the total potential change and of C^{-2} as a function of total potential change.[79] $\varepsilon = 173$. Carrier density is 10^{19} cm^{-3}. (1) $\Delta V_H/(\Delta V_H + \Delta V_{sc})$ vs. $\Delta V_H + \Delta V_{sc}$ calculated by using Eqs. (32)-(34) and (37). (2) As curve 1 but calculated by using Eqs. (37) and (38). (3) C^{-2} vs. $\Delta V_H + \Delta V_{sc}$ calculated by using Eqs. (32)-(34), (37), (39), and (40) and $C_H = 10$ μF cm^{-2}. (4) As curve 3 but $C_H = \infty$.

where C_H and C_{sc}† are the differential capacitance of the Helmholtz layer and the semiconductor, respectively, and the electrode potential of the semiconductor electrode with respect to the flat band potential can be written as

$$V - V_{FB} = \Delta V_{sc} + \Delta V_H \qquad (40)$$

The effect of C_H on the Mott–Schottky plots obtained by using Eqs. (32)-(34), (37), (39), and (40) is shown in Fig. 11. The plots are linear when potential change is large but curved when $(\Delta V_{sc} + \Delta V_H) <$ ca. 0.3 V. The slope of the linear portion of the relation is almost identical whether ΔV_H is neglected or not. Thus, the linearity of the Mott–Schottky plot does not necessarily mean the existence of band-edge pinning.

† C_{sc} can be calculated by differentiation of Eq. (32).

(ii) Effect of Surface States

The argument developed in Section III.2 can be applied to the potential distribution when an external potential is applied to the semiconductor/electrolyte interface.[71] Thus, at a semiconductor with a high density of surface states, the variation of the potential between the semiconductor and the solution leads to a relatively constant ΔV_{sc} and a significant change in ΔV_H. Green[31,83] has shown that if the surface state density becomes very high, the behavior of the semiconductor becomes that of a metal.

IV. DISTRIBUTION OF ENERGY STATES FOR IONS OF REDOX SYSTEM IN SOLUTION

1. Importance of the Energy Levels of Redox Couples

Usually it is thought that electron transfer can take place only between electronic energy states of equal energies, one being occupied and the other vacant.[5,61,84,85] Gurney[86] applied this concept for the electrode kinetics at a metal electrode and obtained the following rate expression:

$$i \propto \int N(E) W(E) D(E) \, dE \quad (41)$$

where $N(E)$, $W(E)$, and $D(E)$ are the number of electrons in the metal, tunneling probability, and the number of acceptor states in the solution, respectively, between E and $E + dE$. Gerischer extended this equation for electrode reactions at semiconductor electrodes in the dark. Thus, the total current, i, is devided into two fractions, namely the conduction band current, i_{CB}, and the valence band current, i_{VB}. These currents can be written as follows[8]:

$$i_{CB} = \vec{i}_{CB} - \overleftarrow{i}_{CB} = \vec{k}_{CB} c_R^S \int_{E_c}^{\infty} N'(E) W(E) D_O(E) \, dE$$

$$- \overleftarrow{k}_{CB} c_O^S \int_{E_c}^{\infty} N(E) W(E) D_R(E) \, dE \quad (42)$$

$$i_{VB} = \overleftarrow{i}_{VB} - \overrightarrow{i}_{VB} = \overleftarrow{k}_{VB} c_R^S \int_0^{E_v} N'E(W) W(E) D_O(E)\, dE$$

$$- \overrightarrow{k}_{VB} c_O^S \int_0^{E_v} N(E) W(E) D_R(E)\, dE \quad (43)$$

where $\overleftarrow{i}_{CB}$ and $\overrightarrow{i}_{CB}$ are the anodic and cathodic component of the conduction band current, the rate constants of which are $\overleftarrow{k}_{CB}$ and $\overrightarrow{k}_{CB}$, respectively. Similar notation is used for the valence band current. $N(E)$, $N'(E)$, $W(E)$, $D_O(E)$, and $D_R(E)$ represent the number of occupied and unoccupied states in the semiconductor, tunneling probability, and the distribution of occupied (reduced) and of unoccupied (oxidized) state of redox couple, respectively, between energy E and $E + dE$. E is taken as zero at the bottom of the valence band, c_R^S and c_O^S are the surface concentrations of reduced and oxidized species, respectively.

In Sections II and III we correlated the electronic energy level of the solid phase, E_F, and that of electrolyte, V_{redox}. This comparison is useful with regards to understanding the potential difference between the two phases and the potential distribution of this interface. However, to describe the kinetics of the electron transfer reactions at the surface, it is necessary to know the energy distributions of occupied and unoccupied levels of the semiconductor and those of the redox couple, as one can see from Eqs. (42) and (43). It is well known[87] that the energy distribution in a semiconductor in the dark is described by Fermi statistics with a function for density of states. The energy distributions of redox couples are, however, still a very controversial issue.[88] There are essentially two major approaches to describe this problem. One approach was originated by Gurney.[86] In this model, the energy distribution is considered to be due to the vibration-rotation interaction between the particular ion and solvent molecules. According to the second model,[61] the energy level of the ion fluctuates around its most probable value due to the thermal fluctuation of solvent dipoles.

2. Gurney's Model

Gurney[86] developed a model for the neutralization of H^+ and Cl^-. Let us first consider the H^+ case. In vacuum the ionization potential,

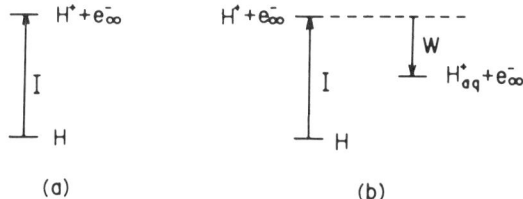

Figure 12. (a) Ionization of H to H^+ in vacuum and electron in vacuum at ∞. Energy required is the ionization potential. (b) Ionization of H to H^+ in water and electron in vacuum at ∞. Energy required is $(I - W)$.

I, is required to complete the following process (Fig. 12a):

$$H \to H^+ + e_\infty^- \tag{44}$$

This step is analogous to that of removing an electron from the Fermi level of the solid phase to the vacuum, the energy of which is given by the work function, Φ. Thus, the ionization potential represents the energy level of electrons of H/H^+ (H as an occupied state and H^+ as an unoccupied state). In an aqueous solution, ions are hydrated and the energy required for

$$H \to H_{aq}^+ + e_\infty^- \tag{45}$$

is smaller than I by the hydration energy, W, and the energy level of electrons of H/H^+ in an aqueous solution is given by $I - W$ (Fig. 12b). In the above consideration, it was assumed that the mutual potential energy of the ion and adjacent water molecule before the neutralization was $-W$ and that after the neutralization it was zero. This is not quite correct. Figure 13 shows the potential energy diagram for this process. The potential energy of the neutral state ion, i.e., H, is represented by a curve such as *abc*. The force between H and H_2O is repulsive at all distances. The curve for the stable vibrational levels of H_3O^+ is represented by *defgh*. The lowest vibration-rotation level of H_3O^+ is *ef* and the energy difference between *ef* and *h* is the hydration energy of positive ion, H^+ in this case, which is in the lowest vibration-rotation level, W_0. The transition representing neutralization of an ion in its lowest vibration-rotation level is on this diagram *fb*.† In accordance with the

---

† The Gerischer model, which will be discussed later, allows the transition at much smaller H^+-H_2O distances.

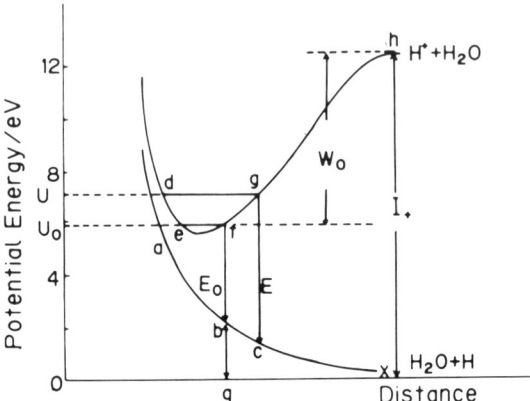

Figure 13. Potential energy diagram of $H^+ + H_2O$ and $H + H_2O$ systems.[86] See text for details.

Franck-Condon principle, bq represents the positive mutual potential energy of the molecular components before they have moved apart. If R_n is the value of this repulsive potential energy resulting from neutralization of an ion in its nth vibration-rotation level, the neutralization energy of H_3O^+ in its nth vibration-rotation level, E_n^+, is given by

$$E_n^+ = I - W_n - R_n \tag{46}$$

Thus, this energy corresponds to the following process:

$$(H - H_2O)^+ + e_\infty^- \to H - H_2O \tag{47}$$

This means that a free electron from infinity is introduced into the solution to occupy the electron state in H_3O^+ in its nth vibration-rotation level without change of the solvation structure (Franck-Condon principle). Therefore, this energy corresponds to the energy of unoccupied states, i.e., H_3O^+.

Similar arguments can be applied to Cl^- with minor modification (Fig. 14). In this case the potential energy curve for $Cl^- - H_2O$ ($klmnp$) lies below the axis and px is the ionization potential of the unhydrated negative ion, Cl^-. Curve $a'b'c'$ represents the potential energy curve for $H_2O - Cl$. The lowest vibration rotation level of $Cl^- - H_2O$ is lm and the energy difference between lm and p is the hydration energy of negative ion, Cl^- in this case,

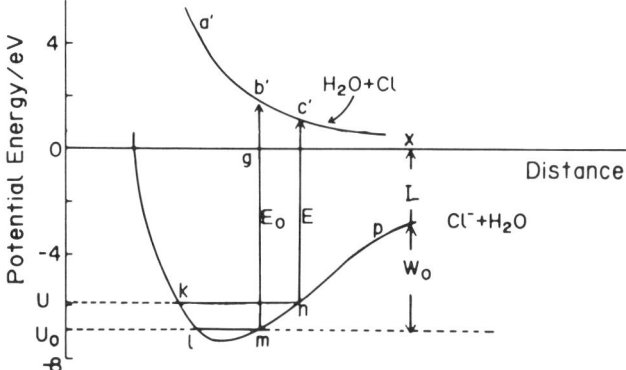

Figure 14. Potential energy diagram of $Cl^- + H_2O$ and $Cl + H_2O$ systems.[86] See text for details.

which is in the lowest vibration-rotation level, W_0. From this figure it is clear that the neutralization energy of Cl^--H_2O in its nth vibration-rotation level, E_n^-, is given by

$$E_n^- = I + W_n + R_n \tag{48}$$

In both cases, if the vibration-rotation energy is given by U, the distribution is given by the Boltzmann distribution:

$$N(U) = N_0 \, e^{(U_0-U)/kT} \tag{49}$$

where U_0 is the vibration-rotation energy of an ion in its ground level and is represented by the energy level of ef in Fig. 13 and by lm in Fig. 14. If E and E_0 are the neutralization energies of an ion, the corresponding vibration-rotation energies of which are U and U_0, respectively, it is clear from Fig. 13 that $(E - E_0)$ is greater than $(U - U_0)$; and, between E and E_0 the same number of levels exists as those between U and U_0, but they are distributed over a larger range of energy. Thus, for H^+

$$N(E) = N_0 \, e^{(E_0-E)/\gamma kT} \tag{50}$$

where $\gamma = |E_0 - E|/(U_0 - U) > 1$. Similarly, the distribution function for Cl^- is given by

$$N(E) = N_0 \, e^{(E-E_0)/\gamma kT} \tag{51}$$

The distribution given by Gurney is shown in Fig. 15, where E_0^+

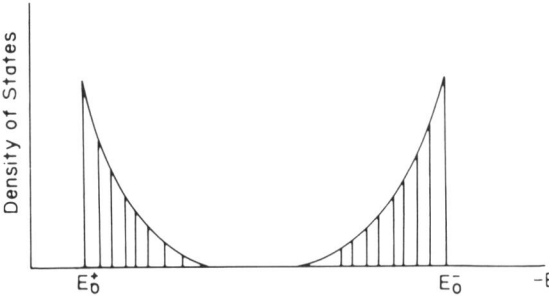

Figure 15. Distribution of density of states as a function of energy (Gurney's model).[86] E_0^+ and E_0^- are the neutralization energy of $H^+ - H_2O$ and $Cl^- - H_2O$, in its ground level, respectively.

and E_0^- represent the neutralization energy of H_3O^+ and Cl^-, respectively, in their ground levels. These distributions allow only the energy levels of $E > E_0$ for positive ions and $E < E_0$ for negative ions. This limitation originates from the fact that Gurney did not allow states in which the ion-water distance is shorter than its ground state. In Fig. 13, only the states at which the H^+-H_2O distance is larger than f are allowed and in Fig. 14, only the states at which the Cl^--H_2O distance is larger than m are allowed. It seems, however, that there is no reason to inhibit the states in which the ion-water distance is shorter than its ground level. Thus, there should be energy levels with $E < E_0$ for a positive ion and with $E > E_0$ for a negative ion. Equation (50) should be used if $E > E_0$ and Eq. (51) should be used if $E < E_0$. The value of γ depends on the shape of the potential energy curve of an ion before and after neutralization. As far as H^+ is concerned, $\gamma < 1$ for $E < E_0$, from Fig. 13, which means that the number of states decreases more quickly when $E < E_0$ than when $E > E_0$. For Cl^-, it is not possible to determine γ from Fig. 14 and the exact shape of the potential energy diagram H_2O-Cl and H_2O-Cl^- must be known to calculate γ.

Bockris[88-96] and his colleagues essentially followed Gurney's model for the hydrogen evolution reaction, but they included the interaction between the electrode and hydrogen. They usually used $\gamma = 1$ at $E > E_0$ and neglected the distribution at $E < E_0$[88-93] as Gurney did or assumed a very sharp drop of the number of states at $E < E_0$.[27,94-96]

3. Gerischer's Model

Gurney treated the neutralization of ions ($H^+ + e^- \to H$ and $Cl^- \to e^- + Cl$), which is a rather complicated process that involves the strong interaction between an electrode and an reaction intermediate such as an adsorbed hydrogen for the hydrogen evolution reaction. Gerischer extended Gurney's treatment to simple redox reactions such as

$$(ox)_{solv} + e_\infty^- \to (red)_{solv} \qquad (52)$$

In this case the potential energy diagram is something like the one shown in Fig. 16, in which I is the ionization energy of the reduced species in vacuum. The corresponding reaction can be written as

$$red \to ox + e_\infty^- \qquad (53)$$

In Fig. 16, $_0r_{ox}$ and $_0r_{red}$ are the reaction coordinates corresponding to the most stable states of the oxidized and reduced forms and $_0U_{ox}$ and $_0U_{red}$ represent the energies corresponding to each form. E is the energy change in the reaction shown in Eq. (52). In other words, E is the energy required to introduce a free electron from infinity into the solution and to occupy the electron state in an oxidized form without changing the solvation structure (Franck-Condon principle) and, thus, gives the energy of the unoccupied states. Consequently, Eq. (52) actually has the same meaning as Eq. (47). The reverse process gives the energy of the occupied states

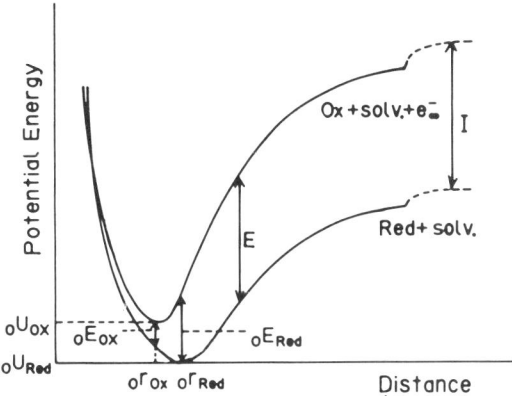

Figure 16. Potential energy diagram of ox + solv. and red + solv. systems.[24]

(reduced form). Gerischer neglected the zero-point energy and represented the vibration-rotation level by a smooth continuous curve, as shown in Fig. 16. Here, $_0E_{ox}$ and $_0E_{red}$ are the energies of unoccupied and occupied states in the most stable state, respectively. He also used the Boltzmann distribution to describe the energy distribution of both forms as Gurney did. The most significant difference between Gurney's treatment and Gerischer's is that Gurney did not allow the configuration where the H^+-H_2O distance is smaller than the ground level as shown before but Gerischer did. The equation given by Gerischer is very similar to that of Gurney:

$$D_{ox}(E) = \frac{g(E)\exp\left(-\frac{U_{ox} - {_0U_{ox}}}{kT}\right)}{\int_{-\infty}^{\infty} g(E)\exp\left(-\frac{U_{ox} - {_0U_{ox}}}{kT}\right) dE} \quad (54)$$

$$D_{red}(E) = \frac{g(E)\exp\left(-\frac{U_{red} - {_0U_{red}}}{kT}\right)}{\int_{-\infty}^{\infty} g(E)\exp\left(-\frac{U_{red} - {_0U_{red}}}{kT}\right) dE} \quad (55)$$

where $g(E)$ is a weighting factor and $D_{ox}(E)$ and $D_{red}(E)$ are the density of states of oxidized and reduced forms, respectively.† To use these equations, the values of $\exp[-(U_{ox} - {_0U_{ox}})/kT]$ and $\exp[-(U_{red} - {_0U_{red}})/kT]$ should be known as functions of E. Figure 17 shows $U_{ox} - {_0U_{ox}}$ and $U_{red} - {_0U_{red}}$ as functions of E. From these, Gerischer found $D_{ox}(E)$ and $D_{red}(E)$ as shown in Fig. 18, which appeared in his original publication.[24] $D_{ox}(E)$ and $D_{red}(E)$ are equivalent to $N(E)$ of H^+ and Cl^-, respectively, of Gurney's treatment. The curves are somewhat asymmetrical, but in his later publications, the curves are more symmetrical and show Gaussian distribution,[5,8] which is also obtained by the continuum model

† Although ΔE instead of E is used in his original paper,[24] ΔE actually means E in Gurney's treatment and Gerischer used E later.[5] Khan commented[88,96] that Gerischer plotted the density of states not as a function of the energy, E, with respect to vacuum but as a function of the difference of energy, ΔE, of the electron in the oxidized and reduced ion. It is, however, clear from the above argument that these two terms have the same meaning.

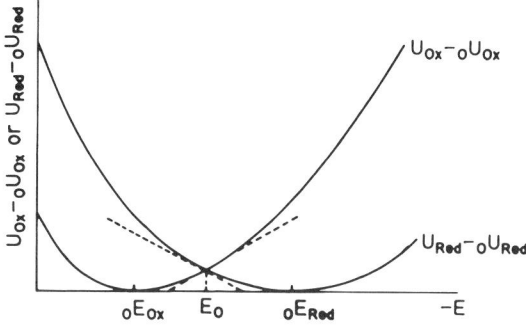

Figure 17. $U_{ox} - {}_0U_{ox}$ and $U_{red} - {}_0U_{red}$ as a function of electron energy, E.[24]

shown below. Since $\gamma(=|{}_0E_{ox} - E|/({}_0U_{ox} - U_{ox})) > 1$ when $E > {}_0E_{ox}$ but <1 when $E < {}_0E_{ox}$, one cannot expect a symmetric distribution. Exact distribution curves can be calculated by knowing the potential energy curves for particular redox couples.

The energy where

$$D_{ox}(E) = D_{red}(E) \tag{56}$$

is, in a sense, equivalent to the Fermi level of the solid state phase at which the occupancy of an electron is 1/2, as in this case. Therefore, Gerischer suggested that this energy be called $E_{F,redox}$.[5,8,23-26,54,55] The significance of $E_{F,redox}$ was discussed in Section II.

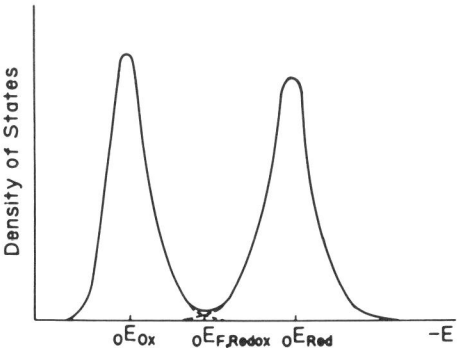

Figure 18. Density of states as a function of electron energy (Gerischer's model).[24]

4. Continuum Solvent Polarization Fluctuation Model

The model of Gurney as well as those of Bockris and of Gerischer assume that an energy level change is caused by the vibration-rotation interaction between the central ion and solvent molecules and describe this change by using the Boltzmann distribution. In the present model, the surrounding solvent is considered to be a continuum dielectric. Marcus,[97-99] Dogonadze,[100,101] Christov,[102] and Levich[103,104] presented the model in this category and the model of Marcus is most often employed in the literature of semiconductor electrochemistry. Here we describe the model presented by Morrison,[61] which is similar to but less rigorous than that of Marcus. Let us consider the following electron transfer reaction:

$$S_0 A^z + e^- \rightarrow S_f A^{z-1} \tag{57}$$

by dividing it into the following three steps:

$$S_0 A^z \rightarrow S_i A^z \tag{58}$$

$$S_i A^z + e^- \rightarrow S_i A^{z-1} \tag{59}$$

$$S_i A^{z-1} \rightarrow S_f A^{z-1} \tag{60}$$

The first step represents the change in polarization or change in dipole configuration around the ion A. The second step is the electron transfer during which the polarization is frozen at the configuration denoted as S_i (Franck–Condon principle). The third step represents the subsequent relaxation of the polarization of the medium to its new equilibrium value. The energy required for the first and third step can be obtained by considering the energy, ΔE_p, necessary to cause fluctuations from the equilibrium polarization of the dielectric around an ion without changing the charge on the ion. A parameter δ is introduced such that the polarization of the dielectric corresponds to $eZ \pm e\delta$, where Z is the actual charge on the ion. A polarization is defined to correspond to a central charge differing from the central charge actually present, eZ, by $\pm e\delta$. It is necessary to know an expression for the energy increase in the polarized dielectric medium when the central charge remains eZ, but the polarization fluctuates from its equilibrium configuration to a new configuration, which would normally correspond to a charge $e(Z \pm \delta)$ on the ion. Marcus[97] and Dogonadze et al.[100]

derived the expression as

$$\Delta E_p = \delta^2 \lambda \qquad (61)$$

where λ is the *reorganization energy* and is given by

$$\lambda = \frac{e^2}{8\pi\varepsilon_0 a}\left(\frac{1}{\kappa_{op}} - \frac{1}{\kappa_s}\right) \qquad (62)$$

where a is the ionic radius and κ_{op} and κ_s are the optical and the static dielectric constant of the medium, respectively. One must note that this is very similar to the Born expression for the solvation energy per ion, W,[105]

$$W = -\frac{e^2 Z^2}{8\pi\varepsilon_0 a}\left(1 - \frac{1}{\kappa_s}\right) \qquad (63)$$

A better expression for λ including changes in the inner sphere is[98]

$$\lambda = \frac{q^2}{8\pi\varepsilon_0 a}\left(\frac{1}{\kappa_{op}} - \frac{1}{\kappa_s}\right) + \sum_j \frac{f_j^{ox} - f_j^{red}}{f_j^{ox} + f_j^{red}} \Delta x_j^2 \qquad (64)$$

where the f's are force constants for the jth bond and Δx_j is the displacement in the bond length. Thus, the energy level shift from its equilibrium value E_0 ($= E_{ox}$ if the ion is the oxidized form) to a value E in the first step is accomplished by a change in polarization δ_i, requiring an energy $\delta_i^2 \lambda$. For the third step, the initial polarization of the surrounding medium corresponds to an ionic charge $e(Z - \delta_i)$ but the central charge is now $e(Z - 1)$. Thus, the change in polarization required to bring the system to its new equilibrium configuration corresponds to a change in central charge of $1 - \delta_i$. Therefore the energy released in the third step is $(1 - \delta_i)^2 \lambda$. The energy change in the second step is given by $E_{cs} - E$, where E_{cs} is the energy of the conduction band edge at the electrode surface.

The total energy change of the three steps which represents the energy required to transfer an electron to an ion in solution with an appropriate change of equilibrium polarization of the medium, Δ, is given by

$$\Delta = \delta_i^2 \lambda - (1 - \delta_i)^2 \lambda + E - E_{cs} \qquad (65)$$

The energy Δ should be independent of intermediate state and, therefore, of E and δ_i. Let us calculate Δ for the equilibrium situation, i.e., $\delta_i = 0$ and $E = E_0$. In this case,

$$\Delta = -\lambda + E_0 - E_{cs} \tag{66}$$

Equations (65) and (66) give

$$-\lambda + E_0 - E_{cs} = \delta_i^2 \lambda - (1 - \delta_i)^2 \lambda + E - E_{cs} \tag{67}$$

Thus,

$$\delta_i = \frac{E_0 - E}{2\lambda} \tag{68}$$

By inserting Eq. (68) into Eq. (61), one obtains

$$\Delta E_p = \frac{(E_0 - E)^2}{4\lambda} \tag{69}$$

This gives the thermal energy required to shift the energy level from E_0 to an arbitrary energy E. The energy distribution function $D(E)$ is given by

$$D(E) \propto \exp(-\Delta E_p / kT) \tag{70}$$

By using Eq. (63)

$$D(E) = \frac{1}{(4\pi\lambda kT)^{1/2}} \exp\left[-\frac{(E_0 - E)^2}{4\lambda kT}\right] \tag{71}$$

where the preexponential term is a normalizing constant. For the oxidized and reduced forms, E_{ox} and E_{red}, respectively, should be used for E_0. This function gives a symmetric curve (Gaussian) with a central energy of E_0. The energy difference between E_{ox} and E_{red} can be obtained by considering the following cycle:

$$S_0 A^z + e^- \rightarrow S_0 A^{z-1} \tag{72}$$

$$S_0 A^{z-1} \rightarrow S_f A^{z-1} \tag{73}$$

$$S_f A^{z-1} \rightarrow S_f A^z + e^- \tag{74}$$

$$S_f A^z \rightarrow S_0 A^z \tag{75}$$

Step (72) represents the electron transfer step from the solid to the unoccupied level of ion in its equilibrium state, $_0E_{ox}$. The energy change for this step is given by $_0E_{ox} - E$, where E is the initial energy level of electron in the solid. Step (73) is the chemical reorganization process with the free energy change ΔG_{red}. Step (74) is the electron transfer from the occupied level of the ion, E_{red}, to the solid phase and the energy released in this process is $E - E_{red}$. The final step (75) is the chemical reorganization to restore the original configuration with a free energy change of ΔG_{ox}. By summing these energies, the energy change around the cycle is given by

$$(_0E_{ox} - E) + \Delta G_{red} + (E - {_0E_{red}}) + \Delta G_{ox} \tag{76}$$

and should be equal to zero. Thus,

$$_0E_{ox} - {_0E_{red}} = -(\Delta G_{red} + \Delta G_{ox}) \tag{77}$$

Since ΔG_{ox} and ΔG_{red} correspond to the free energy change from an energetic state to a stable state, both are negative and, therefore, E_{ox} is always greater than E_{red}. If one neglects entropy effects and assumes that the value of λ is the same for the oxidized and reduced species,

$$\Delta E_p(\delta = 1) = -\Delta G_{ox} = -\Delta G_{red} = \lambda \tag{78}$$

Then,

$$_0E_{ox} - {_0E_{red}} = 2\lambda \tag{79}$$

This equation indicates why, and by how much, the levels are shifted by the polar character of the solvent. By using Eq. (71) and Eq. (79) where $_0E_{ox}$ and $_0E_{red}$ are used instead of E_0, the distribution functions can be drawn as a function of energy. They are similar to those obtained by Gerischer (Figs. 18, 19a).[†]

5. Validity of the Models

The Gaussian distribution is most often used in photoelectrochemistry to describe the energy levels of redox couples and to discuss qualitatively the possibility of the reaction. The major difficulty of this model is that it cannot explain the Tafel behavior

---

[†] Although Gerischer obtained the distribution function based upon Gurney's treatment, the shape of the function is similar to the one based upon the continuum model. One must note that the model of Gerischer is closer to the model of Bockris despite their recent exchange of sharp criticism.[23,28-31,54,55]

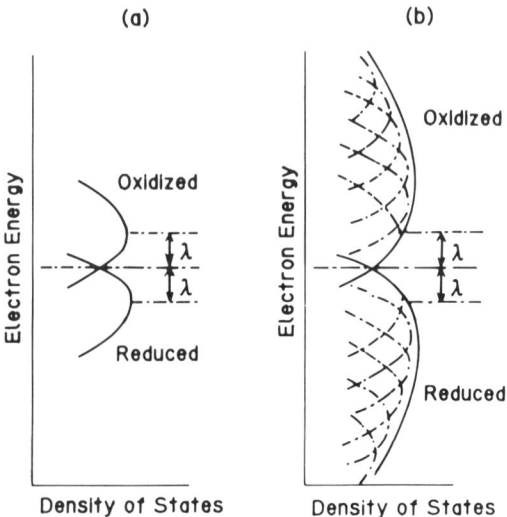

Figure 19. Density of states as a function of electron energy. (a) Solvent fluctuation model; (b) vibronic model.[106,107]

at metal electrodes. It is not clear, however, whether Gurney's model can really explain it or not.

In the present author's view, the Gaussian distribution function based on the solvent fluctuation model, which is developed for a simple redox couple, is used too often even when the basic assumption is not valid. For example, this type of distribution function is often drawn for the hydrogen evolution reaction where the oxidized state is H^+ and reduced state is H_2.[105] Certainly the nature of the solvation is completely different between H^+ and H_2. Moreover, when one considers the kinetics of the hydrogen evolution reaction, one should consider not the energy level of H^+/H_2 but that of $H^+/H(a)$ as Gurney did.

Recently Nakabayashi et al.[106,107] questioned the validity of the Gaussian distribution function based on the fact that the rate of a highly exothermic electron transfer reaction did not agree with the prediction based on this distribution function and proposed an alternate function (vibronic model), which extends to a smaller energy for the oxidized form and to a larger energy for the reduced form (Fig. 19).

V. RATE EXPRESSIONS FOR ELECTRON TRANSFER AT ILLUMINATED SEMICONDUCTOR/ELECTROLYTE INTERFACES

1. Phenomenological Description

Figure 20 is a schematic representation of a p-type semiconductor/electrolyte interface under illumination. Each absorbed photon, the energy of which is larger than the energy gap of the semiconductor, makes an excited electron in the conduction band and a hole in the valence band. Some of the excited electrons reach the semiconductor surface and some recombine with holes before reaching the surface. The electrons which reach the surface would transfer to an acceptor (oxidized form of a redox couple), be trapped by surface states, or recombine with holes in the valence band at the

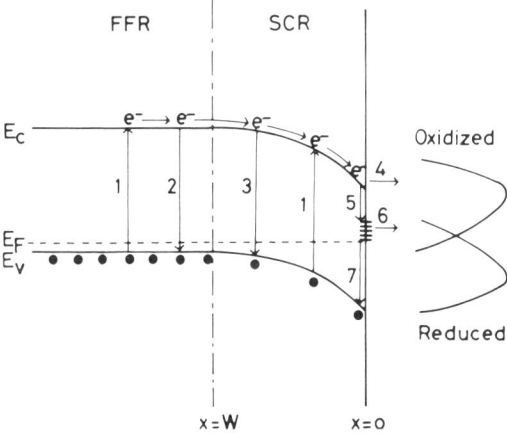

Figure 20. Schematic representation of p-type semiconductor/electrolyte interface under illumination. Semiconductor side is divided into space charge region (SCR) and field free region (FFR) at $x = W$. Numbers in the figure represent the following steps: (1) Excitation of an electron from the valence band to the conduction band, leaving a hole in the valence band. (2) Recombination in the bulk. (3) Recombination in the space charge region. (4) Electron transfer from the conduction band to an oxidized state. (5) Electron capture by a surface state. (6) Electron transfer from the surface state to the oxidized state (electron transfer via surface state). (7) Hole capture by the surface state (surface recombination via surface state). There is also a possibility of direct recombination of a conduction band electron with a valence band hole, although this step is not shown in the figure.

surface directly. The electrons which are trapped by the surface states would either transfer to the acceptor (charge transfer via surface states) or recombine with holes in the valence band (surface recombination via surface states).

To obtain rate expressions for photoelectrochemical reactions, all the steps mentioned above as well as the potential distribution under illumination should be taken into account.

In the early stages of study of photoelectrochemical kinetics, two extreme models were presented. One is the model of Bockris and Uosaki,[93,108,109] in which the charge transfer step was considered to be the rate-determining step and the other is Butler's model,[110] in which the semiconductor/electrolyte interface was considered as a Schottky barrier, i.e., the electrochemical kinetics were neglected and all the potential drop occurred only within the semiconductor.

In this section, these two models are explained first, then the effects of other terms will be considered.

2. The Model of Bockris and Uosaki

Bockris and Uosaki[93,108,109] calculated photocurrent-potential relations mainly for the hydrogen evolution reaction (HER) at p-type semiconductors with the assumption that the following is the rate-determining step:

$$p\text{-S.C.}(e) + H_3O^+ \to p\text{-S.C.}\cdots H\cdots H_2O\dagger \qquad (80)$$

The total current, i, is given by

$$i = i_a - i_c \qquad (81)$$

where i_a and i_c are anodic and cathodic current, which can be separated, respectively, into two terms:

$$i_a = i_{a,\text{CB}} + i_{a,\text{VB}} \qquad (82)$$

$$i_c = i_{c,\text{CB}} + i_{c,\text{VB}} \qquad (83)$$

where $i_{a,\text{CB}}$ is the anodic current passing through the conduction

---

† The final state describes the state immediately after the electron transfer, i.e., before the various atoms and molecules relax into their ground vibration-rotation state.

band; $i_{a,VB}$ is the anodic current passing through the valence band; $i_{c,CB}$ is the cathodic current passing through the conduction band; and $i_{c,VB}$ is the cathodic current passing through the valence band. They expressed each term by modifying Eq. (42) and (43):

$$i_{a,CB} = e\frac{c_R}{c_T} \int_{E_c}^{\infty} N_h(E) W(E) D_R(E)\, dE \qquad (84)$$

$$i_{a,VB} = e\frac{c_R}{c_T} \int_{0}^{E_v} N_h(E) W(E) D_R(E)\, dE \qquad (85)$$

$$i_{c,CB} = e\frac{c_O}{c_T} \int_{E_c}^{\infty} N_e(E) W(E) D_O\, dE \qquad (86)$$

$$i_{c,VB} = e\frac{c_O}{c_T} \int_{0}^{E_v} N_e(E) W(E) D_O(E)\, dE \qquad (87)$$

where $N_h(E)$ is the number of holes with energy E which strike the surface per unit time per unit area; $N_e(E)$ is the number of electrons with energy E which strike the surface per unit time per unit area; c_R and c_O are the numbers of reduced and oxidized species per unit area in the outer Helmholtz plane (OHP), and c_T is the total number of sites per unit area in the OHP. Energy levels are specified to be zero at the bottom of the valence band. They considered only the photoeffect on $i_{c,CB}$ in the case of the HER at p-type semiconductors. $D_O(E)$ is represented by the Boltzmann distribution [Eq. (50)] with the standard enthalpy change for Eq. (80) as the neutralization energy of H_3O^+ in its ground level. $W(E)$ is given by the WKB approximation.[111] In their earlier publications,[93,108] they assumed the energy of excited electrons is conserved, i.e., hot electron injection, which was also recently proposed by Nozik,[112,113] but later they concluded that the electron transfer should take place at the energy of the bottom of the conduction band, i.e., all electrons are thermalized.[109] In this case, there is no need to know the energy distribution of electrons arriving at the surface, but the knowledge of the number of excited electrons in the conduction band arriving at the surface, N_e, is enough to calculate the photocurrent-potential expression. To obtain N_e, they have considered the generation and scattering of minority carriers instead of solving the continuity equation for excess minority carriers. Thus, the number of electrons excited between x and $x + dx$,

$N_e(x)\,dx$, is equal to the number of photons absorbed betweeen x and $x + dx$, $N_{ph}(x)\,dx$, which is given by

$$N_{ph}(x)\,dx = I_0(1 - R_\lambda)\alpha_\lambda \exp(-\alpha_\lambda x)\,dx \qquad (88)$$

where I_0 is the total number of photons of incident light and R_λ and α_λ are the reflectivity and the absorption coefficient of the semiconductor at the wavelength λ. The number of photoexcited electrons, originally expressed for $x = x$ in Eq. (88), decreases to $N_{e,x=x-\Delta x}(x)\,dx$ given by Eq. (89) after traveling Δx:

$$N_{e,x=x-\Delta x}(x)\,dx = N_{e,x}(x)\exp\left[-\frac{\Delta x}{L(x)}\right]dx \qquad (89)$$

where $L(x)$ is the mean free path of electrons at x and is given by[114]

$$L(x) = \frac{2L_D^2}{(L_E^2 + 4L_D^2)^{1/2} - L_E} \qquad (90)$$

where L_D and L_E are the diffusion and the drift length, respectively. Similarly,

$$\begin{aligned}N_{e,x=x-2\Delta x}(x)\,dx &= N_{e,x-\Delta x}(x)\exp\left[-\frac{\Delta x}{L(x-\Delta x)}\right]dx \\ &= N_e(x)\exp\left\{-\left[\frac{1}{L(x)} + \frac{1}{L(x-\Delta x)}\right]\Delta x\right\}dx\end{aligned} \qquad (91)$$

After N steps ($N = x/\Delta x$),

$$\begin{aligned}N_{e,x=0}(x)\,dx &= N_{e,x=\Delta x}\exp\left[-\frac{\Delta x}{L(\Delta x)}\right]dx \\ &= N_e(x)\exp\left\{-\left[\frac{1}{L(x)} + \frac{1}{L(x-\Delta x)}\right.\right. \\ &\qquad\left.\left.+ \cdots + \frac{1}{L(\Delta x)}\right]\Delta x\right\}dx\end{aligned} \qquad (92)$$

$N_{e,x=0}$ gives the number of electrons arriving at the surface per unit area per unit time which are excited between x and $x + dx$. Therefore, the total number of electrons arriving at the surface per unit time per unit area, N_e, is given by

$$N_e = \int_0^\infty N_{e,x=0}(x)\, dx \tag{93}$$

They also took into account the band edge movement due to applied external potential, and their final expression for the photocurrent, i_p, is given by

$$i_p = e\frac{c_O}{c_T} N_e \exp\left\{-\frac{\pi^2 l}{h}[2m(U_{\max} + e\Delta V_H)]^{1/2}\right\}$$
$$\times \exp\{-[\Delta H(e) + e\Delta V_H]/\gamma kT\} \tag{94}$$

where h is the Planck constant, m is the mass of an electron, l is the barrier width, and $U_{\max}$ is the maximum height of the barrier at the flat band potential. The shape of the calculated photocurrent-potential relations (Fig. 21) resembles experimental results but with small efficiencies.[107] Their treatment was quite different from the accepted view of the others at the time of their publication in many respects and was not well accepted. First, the electron transfer process is considered to be rate determining while others believed that the electrochemical reaction rate is so fast that it can be neglected in considering the rate equation. This disagreement was mainly due to the fact that most of the research workers at that time were concerned with photoelectrochemical oxygen evolution reaction (OER) at TiO_2, while Bockris and Uosaki studied HER at p-type semiconductors. The energy of the top of the valence band of TiO_2 is very low and the potential of photogenerated holes at a TiO_2 surface is very much positive compared to the reversible potential of OER, i.e., the overpotential is very high.[116] On the other hand, the energy of the bottom of the conduction band of most of the p-type semiconductors is not too high and the potential of photogenerated electrons at the surface of these semiconductors is not very negative compared to the reversible hydrogen electrode (RHE), i.e., low overpotential.[115] Second, Bockris and Uosaki thought that the band edge energy is not pinned while others believed that band edge pinning occurs, as discussed in Section III. The major deficiency of the treatment by Bockris and Uosaki is that they did not take into account the surface recombination rate, although they mentioned that the electrons which are not accepted by H^+ recombine at the surface with holes in the valence

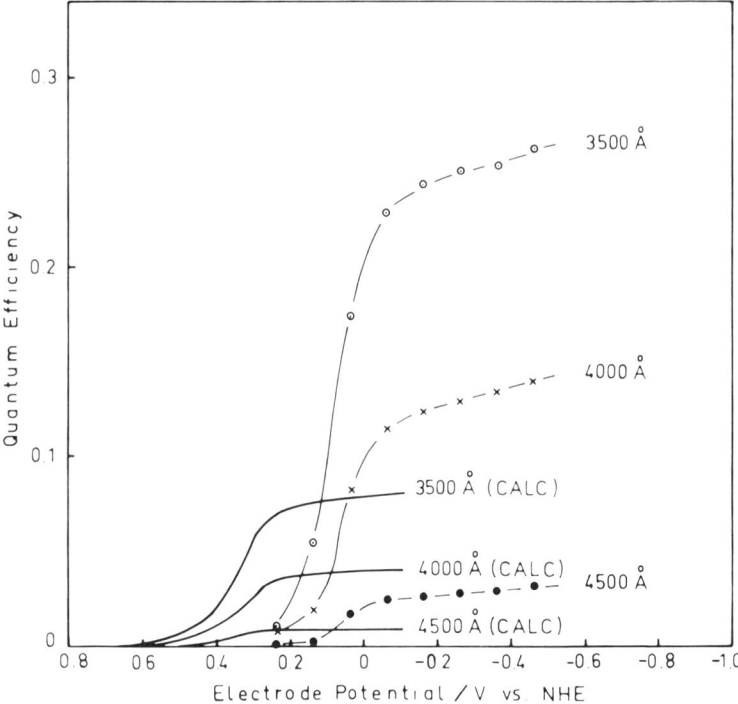

Figure 21. The experimental and calculated [Eq. (94)] quantum efficiency–potential relations of p-GaP in 1 M NaOH.[109]

band.[108,109] Also the calculated photocurrent values are very much affected by the standard enthalpy change for the step shown by Eq. (80) and the shape of tunneling barrier, both of which contained many uncertainties.[109]

3. Butler's Model—Semiconductor/Electrolyte Interface as a Schottky Barrier

A more widely accepted method to obtain the rate expression is to solve the following continuity equation with the appropriate initial and boundary conditions:

$$\frac{\partial p(x)}{\partial t} = g(x) + D\left[\frac{\partial^2 p(x)}{\partial x^2}\right] - \frac{p(x)}{\tau} \qquad (95)$$

where $p(x)$, $g(x)$, and τ are the concentration, the generation rate, and the effective lifetime of the minority carrier, respectively. In this treatment, the energetics of photogenerated electrons or holes were not considered and the charge transfer rate constant was taken as a parameter, if it was included at all. This approach was originally developed by Gärtner for semiconductor/metal junctions[117] where no complication due to slow charge transfer and the potential drop in Helmholtz layer are present. Laser and Bard[118-120] employed a digital simulation technique to solve the differential equation, but they could only obtain results for times very close to the onset of illumination because the minority carrier concentration varies over many orders of magnitude in a small region of space charge layer. Thus, the analytical solutions with some approximations have more often been found in the literature to describe the steady state photocurrent expression. Butler was the first to apply this approach to photoelectrochemical kinetics.[110] He followed Gärtner's treatment, which was developed for a semiconductor/metal junction (Schottky barrier). Butler neglected the charge transfer step with the assumption that the electrochemical reaction rate is very fast. All the applied potential is assumed to drop within the space charge layer. The generation of minority carrier at x and $x + dx$, $g(x)\,dx$, is given by

$$g(x)\,dx = I_0 \alpha_\lambda \exp(-\alpha_\lambda x)\,dx \tag{96}$$

This is similar to Eq. (88) except that the reflection term is neglected, i.e., I_0 represents the number of photons absorbed by the semiconductor. To solve Eq. (95), the semiconductor side is divided into two terms, namely, the space charge region (SCR) where all the potential drop occurs, and the field free region (FFR) (Fig. 20).†
All the minority carriers generated in the space charge region are assumed to reach the semiconductor surface and contribute to the photocurrent, i.e., no recombination in the space charge region and at surface. The photocurrent from this region, i_S, is, therefore, given by

$$i_S = e \int_0^W g(x)\,dx = -eI_0[\exp(\alpha W) - 1] \tag{97}$$

† In a real situation there is no sharp boundary between the space charge region and the field free region, since the electric field extends throughout the semiconductor.

where W is the space charge layer width, which is approximated as

$$W = W_0(V - V_{fb})^{1/2} \qquad (98)$$

with

$$W_0 = (2\varepsilon\varepsilon_0/eN)^{1/2} \qquad (99)$$

where N is the donor or the acceptor concentration of n- or p-type semiconductor, respectively. Here all the applied potential is assumed to drop within the space charge layer. Equation (95) is solved for the field free region with the following boundary conditions:

$$p = p_0 \quad \text{at } x = \infty \qquad (100)$$

and

$$p = 0 \quad \text{at } x = W \qquad (101)$$

where p_0 is the equilibrium concentration of the minority carrier. Since no recombination within the space charge region and at surface is assumed, all the minority carriers arriving at the space charge region from the field free region contribute to the photocurrent. Thus, the photocurrent from this region, i_F, is given by

$$i_F = eI_0 \frac{\alpha L_D}{1 + \alpha L_D} \exp(-\alpha W) + ep_0 \frac{D}{L_D} \qquad (102)$$

where L_D and D are the diffusion length and diffusion coefficient of the minority carrier, respectively. For the wide gap semiconductors, the last term can be neglected, and from Eqs. (97), (98), and (102) the total photocurrent, i_{ph}, is given by

$$i_{ph} = eI_0 \left\{ 1 - \frac{\exp[-\alpha W_0(V - V_{fb})]^{1/2}}{1 + \alpha L_D} \right\} \qquad (103)$$

If $\alpha L_D \ll 1$, the above can be expanded for the region $\alpha W_0(V - V_{fb})^{1/2} \ll 1$ as

$$V - V_{fb} \approx (i_{ph}/\alpha W_0 eI_0)^2 \qquad (104)$$

This relation suggests that a plot of i_{ph}^2 against V would give a linear relation with the intercept of the potential axis as V_{FB}. Near the band edge, α may be expressed as

$$\alpha = A\frac{(h\nu - E_g)^{n/2}}{h\nu} \quad (105)$$

where $h\nu$ is the photon energy, A is a constant, E_g is the band gap energy, and $n = 1$ for direct gap semiconductors and $n = 4$ for indirect gap semiconductors. Under the above assumptions, Eqs. (103) and (105) are combined to give

$$\frac{i_{ph}}{eI_0}h\nu = |L_D + W_0(V - V_{FB})^{1/2}|A(h\nu - E_g)^{n/2} \quad (106)$$

Thus, a plot of $(i_{ph}h\nu)^{2/n}$ against $h\nu$ should be linear with the intercept of the photon energy axis being E_g. These relations are quite often used to determine V_{FB} and E_g without detailed consideration of the assumption made to obtain the equations. Lemasson et al.[121-123] and Salvador[124-127] applied this simple model to determine the minority carrier diffusion length, but used only data obtained with a large bias.

4. Competition between Surface Recombination and Charge Transfer

Although Butler's model was successful in describing the photoelectrochemical kinetics at certain semiconductor electrodes, it is now clear that the assumptions made in his model are valid only for very limited cases.[128-130] The following experimental results suggest that the surface steps control the electrochemical reaction rate under illumination. They are (1) the Tafel-like behavior of the photocurrent-potential relations,[131-133] (2) the nonlinear relation between the photocurrent and the light intensity,[132-134] (3) the photocurrent enhancement by addition of an easily reactive compound,[131,132,135,136] and (4) the photocurrent enhancement by the modification of an electrode surface with metals or metal ions which catalyze the electrochemical reaction[137-146] or passivate the surface recombination.[137,147-150] Thus, only a part of photogenerated minority carriers arriving at the surface transfer to solution and contribute to the photocurrent. Thus, the effect of surface recombination rate and charge transfer rate on the rate expression should be considered.

(i) Wilson's Model

Wilson obtained the photocurrent expression by using the differential equation [Eq. (95)] for minority carriers photogenerated in the field free region.[15,151] He also assumed that all the minority carriers generated in the depletion region reach the semiconductor surface, no recombination within the space charge region and all the applied potential drops within the space charge region. He, however, employed more appropriate boundary conditions and took into account competition between surface recombination and charge transfer. The boundary conditions he used are

$$p(x) \quad \text{finite at } x \to \infty \tag{107}$$

$$p(x) \quad \text{continuous at any place} \tag{108}$$

and

$$-D\frac{dp(x)}{dx} = Sp(x) \quad \text{at } x = W \tag{109}$$

S is the parameter related to the two surface reactions, namely, the recombination of the photogenerated minority carriers with majority carriers through surface states and a charge transfer reaction with the electrolyte. The total minority carrier flux, J_T, to the surface is the sum of the minority carrier flux generated in the space charge region, J_S, plus the flux generated in the field free region, J_F. From Eq. (97), J_S is given by

$$J_S = I_0[1 - \exp(-\alpha W)] \tag{110}$$

By solving Eq. (95) with the boundary conditions given by Eqs. (107)-(109), one obtains

$$J_F = I_0 \frac{L_D}{L_D + D/S} \frac{\alpha L_D}{1 + \alpha L_D} \exp(-\alpha W) \tag{111}$$

This must be compared with Eq. (102), where no surface parameter is included.

Surface recombination is analyzed using Hall-Shockley-Read[152,153] recombination analysis. With the usual approximation, the recombination flux, J_R, is given by

$$J_R = S_R p_s = S_R p_0 \exp(e\Delta V_{sc}/kT) \tag{112}$$

where p_s and p_0 are the density of minority carriers at the surface and at the edge of the space charge region and S_R is the surface recombination parameter, which is given by

$$S_R \approx \int_{E_v}^{E_c} v\sigma_R N_R(E)\, dE \Big/ \left\{ 1 + \exp\left(\frac{e\Delta V_{sc}}{kT}\right) \exp\left(\frac{E - E_F}{kT}\right) \right.$$
$$\left. + R \exp\left(\frac{e\Delta V_{sc}}{kT}\right)\left[1 + \exp\left(-\frac{e\Delta V_{sc}}{kT}\right) \exp\left(\frac{E - E_F}{kT}\right) \right] \right\}$$
(113)

where R is the parameter proportional to the light intensity and was assumed to be zero in the calculation. $N_R(E)$ is the area density of recombination centers at the surface as a function of energy, v is the thermal velocity of minority carriers in the semiconductor, σ_R is the minority carrier capture cross section of the recombination centers, and E_F is the Fermi level at the surface when $\Delta V_{sc} = 0$, i.e., at the flat band potential.

Similarly, the charge transfer flux can be written as

$$J_{CT} = S_{CT} p_s = S_{CT} p_0 \exp(e\Delta V_{sc}) \quad (114)$$

where S_{CT} is the charge transfer parameter for electrons crossing the electrolyte–semiconductor interface and is given by

$$S_{CT} = \int_{E_v}^{E_c} v\sigma_{CT} N_{CT}(E)\, dE \quad (115)$$

where $N_{CT}(E)$ is the area density of reaction centers, σ_{CT} is their cross section for interaction with minority carriers at the surface, and v is the thermal velocity of minority carriers in the semiconductor. According to Wilson, N_{CT} is an intermediate state, the value of which is dependent on its rate of formation from the electrolyte. As far as the current is proportional to the light intensity,† the minority carrier production is the rate-limiting step. In that case S_{CT} can be considered to be independent of current. At high current levels, when the charge transfer rate to the semiconductor becomes comparable to the exchange rate between the intermediate states and the electrolyte, this would no longer be true. Thus, observations

† Wilson stated "it has been observed that the current is proportional to the light intensity"; but there are many examples where the current is independent of light intensity even when the light intensity is relatively weak.[132–134]

at high current densities could give information on this exchange rate.

In the steady state condition, the flux of minority carriers reaching the surface equals the recombination flux plus the charge transfer flux, i.e.,

$$J_T = J_R + J_{CT} = p_0 \exp(e\Delta V_{sc}/kT)(S_R + S_{CT}) \quad (116)$$

Furthermore, from Eqs. (109) and (111), the minority carrier flux at the edge of the depletion region is

$$J_F = I_0 \exp(-\alpha W) \frac{L_D}{L_D + D/S} \frac{\alpha L_D}{\alpha L_D + 1} = Sp_0 \quad (117)$$

from which one obtains

$$p_0 = I_0 \exp(-\alpha W) \frac{L_D}{SL_D + D} \frac{\alpha L_D}{\alpha L_D + 1} \quad (118)$$

From these, S is given by

$$S = \frac{(S_R + S_{CT}) \exp(e\Delta V_{sc}) - \dfrac{1 - \exp(-\alpha W)}{\exp(-\alpha W)[\alpha L_D/(\alpha L_D + 1)]} \dfrac{D}{L_D}}{1 + \dfrac{1 - \exp(-\alpha W)}{\exp(-\alpha W)[\alpha L_D/(\alpha L_D + 1)]}} \quad (119)$$

Finally, an expression for current through the cell, i_{ph}, is obtained by multiplying the total minority carrier flux to the surface by the fraction of the flux causing charge transfer across the electrode-electrolyte interface:

$$i_{ph} = \frac{J_{CT}}{J_{CT} + J_R} eJ_T \quad (120)$$

Thus, the quantum efficiency, η, is given by

$$\eta = \frac{J_{CT}}{J_{CT} + J_R} \frac{J_T}{I_0} \quad (121)$$

$$= \frac{S_{CT}}{S_{CT} + S_R} \left[1 - \exp(-\alpha W) \right.$$

$$\left. + \exp(-\alpha W) \frac{L_D}{L_D + D/S} \frac{\alpha L_D}{\alpha L_D + 1} \right]$$

Thus, Eq. (121) in conjunction with Eqs. (98), (99), (113), and (114) determines the voltage dependence of the normalized current, i.e., quantum efficiency.

The voltage dependence of Eq. (121) is implicit in the space charge region width, W, and in the parameters S, S_{CT} and S_R. The effect of several parameters on the calculated $i_{ph} - V$ relations and brief comparison with the experimental results are shown in Fig. 22. Wilson discussed the validity of the assumption he made, namely, the depletion approximation [Eqs. (98), (99)], the neglect of the field perturbation by photogenerated carriers, and the neglect of the recombination in the depletion region and showed that the model is a good approximation for $N > 10^{16}$ cm^{-3}, $eI_0 < 100$ mA cm^{-2}, and $\Delta V_{sc} <$ about one-half the band gap.

(ii) Other Models

Hall–Schokley–Read statistics were also employed by Reiss,[50] Guibaly et al.,[154-156] and McCann and Haneman[157] to treat surface recombination.

Reiss solved the equation of continuity for both minority and majority carriers and used a modified Butler–Volmer equation to describe the net nonequilibrium rate for charge transfer with exchange current densities of minority and majority carriers as

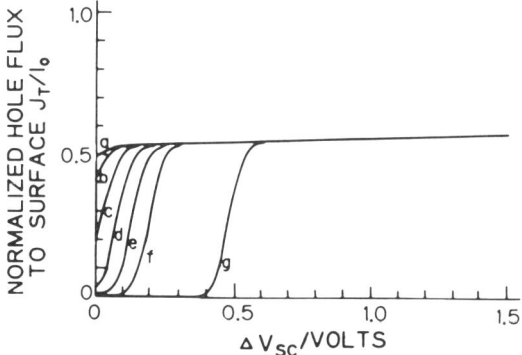

Figure 22. Calculated hole flux to the surface at illuminated n-type semiconductor.[151] The surface reaction parameters ($S = S_R + S_{CT}$) are (a) 10^4, (b) 10^3, (c) 10^2, (d) 10, (e) 1, (f) 0.1, and (g) 10^{-6} cm sec^{-1}. Other parameters used are $L_D = 10^{-4}$ cm, $\alpha = 10^4$ cm^{-1}, and $N = 10^{18}$ cm^{-3}.

parameters.[50] He neglected recombination but avoided the quasiequilibrium condition in the space charge region. Guibaly et al.[154-156] took into account not only charge transfer kinetics and surface recombination but also the recombination in the space charge region and the effect of separation between the quasi-Fermi level of minority carrier and that of majority carrier.

Less vigorous approaches to describe surface recombination were taken by Kelly and Memming,[158] Rajeshwar,[134] and Peter et al.[159,160] These authors considered the possibility of charge transfer via surface states. Thus, as shown in Fig. 20, the minority carriers which are trapped by surface states would either recombine with majority carriers (surface recombination via surface states) or transfer to the electrolyte (charge transfer via surface states or surface state mediated charge transfer). There are many experimental results which support the surface state mediated charge transfer under illumination[134,158-161] and in the dark.[162-164]

Kelly and Memming assumed that all the photogenerated minority carriers arrive at the surface and that the surface state occupancy is determined only by the capture and recombination of the excited minority carriers, and they describe the photocathodic reactions at p-type semiconductors by the following equations[158]:

$$\frac{dn_s}{dt} = J_n - k_n n_s N_t (1 - f) - k_c c_O n_s \qquad (122)$$

$$N_t \frac{df}{dt} = k_s n_s N_t (1 - f) - k_p p_s N_t f - k_{ss} c_O N_t f \qquad (123)$$

where n_s and p_s are the surface concentrations of electrons and holes, k_n and k_p are the rate constants for electron and hole capture by the surface state, k_c and k_s are the rate constant for the electron transfer from conduction band and from the surface states to the oxidized species in solution, the concentration of which is c_O, and N_t and f are the surface state density and the fraction of the surface states occupied by electrons. Then the Faraday flux via conduction band, J_c, and surface states, J_{ss}, are given by

$$J_c = k_c c_O n_s \qquad (124)$$

$$J_{ss} = k_{ss} c_O N_t f \qquad (125)$$

If the surface states communicate better with the semiconductor than with the electrolyte, i.e., $k_p p_s > k_c c_O$, they cause a loss of photogenerated carriers by surface recombination. On the other hand, if the states communicate better with the electrolyte than with the semiconductor, i.e., $k_p p_s < k_c c_O$, they do not contribute to recombination but mediate fast electron transfer. Kobayashi et al.[166] demonstrated that the surface state capacitance disappears in the former situation.

Rajeshwar[134] and Chazalviel,[165] who assumed the only pathway for charge transfer is via surface states, and Peter et al.[159] used similar models.

5. Effects of Recombination in Space Charge Region

So far only the recombination in the bulk and at the surface of the semiconductor are considered and the recombination in the space charge region is neglected. The assumption is valid only if the transit time of the photogenerated minority carriers across the depletion region is less than the minority carrier lifetime.[151,167]

The method of Sah, Noyce, and Shockley[168] is often employed to treat the effect of recombination in the space charge region on the photocurrent-potential characteristics.[154-156,169] According to Reichman, the space charge region recombination flux, J_s^R, is given by[169]

$$J_s^R = \left(\frac{p_s}{p_s^0}\right)^{1/2} \frac{\pi k T n_i W \exp[(\Delta V_{sc}^0 - \Delta V_{sc})/2kT]}{4\tau \Delta V_{sc}} \quad (126)$$

where p_s and p_s^0 are the surface concentration of the minority carrier and that at equilibrium, respectively, ΔV_{sc}^0 is the potential drop in the semiconductor at equilibrium, and n_i is the intrinsic electron density, which is given by

$$p_0 = n_i^2/N \quad (127)$$

Reichman calculated the current-potential characteristics based on the above consideration. The results are shown in Fig. 23. In the calculation, the current due to majority carrier is also taken into account but surface recombination is neglected. Horowitz applied Reichman's treatment to show that a discrepancy observed at p-GaP

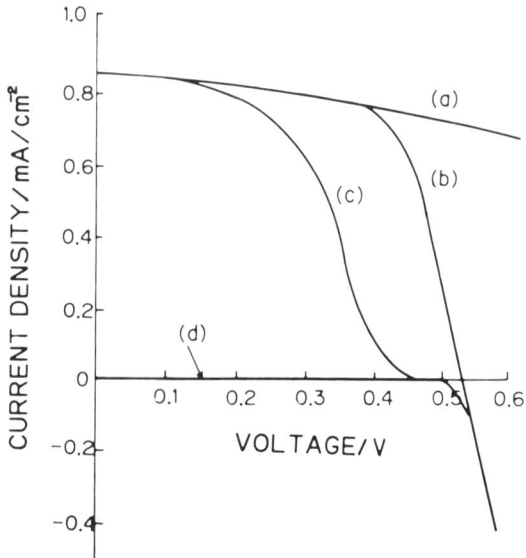

Figure 23. Comparison of calculated current-voltage characteristics of n-type semiconductor-electrolyte junction device under illumination of 1 mA cm^{-2} (equivalent) photon flux.[169] Parameters used are $L_D = 0.5 \times 10^{-4}$ cm, $N = 10^{16}$ cm^{-3}, $\alpha = 3 \times 10^4$ cm^{-1}, $n_i = 10^7$ cm^{-3}, $\varepsilon = 12$, $\Delta V_{sc}^0 = 0.7$ V, $\tau = 10^{-9}$ sec, and the electron and hole exchange current parameters are 10^{-10} mA cm^{-2} and 10^{-5} mA cm^{-2}, respectively. (a) Gärtner's (Butler's) model [Eq. (103)]. (b) Reichman's model considering both electron and hole currents but neglecting the space charge recombination. (c) Reichman's model with the space charge recombination. (d) As (b) but in dark.

between the photocurrent onset potential and the flat band potential determined by a Mott-Schottky plot is due to slow charge transfer across the semiconductor/electrolyte interface and the recombination in the space change region.[105] The discrepancy had been attributed previously by Butler and Ginley to a bulk recombination center.[170] Thus, the recombination in the space charge region can be a quite significant loss mechanism, but it is difficult to justify the neglect of surface recombination.

Khan and Bockris[167] presented a model which accounts for charge transfer kinetics and surface recombination as well as potential drop in the Helmholtz layer. In their paper, they estimated the transit time across the space charge region and the lifetime of the minority carrier and showed that it is reasonable to neglect the recombination in the space charge region if the mobility of the minority carrier is >10 cm² sec⁻¹ V⁻¹. For most of the n-type oxide semiconductors, this requirement is not met and space charge recombination cannot be neglected.

Albery et al.[171] took a quite different approach to treat the space charge recombination. It was assumed that the minority carriers recombine in both bulk and depletion region by first-order kinetics with a rate constant k. This may be true in the field free region, but in the space charge region, traps can be filled or emptied by the field and so the order of kinetics may change. The concentration of majority carriers will be less near the surface because of the band bending. This leads to less recombination. Thus, their treatment should be restricted to the cases of low illumination intensities and small band bending. In their later publication,[172] they further considered the space charge region recombination by assuming the recombination process to be the same as that of Shockely and Read[153] but using the notation of chemical kinetics.

6. Effects of Grain Boundary Recombination

If one considers the practical photoelectrochemical energy conversion devices, it is essential to develop efficient polycrystalline semiconductor electrodes.[173] In polycrystalline semiconductors, there are grain boundaries at which recombination is more effective than in the bulk. These boundaries give rise to space charge regions on both sides which extend into the adjoining crystallites.[174] Recombination at grain boundaries in solid state solar cells has been considered by many workers.[175-177] The method proposed by Ghosh et al. for Si MIS and heterostructure cells[177] was applied to photoelectrochemical cells by McCann and Haneman.[157] Since usually the average minority carrier lifetime due to grain boundary recombination, τ_p^{gb}, is much shorter than that in other regions, the effective average excess minority carrier lifetime, τ_p, is given by

$$\tau_p \approx \tau_p^{gb} \qquad (128)$$

and τ_p^{gb} can be approximated as

$$\tau_p^{gb} = \frac{1}{\sigma_p^{gb} s_p^{gb} \rho_p^{gb}} \qquad (129)$$

where σ_p^{gb} is the mean capture cross section of minority carriers at grain boundary, s_p^{gb} is the mean velocity of minority carrier recombination at a grain boundary, and ρ_p^{gb} is the equivalent density of minority carrier centers in the grain. For a cubic grain,

$$\rho_p^{gb} = \frac{A_s^g}{V^g} N_R^{gb} \tag{130}$$

where A_s^g and V^g are the surface area and volume, respectively, of a grain of average dimension and N_R^{gb} is an effective area density of grain boundary recombination centers. The effects of grain size on the open circuit voltage and the short circuit current for n-GaAs/electrolyte/counter electrode cell are shown in Fig. 24.[157]

VI. CONCLUDING REMARKS

It is now clear that when one considers the photoelectrochemical kinetics at semiconductor/electrolyte interface, at least light absorp-

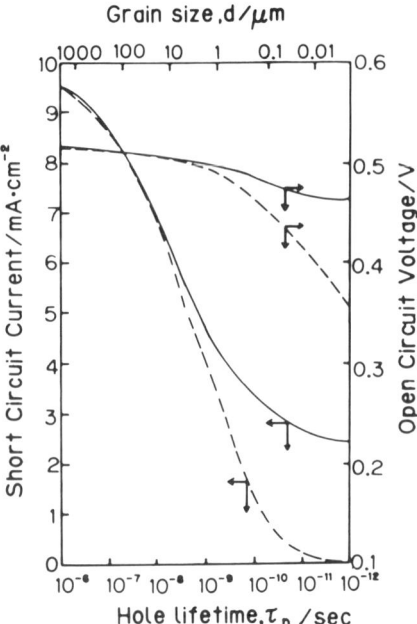

Figure 24. The effects of hole lifetime and grain size on open circuit voltage and short circuit current.[157] Solid curves are without and dashed curves are with space charge region recombination. The approximate relation $\tau_p = 5.21 \times 10^{-6} d$ [177] was assumed to simulate grain boundary recombination effects.

tion, generation of electron-hole pairs, recombination in bulk, space charge region, and at surface and charge transfer steps must be taken into account. Steps within the semiconductor have been treated rigorously but more attention should be given to those at the interface, which are quite different from those at semiconductor/metal interface, The role of surface states must be quantitatively understood—i.e., under what conditions do they act as recombination centers or as mediators for electron transfer? The effect of the surface states on the potential distribution at the semiconductor/electrolyte interface, particularly under illumination,[178] should be investigated. The charge transfer rate constant is in most cases treated as a parameter, but to design a better electrode surface for the particular charge transfer reaction, a quantitative expression which describes the rate constant in terms of the properties of semiconductor and electrolyte is needed.

ACKNOWLEDGMENTS

We would like to thank Professor J. O'M. Bockris for stimulating discussions. The correspondence with Professor S. Trasatti was very helpful to clarify the problem presented in Section II. We are also thankful to Miss M. Yoshida for typing this manuscript in a most expert and efficient manner.

REFERENCES

[1] E. Becquerel, *C. R. Acad. Sci.* **9** (1839) 561.
[2] W. H. Brattain and G. G. B. Garett, *Bell Sys. Tech. J.* **34** (1955) 129.
[3] M. Green, in *Modern Aspects of Electrochemistry*, No. 2, Ed. by J. O'M. Bockris, Butterworths, London, 1959, pp. 343-407.
[4] J. F. Dewald, in *Semiconductors, ACS Monograph*, No. 140, Ed. by N. B. Hannay, Reinhold, New York, 1959, pp. 727-752.
[5] H. Gerischer, in *Advances in Electrochemistry and Electrochemical Engineering*, Vol. 1, Ed. by P. Delahay, Interscience, New York, 1961, pp. 139-232.
[6] P. J. Holmes, Ed., *The Electrochemistry of Semiconductors*, Academic, London, 1962.
[7] V. A. Myamlin and Yu. V. Pleskov, *Electrochemistry of Semiconductors*, Plenum Press, New York, 1967.
[8] H. Gerischer, in *Physical Chemistry: An Advanced Treatise*, Vol. IXA, Ed. by H. Eyring, Academic, New York, 1970, Chap. 5.

[9] S. R. Morrison, *Prog. Surf. Sci.* **1** (1971) 105.
[10] Yu. V. Pleskov, *Prog. Surf. Membrane Sci.* **7** (1973) 57.
[11] A. Fujishima and K. Honka, *Nature* **238** (1972) 37.
[12] B. Parkinson, *Acc. Chem. Res.* **17** (1984) 431 and references therein.
[13] M. Grätzel, in *Modern Aspects of Electrochemistry*, No. 15, Ed. by R. E. White, J. O'M. Bockris, and B. E. Conway, Plenum Press, New York, 1983, pp. 83-165.
[14] K. Rajeshwar, P. Singh, and J. DuBow, *Electrochim. Acta* **23** (1978) 1117.
[15] R. H. Wilson, *CRC Crit. Rev. Solid State Mater. Sci.* **10** (1980) 1.
[16] F. Williams and A. J. Nozik, *Nature* **312** (1984) 21.
[17] A. Sommerfeld and H. Bethe, *Electronentheorie der Metalle*, Springer, Berlin, 1967.
[18] S. Trasatti, in *Comprehensive Treatise of Electrochemistry*, Vol. 1, Ed. by J. O'M. Bockris, B. E. Conway, and E. Yeager, Plenum Press, New York, 1980, pp. 45-81.
[19] C. Kittel, *Elementary Solid State Physics*, Wiley, New York, 1962.
[20] A. G. Milnes and D. L. Feucht, *Heterojunctions and Metal-Semiconductor Junctions*, Academic, New York, 1972.
[21] J. O'M. Bockris and A. K. N. Reddy, *Modern Electrochemistry*, Vol. 2, Plenum Press, New York, 1970, Chap. 7.
[22] S. Trasatti, *J. Electroanal. Chem.* **139** (1982) 1.
[23] H. Gerischer, *Appl. Phys. Lett.* **45** (1984) 913.
[24] H. Gerischer, *Z. Phys. Chem. N. F.* **26** (1960) 223.
[25] H. Gerischer, *Z. Phys. Chem. N. F.* **26** (1960) 325.
[26] H. Gerischer, *Z. Phys. Chem. N. F.* **27** (1960) 48.
[27] S. U. M. Khan and J. O'M. Bockris, in *Photoelectrochemistry: Fundamental Processes and Measurement Techniques*, Ed. by W. L. Wallace, A. J. Nozki, S. K. Deb, and R. H. Wilson, The Electrochemical Society, New Jersey, 1982, pp. 39-60.
[28] J. O'M. Bockris and S. U. M. Khan, *Appl. Phys. Lett.* **42** (1983) 124.
[29] S. U. M. Khan and J. O'M. Bockris, *J. Phys. Chem.* **87** (1983) 2599.
[30] J. O'M. Bockris and S. U. M. Khan, *Appl. Phys. Lett.* **45** (1984) 913.
[31] S. U. M. Khan and J. O'M. Bockris, *J. Phys. Chem.* **89** (1985) 555.
[32] W. P. Gomes and F. Cardon, *Prog. Surf. Sci.* **12** (1982) 155.
[33] J. O'M. Bockris, *Energy Convers.* **10** (1970) 41.
[34] J. O'M. Bockris, *J. Electroanal. Chem.* **36** (1972) 495.
[35] E. Gileadi and G. Stoner, *J. Electroanal. Chem.* **36** (1972) 492.
[36] B. Jakuszewski, *Bull. Soc. Sci. Lett. Lodz, Cl. III* **8** (1957) 1.
[37] B. V. Ershler, *Usp. Khim.* **21** (1952) 237.
[38] J. O'M. Bockris and S. D. Argade, *J. Chem. Phys.* **49** (1968) 5133.
[39] F. Lohmann, *Z. Naturforsch.* **22a** (1967) 843.
[40] R. M. Noyes, *J. Am. Chem. Soc.* **84** (1962) 513.
[41] S. Trasatti, *The Absolute Electrode Potential: An Explanatory Note*, IUPAC Commission I. 3 (Electrochemistry), 1984.
[42] S. Trasatti, in *Advances in Electrochemistry and Electrochemical Engineering*, Vol. 10, Ed. by H. Gerischer and C. W. Tobias, Wiley, New York, 1977, pp. 213-321.
[43] S. Trasatti, *J. Electroanal. Chem.* **52** (1974) 313.
[44] A. Frumkin and B. Damaskin, *J. Electroanal. Chem.* **79** (1977) 259.
[45] J. Randles, *Trans. Faraday Soc.* **52** (1956) 1573.
[46] D. C. Grahame, E. M. Coffin, J. I. Cammings, and M. A. Poth, *J. Am. Chem. Soc.* **74** (1952) 1207.
[47] J. R. Farrell and P. McTigue, *J. Electroanal. Chem.* **139** (1982) 37.
[48] *JANAF Thermodynamical Tables*, 2nd edition, 1971; *JANAF Thermodynamical Tables, 1975, Supplement*.
[49] L. Benjamin and V. Gold, *Trans. Faraday Soc.* **50** (1954) 797.
[50] H. Reiss, *J. Electrochem. Soc.* **125** (1978) 937.

[51] R. Gomer and G. Tryson, *J. Chem. Phys.* **66** (1977) 4413.
[52] W. N. Hansen and D. M. Kolb, *J. Electroanal. Chem.* **100** (1979) 493.
[53] O. Hofmann, K. Doblhofer, and H. Gerischer, *J. Electroanal. Chem.* **161** (1984) 337.
[54] H. Gerischer and W. Ekardt, *Appl. Phys. Lett.* **43** (1983) 393.
[55] D. Scherson, W. Ekardt, and H. Gerischer, *J. Phys. Chem.* **89** (1985) 555.
[56] J. Bardeen, *Phys. Rev.* **71** (1947) 717.
[57] J. I. Pankove, *Optical Processes in Semiconductors*, Prentice-Hall, Englewood Cliffs, New Jersey, 1971, Chap. 14.
[58] C. A. Mead, *Solid State Electron.* **9** (1966) 1023.
[59] D. C. Geppert, A. M. Cowley, and B. V. Dore, *J. Appl. Phys.* **37** (1966) 2458.
[60] W. E. Spicer, P. W. Chye, P. R. Skeath, C. Y. Su, and I. Lindau, *J. Vac. Sci. Technol.* **16** (1979) 1422.
[61] S. R. Morrison, *Electrochemistry at Semiconductor and Oxidized Metal Electrodes*, Plenum Press, New York, 1980, Chap. 1.
[62] S. Fonash, M. Rose, K. Corby, and J. Jordan, in *Photoelectrochemistry: Fundamental Processes and Measurement Techniques*, Ed. by W. L. Wallace, A. J. Nozik, S. K. Deb, and R. H. Wilson, The Electrochemical Society, New Jersey, 1982, pp. 14-38.
[63] C. M. Gronet, N. S. Lewis, G. W. Cogan, J. F. Gibbons, and G. R. Moddel, *J. Electrochem. Soc.* **131** (1984) 2873.
[64] P. A. Kohl and A. J. Bard, *J. Electrochem. Soc.* **126** (1979) 59.
[65] A. B. Bocarsly, E. G. Walton, M. G. Bradley, and M. S. Wrighton, *J. Electroanal. Chem.* **100** (1979) 283.
[66] F.-R. Fan and A. J. Bard, *J. Am. Chem. Soc.* **102** (1980) 3677.
[67] A. B. Bocarsly, D. C. Bookbinder, R. N. Diminey, N. S. Lewis, and M. S. Wrighton, *J. Am. Chem. Soc.* **102** (1980) 3683.
[68] F.-R. Fan, H. S. White, B. Wheeler, and A. J. Bard, *J. Am. Chem. Soc.* **102** (1980) 5142.
[69] R. E. Malpas, K. Itaya, and A. J. Bard, *J. Am. Chem. Soc.* **103** (1981) 1622.
[70] S. Tanaka, J. A. Bruce, and M. S. Wrighton, *J. Phys. Chem.* **85** (1981) 3778.
[71] A. J. Bard, A. B. Bocarsly, F.-R. Fan, E. G. Walton, and M. S. Wrighton, *J. Am. Chem. Soc.* **102** (1980) 3671.
[72] R. H. Kingston and S. F. Neustadter, *J. Appl. Phys.* **26** (1955) 718.
[73] H. Gerischer, *Faraday Dis. Chem. Soc.* **70** (1980) 98.
[74] A. J. Nozik, *Faraday Dis. Chem. Soc.* **70** (1980) 102.
[75] D. J. Benard and P. Handler, *Surf. Sci.* **40** (1973) 141.
[76] W. Kautek and H. Gerischer, *Ber. Bunsenges, Phys. Chem.* **84** (1980) 645.
[77] C. D. Jaeger, H. Gerischer, and W. Kautek, *Ber. Bunsenges. Phys. Chem.* **86** (1982) 20.
[78] K. Uosaki and H. Kita, *J. Phys. Chem.* **88** (1984) 4197.
[79] K. Uosaki and H. Kita, *J. Electrochem. Soc.* **130** (1983) 895.
[80] H. Gerischer, *J. Electrochem. Soc.* **131** (1984) 2452.
[81] R. De Gryse, W. P. Gomes, and F. Cardon, *J. Electrochem. Soc.* **122** (1975) 711.
[82] J. A. Turner and A. J. Nozik, *Appl. Phys. Lett.* **41** (1982) 101.
[83] M. Green, *J. Chem. Phys.* **31** (1959) 200.
[84] J. O'M. Bockris and S. U. M. Khan, *Quantum Electrochemistry*, Plenum Press, New York, 1979, Chap. 8.
[85] R. Memming, in *Comprehensive Treatise of Electrochemistry*, Vol. 7, Ed. by B. E. Conway, J. O'M. Bockris, E. Yeager, S. U. M. Khan, and R. E. White, Plenum Press, New York, 1983, pp. 529-592.
[86] R. W. Gurney, *Proc. R. Soc. London* **A134** (1932) 137.

[87] C. Kittel, *Introduction to Solid State Physics*, 5th Edition, John Wiley, New York, 1976, Chap. 8.
[88] S. U. M. Khan and J. O'M. Bockris, in *Comprehensive Treatise of Electrochemistry*, Vol. 7, Ed. by B. E. Conway, J. O'M. Bockris, E. Yeager, S. U. M. Khan, and R. E. White, Plenum Press, New York, 1983, pp. 44-86.
[89] J. O'M. Bockris and D. B. Matthews, *Proc. Soc. London* **A292** (1966) 479.
[90] D. B. Matthews and J. O'M. Bockris, *J. Chem. Phys.* **48** (1968) 1989.
[91] J. O'M. Bockris and A. K. N. Reddy, *Modern Electrochemistry*, Vol. 2, Plenum Press, New York, 1970, Chap. 8.
[92] D. B. Matthews and J. O'M. Bockris, in *Modern Aspects of Electrochemistry*, No. 6, Ed. by J. O'M. Bockris and B. E. Conway, Plenum Press, New York, 1971, Chap. 4.
[93] J. O'M. Bockris and K. Uosaki, in *Solid State Chemistry of Energy Conversion and Storage, Advances in Chemistry Series*, No. 163, Ed. by J. B. Goodenough and M. S. Whittingham, American Chemical Society, Washington, D.C., 1977, pp. 33-70.
[94] J. O'M. Bockris and S. U. M. Khan, *Quantum Electrochemistry*, Plenum Press, New York, 1979, Chap. 13.
[95] S. U. M. Khan and J. O'M. Bockris, in *Modern Aspects of Electrochemistry*, No. 14, Ed. by J. O'M. Bockris, B. E. Conway, and R. E. White, Plenum Press, New York, 1982, pp. 151-193.
[96] S. U. M. Khan, in *Modern Aspects of Electrochemistry*, No. 15, Ed. by R. E. White, J. O'M. Bockris, and B. E. Conway, Plenum Press, New York, 1983, pp. 305-350.
[97] R. A. Marcus, *J. Chem. Phys.* **24** (1956) 966.
[98] R. A. Marcus, *Can. J. Chem.* **37** (1959) 155.
[99] R. A. Marcus, *J. Chem. Phys.* **43** (1965) 43.
[100] R. R. Dogonadze, A. M. Kuznetsov, and Yu. A. Chizmadzev, *Russ. J. Phys. Chem.* **38** (1964) 652.
[101] R. R. Dogonadze and A. M. Kuznetsov, *J. Electroanal. Chem.* **65** (1975) 545.
[102] S. G. Christov, *Ber. Bunsenges, Phys. Chem.* **79** (1975) 357.
[103] V. G. Levich, in *Advances in Electrochemistry and Electrochemical Engineering*, Vol. 4, Ed. by P. Delehay, Interscience Publishers, New York, 1966, Chap. 5.
[104] V. G. Levich, in *Physical Chemistry: An Advanced Treatise in Physical Chemistry*, Vol. IX 13, Ed. by H. Eyring, Academic, New York, 1970, Chap. 12.
[105] G. Horowitz, *Appl. Phys. Lett.* **40** (1982) 409.
[106] S. Nakabayashi, A. Fujishima, and K. Honda, *J. Phys. Chem.* **87** (1983) 3487.
[107] S. Nakabayashi, K. Itoh, A. Fujishima, and K. Honda, *J. Phys. Chem.* **87** (1983) 5301.
[108] J. O'M. Bockris and K. Uosaki, *Int. J. Hydrogen Energy* **2** (1977) 123.
[109] J. O'M. Bockris and K. Uosaki, *J. Electrochem. Soc.* **125** (1978) 223.
[110] M. A. Butler, *J. Appl. Phys.* **48** (1977) 1914.
[111] L. Pauling and E. B. Wilson, Jr., *Introduction to Quantum Mechanics with Applications to Chemistry*, McGraw-Hill, New York, 1935, Chap. 7.
[112] R. Ross and A. J. Nozik, *J. Appl. Phys.* **53** (1982) 3813.
[113] G. Cooper, J. A. Turner, B. A. Parkinson, and A. J. Nozik, *J. Appl. Phys.* **54** (1983) 6463.
[114] S. M. Ryvkin, *Photoelectric Effects in Semiconductors*, English Edition, Consultants Bureau, New York, 1964, Chap. XIII.
[115] J. O'M. Bockris and K. Uosaki, *J. Electrochem. Soc.* **124** (1977) 1348.
[116] H. Gerischer, in *Semiconductor Liquid-Junction Solar Cells*, Ed. by A. Heller, The Electrochemical Society, New Jersey, 1977, pp. 1-19.

Theoretical Aspects of Semiconductor Electrochemistry

[117] W. W. Gärtner, *Phys. Rev.* **116** (1959) 84.
[118] D. Laser and A. J. Bard, *J. Electrochem. Soc.* **123** (1976) 1828.
[119] D. Laser and A. J. Bard, *J. Electrochem. Soc.* **123** (1976) 1833.
[120] D. Laser and A. J. Bard, *J. Electrochem. Soc.* **123** (1976) 1837.
[121] P. Lemasson, A. Etcheberry, and J. Gautron, *Electrochim. Acta* **27** (1982) 607.
[122] A. Etcheberry, M. Etman, B. Fotouhi, J. Gautron, J. L. Sculfort, and P. Lemasson, *J. Appl. Phys.* **53** (1982) 8867.
[123] J. L. Sculfort, R. Triboulet, and P. Lemasson, *J. Electrochem. Soc.* **131** (1984) 209.
[124] P. Salvador, *Solar Energy Mater.* **2** (1980) 413.
[125] P. Salvador, *Mater. Res. Bull.* **15** (1980) 413.
[126] P. Salvador, *Solar Energy Mater.* **6** (1982) 241.
[127] P. Salvador, *J. Appl. Phys.* **55** (1984) 2977.
[128] J. O'M. Bockris, *Faraday Dis. Chem. Soc.* **70** (1980) 112.
[129] M. A. Butler, *Faraday Dis. Chem. Soc.* **70** (1980) 116.
[130] K. Uosaki and H. Kita, in *Photoelectrochemistry: Fundamental Processes and Measurement Techniques*, Ed. by W. L. Wallace, A. J. Nozik, S. K. Deb, and R. H. Wilson, The Electrochemical Society, New Jersey, 1982, pp. 392-400.
[131] K. Uosaki and H. Kita, *J. Electrochem. Soc.* **128** (1981) 2153.
[132] K. Uosaki and H. Kita, *Solar Energy Mater.* **7** (1983) 421.
[133] K. Uosaki, S. Kaneko, H. Kita, and A. Chevy, *Bull. Chem. Soc. Jpn.* **59** (1986) 599.
[134] K. Rajeshwar, *J. Electrochem. Soc.* **129** (1982) 1003.
[135] B. Reichmann, F.-R. F. Fan, and A. J. Bard, *J. Electrochem. Soc.* **127** (1980) 333.
[136] J. O'M. Bockris, K. Uosaki, and H. Kita, *J. Appl. Phys.* **52** (1981) 808.
[137] Y. Nakato, S. Tonomura, and H. Tsubomura, *Ber. Bunsenges. Phys. Chem.* **80** (1976) 1289.
[138] W. Kautek, J. Gobrecht, and H. Gerischer, *Ber. Bunsenges. Phys. Chem.* **84** (1980) 1034.
[139] A. Heller and R. G. Vadimsky, *Phys. Rev. Lett.* **46** (1981) 1153.
[140] A. Heller, *Acc. Chem. Res.* **14** (1981) 154.
[141] R. N. Dominey, N. S. Lewis, J. A. Bruce, D. C. Bookbinder, and M. S. Wrighton, *J. Am. Chem. Soc.* **104** (1982) 467.
[142] F.-R. F. Fan, R. G. Keil, and A. J. Bard, *J. Am. Chem. Soc.* **105** (1983) 220.
[143] M. A. Butler and D. S. Ginley, *Appl. Phys. Lett.* **42** (1983) 582.
[144] K. Uosaki and H. Kita, *Chem. Lett.* (1984) 301.
[145] M. Szklarczyk and J. O'M. Bockris, *J. Phys. Chem.* **88** (1984) 5241.
[146] C. R. Cabrera and H. D. Abruña, *J. Phys. Chem.* **89** (1985) 1279.
[147] B. A. Parkinson, A. Heller, and B. Miller, *Appl. Phys. Lett.* **33** (1978) 521.
[148] B. A. Parkinson, A. Heller, and B. Miller, *J. Electrochem. Soc.* **126** (1979) 954.
[149] R. J. Nelson, J. S. Williams, H. J. Leamy, B. Miller, H. J. Casey, Jr., B. A. Parkinson, and A. Heller, *Appl. Phys. Lett.* **36** (1980) 76.
[150] M. P. Dare-Edwards, A. Hamnett, and J. B. Goodenough, *J. Electroanal. Chem.* **119** (1981) 109.
[151] R. H. Wilson, *J. Appl. Phys.* **48** (1977) 4297.
[152] R. N. Hall, *Phys. Rev.* **87** (1952) 387.
[153] W. Schokley and W. T. Read, *Phys. Rev.* **87** (1952) 835.
[154] F. E. Guibaly, K. Colbow, and B. L. Funt, *J. Appl. Phys.* **52** (1981) 3480.
[155] F. E. Guibaly and K. Colbow, *J. Appl. Phys.* **53** (1982) 1737.
[156] F. E. Guibaly and K. Colbow, *J. Appl. Phys.* **54** (1983) 6488.
[157] J. F. McCann and D. Haneman, *J. Electrochem. Soc.* **129** (1982) 1134.
[158] J. J. Kelly and R. Memming, *J. Electrochem. Soc.* **129** (1982) 730.
[159] L. M. Peter, J. Li, and R. Peat, *J. Electroanal. Chem.* **165** (1984) 29.
[160] J. Li, R. Peat, and L. M. Peter, *J. Electroanal. Chem.* **165** (1984) 41.

[161] R. Haak, D. Tench, and M. Russak, *J. Electrochem. Soc.* **131** (1984) 2709.
[162] J. Vandermolen, W. P. Gomes, and F. Cardon, *J. Electrochem. Soc.* **127** (1980) 324.
[163] P. Salvador and C. Gutierrez, *Surf. Sci.* **124** (1983) 398.
[164] P. Salvador and C. Gutierrez, *J. Electrochem. Soc.* **131** (1984) 326.
[165] J.-N. Chazalviel, *J. Electrochem. Soc.* **129** (1982) 963.
[166] K. Kobayashi, M. Takata, S. Okamoto, and M. Sukigara, *J. Electroanal. Chem.* **185** (1985) 47.
[167] S. U. M. Khan and J. O'M. Bockris, *J. Phys. Chem.* **88** (1984) 2504.
[168] C. T. Sah, R. N. Noyce, and W. Shockley, *Proc. IRE* **45** (1957) 1228.
[169] J. Reichman, *Appl. Phys. Lett.* **36** (1980) 574.
[170] M. A. Butler and D. S. Ginley, *J. Electrochem. Soc.* **127** (1980) 1273.
[171] W. J. Albery, P. N. Bartlett, A. Hamnett, and M. P. Dare-Edwards, *J. Electrochem. Soc.* **128** (1981) 1492.
[172] W. J. Albery and P. N. Bartlett, *J. Electrochem. Soc.* **130** (1983) 1699.
[173] G. Fulop, M. Doty, P. Meyers, J. Betz, and C. H. Liu, *Appl. Phys. Lett.* **40** (1982) 327.
[174] J. Seto, *J. Appl. Phys.* **46** (1975) 5247.
[175] H. C. Card and E. S. Yang, *IEEE Trans. Electron Devices* **ED-24** (1977) 397.
[176] L. L. Kazmerski, *Solid State Electron.* **21** (1978) 1545.
[177] A. K. Ghosh, C. Fishman, and T. Feng, *J. Appl. Phys.* **51** (1980) 446.
[178] P. Alongue and H. Cachet, *J. Electrochem. Soc.* **132** (1985) 45.

2

Photoelectrochemical Devices for Solar Energy Conversion

Mark E. Orazem

Department of Chemical Engineering, University of Virginia, Charlottesville, Virginia 22901

John Newman

Materials and Molecular Research Division, Lawrence Berkeley Laboratory, and Department of Chemical Engineering, University of California, Berkeley, California 94720

Photoelectrochemical cells are distinguished by the use of a semiconductor-electrolyte interface to create the necessary junction for use as a photovoltaic device.[1-8] This chapter presents a description of this device from an electrochemical engineering viewpoint. The traditional chemical engineering fundamentals of transport phenomena, reaction kinetics, thermodynamics, and system design provide a useful foundation for the study of semiconducting devices. The motivation for the study of photoelectrochemical cells is discussed, and a physical description of the cell features is presented. A tutorial on the mechanism of cell operation is presented which includes descriptions of the phenomena of band-bending and straightening, the effect of interfacial phenomena, and current flow. Mathematical relationships are developed which describe this system, and the influence of cell design is discussed.

Many review papers, including recent chapters in this series, cover the physics and chemistry of photoelectrochemical cells (see, e.g., Refs. 9-19). The reader is referred to these for a historical perspective of the development of the field of photoelectrochemistry as applied to photovoltaic devices. An overview of the analytic and numerical models that have been developed for photoelectrochemical cells is presented in Section III.2. This chapter is further distinguished by an emphasis on quantitative design and optimization of large-scale photoelectrochemical cells and by a mathematical description that accounts for the influence of the nonideal behavior associated with large electron and hole concentrations in the semiconductor. This review also provides a contrast between a physical description of electrons in terms of energies and statistical distributions and a chemical description in terms of concentrations and activity coefficients.

I. SEMICONDUCTOR ELECTRODES

Semiconductors are characterized by the difference in energy between valence and conduction-band electrons. Electrons can be transferred from the valence band to the conduction band by absorption of a photon with energy greater than or equal to the transition or band-gap energy. When the electron moves into the higher energy level, it leaves behind a vacancy in the valence band, or *hole*. Both the negatively charged electrons and the positively charged holes are mobile and can serve as charge carriers (see, e.g., Refs. 20-24). In the presence of a potential gradient (or electric field), electrons and holes tend to migrate in opposite directions and can result in a net flow of electrical current. In the absence of a potential gradient, electron–hole pairs produced by illumination recombine with no net flow of electrical current. Photovoltaic devices therefore require an equilibrium potential gradient in the illuminated region of the semiconductor.

A potential gradient can be created by forming an interface or junction with a semiconducting material. Metal–semiconductor, p–n semiconductor, and semiconductor–electrolyte interfaces have been used in the construction of photovoltaic cells.[25-27] The interface in a p–n junction photovoltaic cell can be constructed by doping the surface of an n or p-type semiconductor with atoms that invert the semiconductor type. These atoms are then thermally diffused into the host semiconductor to an optimal depth. Diffusion rates in grain boundaries greatly exceed those in the bulk crystal; thus the need for a distinct boundary limits this technique to single-crystal host semiconductors. The junction between an electrolyte and a semiconductor, in contrast to the thermally diffused p-n junction, is formed spontaneously when the semiconductor is immersed in the electrolyte. The doping and diffusion processes are not needed, and polycrystalline semiconductors can be used.[28] Vapor or plasma deposition of thin metallic or semiconducting films also allows construction of solar cells with polycrystalline materials. These solid-state-junction photovoltaic devices have some of the advantages attributed to photoelectrochemical cells and avoid the associated corrosion problems.[29]

1. Physical Description

The principal elements of the liquid-junction photovoltaic cell, as shown in Fig. 1, are the counterelectrode, the electrolyte, the semiconductor-electrolyte interface, and the semiconductor. The distribution of charged species (ionic species in the electrolyte and electrons and holes in the semiconductor) is altered by the semiconductor-electrolyte interface, and an equilibrium potential gradient is formed in the semiconductor. The interfacial region may be

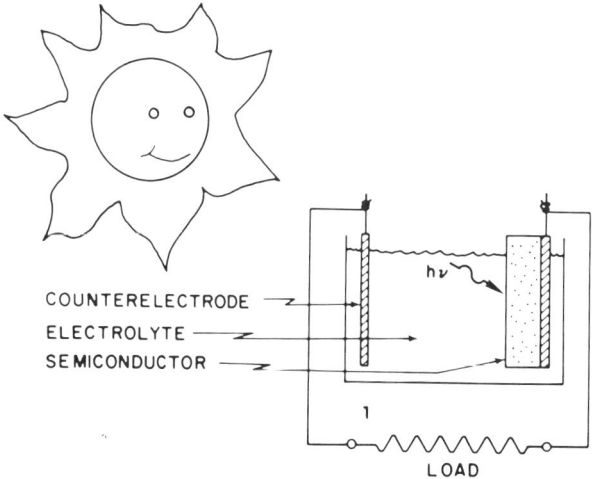

Figure 1. The photoelectrochemical cell.

associated with adsorption of charged species or with surface sites. The charge is distributed such that the interface taken as a whole is still electrically neutral. Sunlight is absorbed within the semiconductor and causes generation of electron-hole pairs which are separated by the potential gradient. This separation leads to concentration and potential driving forces for electrochemical reactions at the semiconductor-electrolyte interface. The electrochemical reactions allow passage of electrical current through the cell.

The semiconductor and the electrolyte phases are conveniently characterized through macroscopic relations. A microscopic model is required for the interface between the bulk phases. This model can be arbitrarily complex but is restricted by the requirement that thermodynamic relationships among the bulk phases hold. A convenient model for the interfacial region is represented in Fig. 2. The interface is represented by four planes, inner and outer Helmholtz planes on the electrolyte side of the interface and inner and outer surface states on the semiconductor side. The outer Helmholtz plane (OHP) is the plane of closest approach for (hydrated) ions associated with the bulk solution. The inner Helmholtz plane (IHP) passes through the center of ions specifically adsorbed on the semiconductor surface. The outer surface state (OSS) represents the plane of closest approach for electrons (and holes) associated with the bulk of the semiconductor. The inner surface state (ISS) is a plane of surface sites for adsorbed electrons. If surface sites are neglected, the ISS and the OSS are coincident.

This model of the semiconductor-electrolyte interface is an application of the classical Stern-Gouy-Chapman diffuse-double-layer theory[30-33] to the semiconductor and the electrolyte sides of the interface. Charge adsorbed onto the IHP and the ISS planes is balanced by charge in the diffuse region of the electrolyte and the space-charge region of the semiconductor. The net charge of the interface, including surface planes and diffuse and space-charge regions, is equal to zero. Within a given model, reactions may be written to relate concentrations and potentials at interfacial planes. Interfacial sites or energy levels for electrons or holes can be included at the ISS. Interfacial reactions may thus include adsorption of ionic species from the OHP to the IHP, adsorption of electrons from the OSS to sites of specified energy at the ISS, surface recombination through ISS sites, and direct surface re-

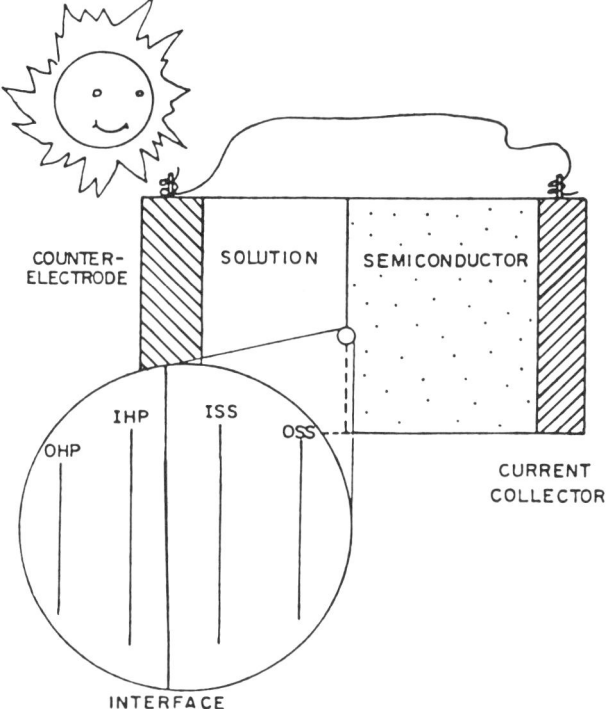

Figure 2. Physical model of the semiconductor-electrolyte interface in a photoelectrochemical cell.

combination. Emission of electrons from the semiconductor may take place by electron transfer from surface sites at the ISS to adsorbed ions (trapping and subsequent emission), transfer of OSS electrons to adsorbed ions (thermionic emission), transfer of electrons in the space charge region to adsorbed ions (thermally enhanced field emission or direct tunneling), transfer of electrons in the neutral region to adsorbed ions (field emission or direct tunneling), and transfer of electrons in the neutral region to adsorbed ions through defects in the bulk (multistep tunneling). Such reactions may also involve ions associated with the OSS or with the diffuse part of the double layer. These reactions may also take place in the reverse direction. These reaction mechanisms are described by Fonash.[34]

2. The Mechanism of Cell Operation

Formation of an interface perturbs the potential distribution in the semiconductor, and this perturbation creates the junction necessary for the photovoltaic effect in solar cells. Examination of the potential distribution in the liquid-junction cell therefore provides insight into the forces driving the cell. The following discussion is based on the numerical solution of the equations governing a n-GaAs photoanode with a selenium redox couple (see Refs. 35-37). The governing equations are presented in Section II. Interfacial reactions were included but were assumed to be sufficiently fast that the operation of the cell was limited by the transport and generation of electrons and holes in the semiconductor. The potential distribution is presented in Fig. 3. In the dark, at open circuit (curve a), the system is equilibrated. The potential is nearly constant throughout the solution and the interfacial planes (OHP, IHP, ISS, and OSS). The potential varies in the semiconductor in response

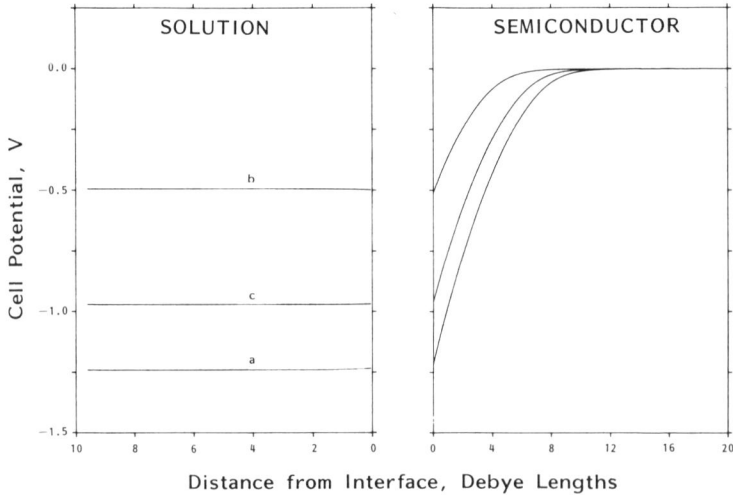

Figure 3. Potential distribution for an n-GaAs rotating disk electrode with no interfacial kinetic limitations. Curve a, open circuit in the dark; curve b, open circuit under 882 W/m² illumination; and curve c, near short circuit ($i = -23.1$ mA/cm²) under illumination. The electrolytic Debye length is 0.2 nm, and the semiconductor Debye length is 70 nm.

to charge distributed in the semiconductor. This variation of electrical potential in the equilibrated semiconductor is termed *band-bending*. The *flat-band potential* is the potential that would need to be applied in order to achieve uniform potential in the semiconductor.

Under illumination at open circuit (curve b), the concentrations of electrons and holes increase, and the variation of potential in the semiconductor decreases. The decrease in potential variation in response to illumination is referred to as the *straightening of the bands*. The charges of holes and electrons, generated by the illumination, tend to go in opposite directions under the influence of the electric field. Their accumulation, at open circuit, at various locations creates an electric field which tends to cancel that existing in the dark and leads to this straightening of the bands. The difference between the potential in the dark and under illumination represents a driving force for flow of electrical current. The potential distribution near the short-circuit condition (curve c) approaches the equilibrium distribution. Short circuit is defined as the condition of a zero cell potential. The description presented here neglects the influence of electrolyte resistance and counterelectrode kinetic and mass-transfer limitations in determining the condition of short circuit. The inclusion of these effects cause the short-circuit potential distributions described here to occur at negative cell potentials. All the variation in potential (from open circuit in the dark to open circuit under illumination to short circuit under illumination) takes place in the semiconductor. The potential drop across interfacial planes is comparatively small and invariant.

All potentials given in Fig. 3 are referenced to the potential at the interface between the semiconductor and the current collector. This choice of reference potential is arbitrary, and is used here to emphasize the degree of band bending and straightening in the semiconductor. A number of researchers (see, e.g., Refs. 17 and 19) have reported that the potential of the solution is independent of current and illumination intensity when referenced to an external quantity such as the Fermi energy of an electron in vacuum. This concept does not have strict thermodynamic validity because it depends upon the calculation of individual ionic activity coefficients[38]; however, it has proved useful for the prediction of the interaction among semiconductors and a variety of redox systems. Memming,[17,39] for example, has provided a chart of the relative energies for redox systems as compared to the valence and conduction band edge energies for several semiconductors. Concentration distributions of holes and electrons in the semiconductor are presented in Fig. 4 for a system with no interfacial kinetic limitations. Under equilibrium conditions the concentration of holes (curve a) is essentially zero in the bulk of the semiconductor and increases near the negatively charged interface. Conduction electrons are depleted near the interface and reach a value close to unity at the current collector, where the concentrations are scaled by the dopant concentration ($N_d - N_a$), and N_d and N_a represent the concentrations of ionized electron donors and acceptors, respectively. The equilibrated semiconductor of Fig. 4 can therefore be described as having an inversion region extending from the semiconductor-electrolyte interface to three Debye lengths from the interface [see Eq. (20) for a mathematical definition of the Debye length], a depletion region extending from three to eight Debye

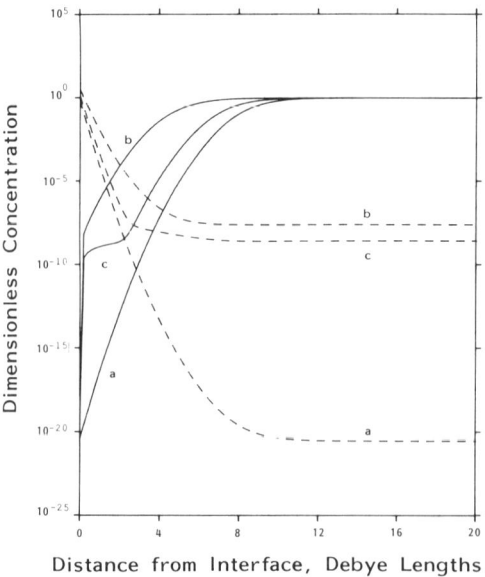

Figure 4. Concentration distribution for an n-GaAs rotating disk electrode with no interfacial kinetic limitations. Dashed lines, holes; solid lines, electrons; curve a, open circuit in the dark; curve b, open circuit under 882 W/m^2 illumination; and curve c, near short circuit ($i = -23.1$ mA/cm^2) under illumination. Concentrations are made dimensionless by the net dopant concentration $N_d - N_a$.

lengths, and a neutral region adjacent to the current collector. An inversion region is usually defined as the region for which the dimensionless minority carrier concentration is greater than one. The definition used here is for a region where the minority carrier concentration is greater than the majority carrier concentration.

The semiconductor described by Fig. 4 has a net positive charge which is balanced by charge associated with the diffuse region of the electrolyte and with the interface such that system electroneutrality is maintained. The potential gradient, the driving force for migration of charged species, is balanced by the concentration gradient, which drives diffusion. Thus, the net flux of each species in the semiconductor is equal to zero at equilibrium.

Illumination under open-circuit conditions produces electron-hole pairs, which are separated by the potential gradient (see Fig. 3). The concentration of holes increases near the interface, and the concentration of electrons increases near the current collector (curve b in Fig. 4). Under steady-state conditions, the rate of generation of electron-hole pairs is balanced by the rate of homogeneous and interfacial recombination. As the system without kinetic limitations approaches short circuit (curve c in Fig. 4), the concentrations of holes and electrons approach the equilibrium distributions.

The potential and concentration distributions described for the system with no kinetic limitations to interfacial reactions are constrained by the rates of generation and mass transfer in the semiconductor. More generally, kinetic limitations to interfacial reactions are compensated by the increased interfacial potential and concentration driving forces required to allow passage of electrical current. In contrast to the results shown as curve c in Fig. 4, the surface concentration of holes under kinetic limitations to interfacial reactions can increase with increasing current density.[37] The presence of these limitations may be inferred from experimental data by inflection points in the current-potential curve.

II. MATHEMATICAL DESCRIPTION

Macroscopic transport equations are commonly used to describe the semiconductor and the electrolyte in the liquid-junction cell. A microscopic model of the semiconductor-electrolyte interface couples the equations governing the macroscopic systems.

1. Semiconductor

The electrochemical potential of a given species can arbitrarily be separated into terms representing a reference state, a chemical contribution, and an electrical contribution:

$$\mu_i = \mu_i^\Theta + RT \ln(c_i f_i) + z_i F \Phi \tag{1}$$

where Φ is a potential which characterizes the electrical state of the phase and can be arbitrarily defined. The potential used here is the electrostatic potential which is obtained through integration of Poisson's equation.[40] Equation (1) defines the activity coefficient, f_i.

The flux density of an individual species within the semiconductor is driven by a gradient of electrochemical potential:

$$\mathbf{N}_i = -c_i u_i \nabla \mu_i \tag{2}$$

This can be written for electrons and holes in terms of concentration and potential gradients (see, e.g., Ref. 10 or 33). The flux density of holes $\mathbf{N}_h$ is therefore given by

$$\mathbf{N}_h = \frac{u_h RT}{f_h} \nabla(pf_h) - u_h pF \nabla \Phi \tag{3}$$

and a similar result is obtained for the flux density of electrons. The concentrations of electrons and holes are represented by n and p, respectively, and the mobilities u_i are related to the diffusivities D_i by the Nernst-Einstein equation, i.e.,

$$D_i = RT u_i \tag{4}$$

This equation is appropriate for both dilute and concentrated solutions.[31] Nonidealities associated with more concentrated solutions are incorporated within the activity coefficient.

Equation (3) can be simplified through the assumption of constant activity coefficients. Under the assumption of constant activity coefficients, Eq. (3) is in harmony with a Boltzmann distribution of electrons and holes. Such an approach is valid for p/N_v and n/N_c less than 0.1. At higher concentrations, Fermi-Dirac statistics must be used to account for the distribution of electrons and holes as functions of energy. These effects can be treated by introduction of concentration-dependent activity coefficients for electrons and holes such as those originally presented by Rosenberg,[41-46] i.e.,

$$f_e = \frac{\exp(\eta_e)}{F_{1/2}(\eta_e)} \tag{5a}$$

and

$$f_h = \frac{\exp(\eta_h)}{F_{1/2}(\eta_h)} \tag{5b}$$

respectively, where

$$\eta_e = \frac{E_f - E_c}{kT} \tag{5c}$$

$$\eta_h = \frac{E_v - E_f}{kT} \tag{5d}$$

E_f is the Fermi energy, E_c and E_v are the energies of the conduction and valence band edges, respectively, and $F_{1/2}(\eta)$ is the Fermi integral of order one-half for the argument η. The activity coefficients approach values of unity at dilute carrier concentration because the value of $F_{1/2}(\eta)$ approaches $\exp(\eta)$ at dilute carrier concentrations. The concentration dependency of Eqs. (5a)-(5d) can be obtained explicitly through analytic expressions relating $\exp(\eta)$ to $F_{1/2}(\eta)$.[47-50]

Equation (5) is restricted by the assumption that the energies of the band edges are independent of carrier concentration. The expressions

$$f_e = \frac{\exp(\eta_e)}{F_{1/2}(\eta_e)} \exp(-\Delta E'_c/kT) \tag{6a}$$

and

$$f_h = \frac{\exp(\eta_h)}{F_{1/2}(\eta_h)} \exp(-\Delta E'_v/kT) \tag{6b}$$

account for the interactions among electrons and holes which cause shifts in the conduction and valence band-edge energies, $\Delta E'_c$ and $\Delta E'_v$, respectively. An additional shift in band energies is associated with large concentrations of dopant species. These effects are included within material balances which couple transport and kinetic expressions. This separation of the influence of dopant and carrier concentrations is necessary in regions where the assumption of electroneutrality is not valid. Several reviews of the influence of carrier and dopant concentrations are available (see, e.g., Refs. 51–57).

The thermodynamic consistency of the expressions used for electron and hole activity coefficients can be evaluated by application of the second cross derivative of the Gibbs function,[58] i.e.,

$$\left(\frac{\partial \mu_e}{\partial p}\right)_{T,P,n} = \left(\frac{\partial \mu_h}{\partial n}\right)_{T,P,p} \tag{7}$$

Equation (7) is properly expressed in terms of mole numbers. Under the assumption that the concentrations are sufficiently dilute to allow lattice expansion to be ignored, the thermodynamic relationship can be expressed in terms of concentrations. Application to Eq. (6) yields[59]

$$\left(\frac{\partial \Delta E'_c}{\partial p}\right)_{T,P,n} = \left(\frac{\partial \Delta E'_v}{\partial n}\right)_{T,P,p} \tag{8}$$

Equation (8) constrains the choice of expressions used to account for the influence of carrier interactions on shifts of the conduction and valence band-edge energies.

Experimental results have been used to obtain averaged activity coefficients.[60] Another approach toward characterization of degenerate semiconductors has been to include the nonidealities associated with degeneracy within a modified Nernst–Einstein relationship.[61–64] The modified Nernst–Einstein relationship is given by[65]

$$D_i = RTu_i \frac{F_{1/2}(\eta_i)}{\frac{\partial}{\partial \eta_i} F_{1/2}(\eta_i)} \tag{9}$$

This approach is related to the activity coefficient used in the above development by[43]

$$\frac{F_{1/2}(\eta_i)}{\frac{\partial}{\partial \eta_i} F_{1/2}(\eta_i)} = \left(1 + \frac{\partial \ln f_i}{\partial \ln c_i} - \frac{\partial \Delta E'_i}{\partial \ln c_i}\right) \tag{10}$$

The validity of the Nernst–Einstein relation rests on the fact that the driving force for both migration and diffusion is the gradient of the electrochemical potential, and the decomposition of this into chemical and electrical contributions is arbitrary and without basic physical significance.[38] Correction of the Nernst–Einstein relationship to account for nonideal behavior represents a decomposition of the electrochemical potential gradient such that the diffusional flux density is proportional to the gradient of concentration, not activity as given in Eq. (3). The approach represented by Eqs. (1)-(8) allows separation of the influence of nonideal behavior from transport properties. The principal advantage of this approach is that the activity coefficients presented in Eqs. (5) and (6) can also be employed within the framework of the transport theory for concentrated solutions.[33,66]

Homogeneous reaction takes place in the semiconductor; thus a material balance for a given species, say holes, yields

$$\nabla \cdot \mathbf{N}_h = R_h \tag{11}$$

where R_h is the net rate of production of holes under steady-state conditions. The rate of production of holes is, by stoichiometry, equal to the rate of production of electrons and is governed by three concurrent processes: generation by absorption of light, generation by absorption of heat, and recombination of electrons and holes (i.e., transfer of an electron from the conduction band to the valence band):

$$R_h = G_L + G_{th} - R_{rec} \tag{12}$$

Mathematical models of the homogeneous recombination process have been developed which incorporate single-step electron transfer from one energy level to another. They differ in the assumption of the presence or absence of impurities which allow electrons to have energies between the conduction and valence-band energies.[67-69]

Band-to-band kinetic models (presented in Fig. 5) allow electrons to have only valence or conduction-band energies. Absorption of the appropriate amount of thermal or electromagnetic energy creates an electron–hole pair; recombination of an electron and a hole releases energy in the form of heat or light. The band-to-band model yields

$$R_h = \eta m q_0 \, e^{-my} - k_{rec}(np - n_i^2) \tag{13}$$

where η is the fraction of incident photons with energy greater

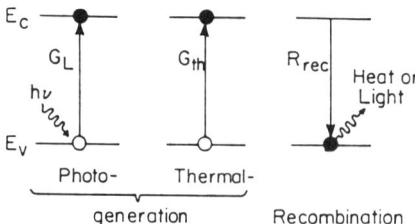

Individual Reaction Rates

$G_{th} = k_{th}(N_c - n)(N_v - p)$

$R_{rec} = k_{rec}\, np$

Figure 5. Schematic representation of band-to-band recombination kinetics in the semiconductor.

than the band gap energy, m is the absorption coefficient, q_0 is the incident solar flux, and n_i is the intrinsic concentration,

$$n_i = \left[\frac{k_{th}(N_c - n)(N_v - p)}{k_{rec}}\right]^{1/2} \quad (14)$$

The intrinsic concentration is written in terms of N_c and N_v, the number of available conduction and valence-band sites, respectively, and k_{th} and k_{rec}, thermal generation and recombination rate constants. Under equilibrium conditions, the rate of thermal generation is equal to the rate of recombination, and $np = n_i^2$.

Most semiconducting materials contain within their lattice structure impurities or imperfections which may be described as fixed sites with valence-band electron energies within the semiconductor band gap. The trap-kinetics model allows recombination to occur through these sites (see Fig. 6). Absorbed radiation drives an electron from the valence band to the conduction band, and all recombination and thermal generation reactions are assumed to occur though trap sites. This model results in

$$R_h = \eta m q_0\, e^{-my} - \frac{N_t k_2(np - n_i^2)}{\dfrac{k_1(N_v - p) + k_3(N_c - n)}{k_4} + \dfrac{k_2}{k_4}p + n} \quad (15)$$

where k_1, k_2, k_3, and k_4 are the rate constants for reactions 1–4,

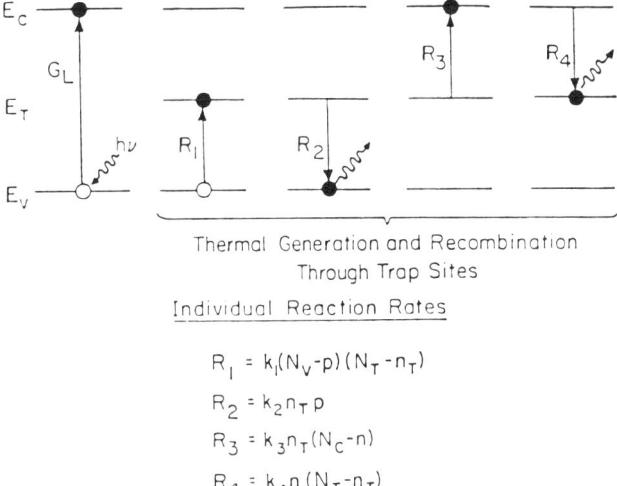

Figure 6. Schematic representation of single-trap recombination kinetics in the semiconductor (Ref. 36). Reprinted by permission of the publisher, the Electrochemical Society, Inc.

respectively, shown in Fig. 6. The intrinsic concentration is given by

$$n_i = \left[\frac{k_1 k_3 (N_c - n)(N_v - p)}{k_2 k_4}\right]^{1/2} \quad (16)$$

The expressions for the intrinsic concentration [Eqs. (13) and (16)] are consistent with the expression derived through statistical-mechanical models, e.g.,

$$np = n_i^2 = (N_c - n)(N_v - p)\, e^{-E_g/kT}$$

The intrinsic concentration can be considered to be a constant for a given semiconductor only if the ratios n/N_c and p/N_v are negligibly small compared to unity. The intrinsic concentration is related to the nondegenerate limit $n_{\lim}$ and the activity coefficients of Eqs. (5) and (6) by

$$n_i = n_{\lim}\left(\frac{1}{f_e f_h}\right)^{1/2} \exp(\Delta E_g/kT) \quad (17)$$

where ΔE_g represents the narrowing of the band gap associated with large dopant concentrations. The assumption of a constant intrinsic concentration is consistent with the assumption of unity activity coefficients for electrons and holes. The value of the intrinsic concentration derived from statistical-mechanical arguments serves as a relationship among the kinetic parameters in Eqs. (14) and (16). The divergence of the current is zero at steady state; therefore the fluxes of holes and electrons are related by

$$\nabla \cdot \mathbf{N}_e - \nabla \cdot \mathbf{N}_h = 0 \qquad (18)$$

A material balance on electrons, analogous to Eq. (11), could be used to replace Eq. (18). Numerical computational accuracy is enhanced by coupling conservation of the minority carrier with Eq. (18).

Poisson's equation,

$$\nabla^2 \Phi = -\frac{F}{\varepsilon_{sc}}[p - n + (N_d - N_a)] \qquad (19)$$

relates the potential to the charge distribution. The Debye length,

$$\lambda_{sc} = [\varepsilon_{sc} RT/F^2(N_d - N_a)]^{1/2} \qquad (20)$$

characterizes the distance over which the potential varies in the semiconductor. It typically has a value of 10 to 200 nm.

The degree of ionization of donors or acceptors is dependent upon the concentrations of charged species within the semiconductor and upon the temperature. Complete ionization is frequently assumed, and this assumption is reasonable at room temperatures. Gerischer[70] presents development of these equations under the condition of incomplete dopant ionization.

2. Electrolyte

The equations which govern the electrolyte are similar to those which govern the semiconductor with the exceptions that homogeneous reactions can frequently be neglected and that convective transport of ionic species may be important. For dilute electrolytic solutions (less than 3 M) the flux density of an ionic species can be expressed in terms of migrational, diffusional, and convective components, i.e.,

$$\mathbf{N}_i = -z_i u_i c_i F \nabla \Phi - D_i \nabla c_i + c_i \mathbf{v} \qquad (21)$$

This relationship fails for concentrated solutions owing to the neglect of the contributions of activity coefficients [see Eq. (3)] and to the neglect of ion–ion interactions.[33] Under the assumptions of steady state and that homogeneous reactions can be neglected, Eq. (11) can be expressed as

$$\nabla \cdot \mathbf{N}_i = 0 \qquad (22)$$

The expression of Poisson's equation is analogous to Eq. (19), i.e.,

$$\nabla^2 \Phi = -\frac{F}{\varepsilon_{\text{sol}}} \sum_i z_i c_i \qquad (23)$$

Electroneutrality of the electrolyte is not assumed here because the charge held within the diffuse region near the interface can be significant, and this charge contributes to balancing the charge held within the space-charge region of the semiconductor. The Debye length in the solution is given by

$$\lambda_{\text{sol}} = \left(\frac{\varepsilon_{\text{sol}} RT}{F^2 \sum_i z_i c_{i,\infty}} \right)^{1/2} \qquad (24)$$

and typically has a value of 0.1 to 1 nm.

The charge held within the diffuse part of the double layer approaches zero far from the interface. The distance over which this charge approaches zero is scaled by the Debye length. At distances greater than 20 Debye lengths, Eq. (23) can be replaced by the requirement of electroneutrality:

$$\sum_i z_i c_i = 0 \qquad (25)$$

The characteristic scaling distance for the diffusion layer (see Ref. 33) is much larger than the Debye length; therefore efficient numerical treatment of the influence of electrolytic mass transfer on the performance of photoelectrochemical cells requires a change of scale at the inner limit of the diffusion layer (or the outer limit of the diffuse part of the double layer).

The relationship presented above are sufficient to describe the electrolytic solution. An additional relationship yields the current density as a function of the ionic fluxes,

$$i = F \sum_i z_i \mathbf{N}_i \qquad (26)$$

Within the semiconductor, this can be regarded to be an integrated form of Eq. (18).

3. Semiconductor–Electrolyte Interface

The interfacial reactions described in Section I.1 are driven by departure from equilibrium. For a general interfacial reaction given by[33]

$$\sum_i s_i M_i^{z_i} \to 0 \tag{27}$$

the condition of equilibrium is given by

$$\sum_i s_i \mu_i = 0 \tag{28}$$

Here, s_i is the stoichiometric coefficient of species i, M_i is a symbol for the chemical formula of species i, and μ_i is the electrochemical potential of species i. Electrons at a given energy level (or holes) are included explicitly as a reactant.

The rate of reaction l at the interface is given by

$$r_l = k_{f,l} \exp\left[\frac{(1-\beta_l)nF\Delta\Phi_l}{RT}\right] \prod_i c_i^{p_{i,l}}$$
$$- k_{b,l} \exp\left(\frac{-\beta_l nF\Delta\Phi_l}{RT}\right) \prod_i c_i^{q_{i,l}} \tag{29}$$

where β_l is a symmetry factor (usually assumed to be equal to 1/2), $k_{f,l}$ and $k_{b,l}$ are forward and backward reaction rate constants, respectively, n is the number of electrons transferred, and $\Delta\Phi_l$ is the potential driving force for the given reaction, l. The potential driving force enters into reactions involving charge transfer from locations of one potential to locations of another.

The reaction orders for a given species i in the forward and reverse directions are $p_{i,l}$ and $q_{i,l}$, respectively. They are determined from the stoichiometric coefficients, $s_{i,l}$:

For $s_{i,l} = 0$: $p_{i,l} = 0$ and $q_{i,l} = 0$

For $s_{i,l} > 0$: $p_{i,l} = s_{i,l}$ and $q_{i,l} = 0$

For $s_{i,l} < 0$: $p_{i,l} = 0$ and $q_{i,l} = -s_{i,l}$

The reaction rates are written in terms of the equilibrium constants as

$$r_l = k_{b,l} \left\{ K_l \exp\left[\frac{(1-\beta_l)F\Delta\Phi_l}{RT}\right] \prod_i c_i^{p_{i,l}} \\ - \exp\left(\frac{-\beta_l F\Delta\Phi_l}{RT}\right) \prod_i c_i^{q_{i,l}} \right\} \quad (30)$$

The equilibrium constant used here is the ratio of the forward and backward rate constants:

$$K_l = \frac{k_{f,l}}{k_{b,l}} \quad (31)$$

The equilibrium constants can be related to equilibrium interfacial concentrations and potentials, i.e.,

$$K_l = \exp\left(-\frac{F}{RT}\Delta\Phi_l\right) \prod_i c_i^{-s_{i,l}} \quad (32)$$

Through Eq. (30), the equilibrium constants can be related to electron-site and Fermi energies. Within parametric studies, it is convenient to allow one independent rate constant to be characteristic of each group of interfacial reactions. For example, adsorption reactions (IHP-OHP) might have individual rate constants for each reaction l related to the characteristic rate constant by

$$k_{b,l} = k_j^0 K_l^{-1/2} \quad (33a)$$

and

$$k_{f,l} = k_j^0 K_l^{1/2} \quad (33b)$$

where k_j^0 is the preexponential part of the rate constant, with a characteristic value for a given reaction type j, and β was given a value of one half. The value for the equilibrium constant K_l incorporates the energy levels of interfacial sites and associated energies of adsorption. These equations are consistent with Eqs. (30) and (31).

The general approach described above can be applied to an arbitrarily complex interfacial reaction scheme. Concentrations of reacting species are related to surface reactions by material balances which, under steady-state conditions, are expressed by continuity of flux at the OSS and the OHP; i.e.,

$$\mathbf{N}_e \bigg|_{OSS} = \sum_l - s_{e,l} r_{l,OSS} \quad (34a)$$

$$\mathbf{N}_h \bigg|_{OSS} = \sum_l - s_{h,l} r_{l,OSS} \quad (34b)$$

and

$$\mathbf{N}_i \bigg|_{OHP} = \sum_l - s_{i,l} r_{l,OHP} \quad (34c)$$

Material balances are also written for each adsorbed species i at the ISS and the IHP, i.e.,

$$\sum_l s_{i,l} r_{l,ISS} = 0 \quad (35a)$$

and

$$\sum_l s_{i,l} r_{l,IHP} = 0 \quad (35b)$$

These equations apply only if surface states are involved within the microscopic model of the interface.

Gauss's law can be applied to the region between the OSS and ISS:

$$\Phi_{OSS} - \Phi_{ISS} = \frac{\delta_1}{\varepsilon_{sc}} \left[\frac{\varepsilon_1}{\delta_2} (\Phi_{ISS} - \Phi_{IHP}) + F \sum_{ISS} z_i \Gamma_i \right] \quad (36a)$$

and between the ISS and the IHP:

$$\Phi_{ISS} - \Phi_{IHP} = \frac{\delta_2}{\varepsilon_2} \left[\frac{\varepsilon_{sol}}{\delta_3} (\Phi_{IHP} - \Phi_{OHP}) + F \sum_{IHP} z_i \Gamma_i \right] \quad (36b)$$

where Γ_i are the surface concentrations of charged species located at the respective interfacial planes.

The approach described above allows description of interfacial reactions in terms of individual single-step processes. Frequently, reactions are described by a single rate expression. The rate of a charge-transfer reaction, for example, can be expressed through the Marcus-Gerischer theory by

$$r = k_{red}c_{red} \int_{-\infty}^{+\infty} \kappa(E)\rho(E)[1-f(E)]D_{red}(E) \, dE$$
$$- k_{ox}c_{ox} \int_{-\infty}^{+\infty} \kappa(E)\rho(E)f(E)D_{ox}(E) \, dE \tag{37}$$

where k_{ox} and k_{red} are rate constants, $\kappa(E)$ is an energy-dependent transmission or rate constant, $\rho(E)[1-f(E)]$ is the distribution of unoccupied electron states in the electrode, $\rho(E)f(E)$ is the distribution of occupied states in the electrode, and D_{ox} and D_{red} are the distributions of occupied and unoccupied states, respectively, for electrons associated with the ionic species.[10,17,70] This may be regarded to be a form of Eq. (29) integrated over all electron energy levels. Within this approach, the occupancy of electron states $f(E)$ is given by the Fermi-Dirac distribution, and the energy states of electrons associated with ionic species are distributed according to

$$D_{red}(E) = \exp\left[-\frac{(E - E_{F,el} - \lambda)^2}{4kT\lambda}\right] \tag{38a}$$

and

$$D_{ox}(E) = \exp\left[-\frac{(E - E_{F,el} + \lambda)^2}{4kT\lambda}\right] \tag{38b}$$

where λ is called the rearrangement or reorientation energy. This term is used to relate the energy of electrons in the semiconductor to the energy of electrons associated with the ionic species. At equilibrium, the Fermi energy, or electrochemical potential, of electrons in the semiconductor is equal to the Fermi energy of electrons associated with the ionic species. This requirement specifies the concentrations of oxidized and reduced species at the semiconductor surface.

A kinetic argument can be used instead to establish the equilibrium distribution of ions at the semiconductor surface. The rate of adsorption of a species i is given according to Eq. (29) as

$$r_l = k_{f,l} \exp\left[\frac{-z_i(1-\beta_l)F\Delta\Phi_l}{RT}\right]\Gamma_i$$
$$- k_{b,l} \exp\left(\frac{z_i\beta_l F\Delta\Phi_l}{RT}\right)c_i\left(\Gamma_{IHP} - \sum_k \Gamma_k\right) \tag{39}$$

Under equilibrium conditions the reaction rate is zero, and the fractional occupation of the inner Helmholtz plane is given by

$$\frac{\Gamma_i}{\Gamma_{IHP}} = \frac{c_i e^{-\Delta E_i/RT}}{1 + \sum_k c_k e_{-\Delta E_k/RT}} \quad (40)$$

where

$$\Delta E_i = -z_i F \Delta \Phi_l + RT \ln\left(\frac{k_{f,l}}{k_{b,l}}\right) \quad (41)$$

The fractional occupation by a single species i corresponds to that given by the Langmuir adsorption isotherm (see, e.g., Delahay[71]) for which ΔE_i is the "standard" free energy of adsorption. The "standard" free energy of adsorption is therefore a function of $\Delta \Phi_l$, the equilibrium potential difference between the inner and outer Helmholtz planes.

The equilibrium constants given in Eq. (32) couple the equilibrium concentrations of electrons, obtained as functions of Fermi energy, and the equilibrium concentrations of adsorbed ions, obtained as functions of concentration and free energies of adsorption.

4. Boundary Conditions

The boundary conditions are specified by the microscopic model of the various interfaces included within the photoelectrochemical cell. A metal–semiconductor interface, for example, can be described in a manner similar to that presented in the preceding section. Consider a semiconducting electrode bounded at one end by the electrolyte and at the other end by a metallic current collector. The boundary conditions at the semiconductor–electrolyte interface are incorporated into the model of the interface. Appropriate boundary conditions at the semiconductor–current collector interface are that the potential is zero, the potential derivative is equal to a constant, determined by the charge assumed to be located at the semiconductor–current collector interface, and all the current is carried by electrons (the flux of holes is zero). These conditions are consistent with a *selective ohmic contact*.[34] The boundary conditions in the electrolytic solution may be set a fixed distance from the interface. If this location is considered to be the other limit of the diffusion layer, the appropriate boundary conditions are that the potential gradient is continuous and that all concentrations have their bulk value. The electrostatic potential can be set to an arbitrary value at this point.

It is common to treat the semiconductor–electrolyte interface in terms of charge and current density boundary conditions. The total charge held within the electrolytic solution and the interfacial states, which balances the charge held in the semiconductor, is assumed to be constant. This provides a derivative boundary condition for the potential at the interface. The fluxes of electrons and holes are constrained by kinetic expressions at the interface. The assumption that the charge is constant in the space charge region is valid in the absence of kinetic and mass-transfer limitations to the electrochemical reactions. Treatment of the influence of kinetic or mass transfer limitations requires solution of the equations governing the coupled phenomena associated with the semiconductor, the electrolyte, and the semiconductor–electrolyte interface.

5. Counterelectrode

In the region sufficiently far from the interface that electroneutrality holds and under the assumptions that the concentration is uniform and that the solution adjacent to the electrodes may be treated as equipotential surfaces, the potential distribution can be obtained through solution of Laplace's equation, $\nabla^2 \Phi = 0$, and is a function of current density. The potential drop in the region between the counterelectrode and the outer limit of the diffusion layer is given by

$$V_{IR} = \frac{Li}{\kappa} \qquad (42)$$

where κ is the solution conductivity and L is the distance between the counterelectrode and the outer edge of the diffusion layer. The counterelectrode was assumed here to be in a configuration parallel to the semiconductor. Relaxation of this assumption will be discussed in Section III.3. The conductivity of dilute solutions can be related to ionic mobilities and concentrations by

$$\kappa = F^2 \sum_i z_i^2 u_i c_i \qquad (43)$$

or is obtained from experimental measurements for a given electrolyte.

The potential drop across the counterelectrode-electrolyte interface is given by

$$V_{CE} = V_{CE}^0 + \eta_{CE} \tag{44}$$

where V_{CE}^0 is the equilibrium potential drop across the interface and η_{CE} is the total counterelectrode reaction overpotential. The total overpotential is related to the current density through the Butler-Volmer reaction model[33,71]

$$i = i_0 \left\{ \left(1 - \frac{i}{i_{c,\lim}}\right) \exp\left[\frac{(1-\beta)nF}{RT} \eta_{CE}\right] \right.$$
$$\left. - \left(1 + \frac{i}{i_{a,\lim}}\right) \exp\left(-\frac{\beta nF}{RT} \eta_{CE}\right) \right\} \tag{45}$$

where i_0 is the exchange current density associated with the bulk concentrations of reactants, $i_{k,\lim}$ is the diffusion-limited current density associated with species k, and n is the number of electrons transferred in the counterelectrode reaction.

III. PHOTOELECTROCHEMICAL CELL DESIGN

The liquid-junction photovoltaic cell has the advantages that the junction between electrolytic solution and semiconductor is formed easily and that polycrystalline semiconductors can be used. The principal disadvantage is that the semiconductor electrode tends to corrode under illumination. The electrochemical nature of the cell allows both production of electricity and generation of chemical products which can be separated, stored, and recombined to recover the stored energy. Liquid-junction cells also have the advantages that are attributed to other photovoltaic devices. Photovoltaic power plants can provide local generation of power on a small scale. The efficiency and cost of solar cells is independent of scale, and overall efficiency is improved by locating the power plant next to the load.[72]

The design of a liquid-junction photovoltaic cell requires selection of an appropriate semiconductor-electrolyte combination and design of an efficient cell configuration. The selection of a semiconductor is based upon the band gap, which provides an upper limit to the conversion efficiency of the device, and the choice of electrolyte is governed by the need to limit corrosion and by the requirement that interfacial reaction rates be fast.

1. Choice of Materials

The performance of photoelectrochemical cells constructed with thin-film or polycrystalline semiconductors is strongly dependent upon the method of film formation and upon the surface preparation. Conversion efficiencies (incident solar illumination to electrical power) of 3%-6.5% have been reported for cells using thin-film n-CdSe electrodes (see Refs. 73-81), and a conversion efficiency of 5.1% has been achieved with a pressure-sintered polycrystalline CdSe photoelectrode.[82] A comparable efficiency of 8.1% has been reported for a single-crystal CdSe electrode.[83] Conversion efficiencies of 0.038%-0.3% have been reported for polycrystalline CdS films[84-86] as compared to 1.3% for single-crystal CdS.[87] Conversion efficiencies of up to 7.3% have been reported for polycrystalline n-GaAs films[88 90]; 12% has been reported for the single crystal n-GaAs.[91,92] For descriptions of other semiconductor electrolyte combinations see Refs. 93-104. The lower efficiency of the polycrystalline thin-film semiconductors, as compared to the single-crystal counterparts, is expected to be compensated by their lower cost. Reviews of the materials aspects of photoelectrochemical cells are given in Refs. 105-107.

(i) Band Gap

An upper limit to the efficiency of photovoltaic devices can be established, based upon the band gap and the solar spectrum, without consideration of cell configuration. This ultimate efficiency is given by[11]

$$\eta_{\text{ult}} = \frac{E_g \int_{E_g}^{\infty} N(E)\, dE}{\int_0^{\infty} E N(E)\, dE} \qquad (46)$$

where E_g is the semiconductor band-gap energy, E is the photon energy, and $N(E)$ is the number density of incident photons with energy E. The fraction of the power in the solar spectrum that can be converted to electrical power is a function of the band gap of the semiconductor. Photons with energy less than the band gap cannot produce electron–hole pairs. Photons with energy greater than the band gap yield only the band gap energy.[108–110]

The ultimate efficiency of Eq. (46) represents an upper limit to conversion of solar energy[10,111,112]; factors such as reflection and absorption losses of sunlight, kinetic and mass transfer limitations, and recombination will reduce the efficiency. These effects are included in Section III.3. A band gap between 1.0 and 1.5 eV is generally considered to be appropriate for efficient conversion of solar energy.

(ii) Corrosion

The application of liquid-junction technology to photovoltaic power conversion is limited by problems associated with the semiconductor–electrolyte interface. Primary among these problems is corrosion. Efficient conversion of solar energy requires a band gap between 1.0 and 1.5 eV, and most semiconductors near this band gap corrode readily under illumination. Semiconductors with large band gaps (4–5 eV) tend to be more stable but cannot convert most of the solar spectrum.

Among the approaches taken to solve this problem, the most successful concern the matching of an electolyte to the semiconductor. The rate of corrosion is reduced if the semiconductor is in equilibrium with the corrosion products. The rate of corrosion can also be reduced by using a redox couple which oxidizes easily. The oxidation of the redox couple Se_{x+1}^{2-}/Se_x^{2-} for example, has been shown to compete successfully with photocorrosion reactions for holes in n-type GaAs electrodes.[28,113–116]

p-type semiconductors used as cathodes are more stable than the more common and generally more efficient n-type semiconducting anodes. The inefficiency of p-type photocathodes has been attributed to the presence of surface states near the valence band energy. A stable p-type photocathode has been developed, however, with a solar-energy-conversion efficiency of 11.5%.[117] Protective films have been proposed to be a solution to electrode corrosion. The electrode, in this case, would be a small band-gap semiconductor covered by a film composed of either a more stable large band-gap semiconductor,[118] a conductive polymer, or a metal. A large Schottky barrier is frequently present at such semiconductor-metal and semiconductor-semiconductor interfaces which blocks the transfer of holes from the semiconductor to the electrolyte. In cases where the photocurrent is not blocked, corrosion can take place between the semiconductor and the protective film.[40,119] Menezes et al.[120] discuss the difficulties in avoiding absorptive losses in the metal film while maintaining sufficient integrity to serve the semiconductor corrosion protection function. Frese et al.[121] have, however, reported a measurable improvement in the stability of GaAs with less than a monolayer gold metal coverage. Thin conductive poly-pyrrole films appear to be successful in inhibiting corrosion in some electrolytes.[122-128] In addition, insulating polymer films deposited on grain boundaries can improve the performance of polycrystalline semiconductors by reducing surface recombination rates.[129,130] (For more complete reviews of the corrosion of semiconducting electrodes, see Refs. 18, 19, 28, 131, and 132.)

2. Solution of the Governing Equations

Quantitative optimization or prediction of the performance of photoelectrochemical cell configurations requires solution of the macroscopic transport equations for the bulk phases coupled with the equations associated with the microscopic models of the interfacial regions. Coupled phenomena govern the system, and the equations describing their interaction cannot, in general, be solved analytically. Two approaches have been taken in developing a mathematical model of the liquid-junction photovoltaic cell: approximate analytic solution of the governing equations and numerical solution.

(i) Analytic Approach

The semiconductor electrode is typically divided into three regions. Surface-charge and electron and hole-flux boundary conditions model the semiconductor-electrolyte interface. The region adjacent to the interface is assumed to be a depletion layer, in which electron and hole concentrations are negligible. The potential variation is therefore independent of hole and electron concentration in this region. Far from the interface a neutral region is defined in which the potential is constant; here electron and hole fluxes are driven only by diffusion. Current-potential relationships are derived in the analytic approach by invoking assumptions appropriate to each region.

Integration of Poisson's equation in the depletion layer, for example, results in a depletion layer thickness W in terms of the voltage drop V across the layer:

$$W = \left[\frac{2\varepsilon V}{F(N_d - N_a)}\right]^{1/2} \tag{47a}$$

This can also be written in terms of the charge q held within the space-charge region

$$W = \frac{q}{F(N_d - N_a)} \tag{47b}$$

or

$$V = \frac{q^2}{2\varepsilon F(N_d - N_a)} \tag{47c}$$

The depletion layer thickness is, as shown in Fig. 3, a function of illumination intensity. The assumption that the semiconductor can be separated into depletion and neutral regions restricts the voltage drop V to values high enough to deplete the majority carriers (electrons in an n-type semiconductor) in a region adjacent to the interface but small enough to avoid formation of an inversion layer (in which the concentration of minority carriers is significant). This assumption is not appropriate under many operating conditions for which the liquid-junction cell may be practical. (Figure 4, for example, shows an inversion layer adjacent to the solution interface.)

Analytic models of photoelectrochemical devices closely resemble models of solid-state solar cells (see, e.g., Refs. 133-145). Several analytic current-voltage relationships have been derived which use the general approach described above and differ in their treatment of surface reactions and recombination within the depletion and neutral layers. The model of Gärtner,[146] developed for a p-n junction device, is commonly used in the analysis of photoelectrochemical devices.[147-149] Recombination and thermal generation of carriers was neglected in the interfacial and space-charge regions, but was included in the neutral region. The influence of photogenerated carriers on the potential distribution is not treated explicitly; therefore this model applies only in the region where the potential distribution approaches the equilibrium condition. Gärtner's model can be used near the short circuit condition but not at open circuit under illumination (see Fig. 3). Wilson[150,151] included recombination at the semiconductor-electrolyte interface and in the neutral region, but neglected recombination in the space-charge region. The perturbation of the potential distribution by photogenerated carriers was also neglected. Albery *et al.*[152-154] extended the model of Gärtner by including recombination of holes and electrons in the depletion layer. Reichman[155,156] presented a model which included recombination in the depletion region and kinetic limitation at the interface. Reiss[157] presented models for various cases, including within the model the potential drop across the electrolyte double layer, surface recombination, and surface kinetic limitations. The semiconductor was divided into depletion and neutral regions, and the effect of illumination on cell potential was included as an additive photovoltage. Ahlgren[158] incorporated a Butler-Volmer reaction rate expression into the boundary conditions at the semiconductor-electrolyte interface. McCann and Haneman[159] included enhanced recombination associated with grain boundaries within the bulk of the semiconductor. The photovoltage was included in the calculation of the depletion region width. McCann *et al.*[160] used an analytic model to calculate the current-voltage characteristics of front and back-wall-illuminated liquid-junction cells.

Surface states and crystal imperfections have been found to play an important role in charge-transfer and redox reactions at the semiconductor-electrolyte interface (see Refs. 161-173). Mathematical and conceptual relationships have been developed which describe electrochemical reactions at the semiconductor-electrolyte interface in terms of surface states and potentials (see, e.g., Refs. 17, 71, and 174-182). Electrochemical reaction via surface states has been included within an analytic model,[183] but this model is still limited by the restrictions described above.

Equivalent circuit models of the liquid-junction cell have also been presented.[184-186] These models are useful in the analysis of impedance response measurements.

(ii) Numerical Approach

Use of a digital computer in the numerical solution of the equations governing the liquid-junction cell eliminates the need for restrictive assumptions. This approach has been used in the modeling of solid-state devices.[187-196] Laser and Bard[197-200] developed a computer program which was used to calculate open-circuit photopotentials, the transient behavior of the system following charged injection, and the time dependence of photocurrents in liquid-junction cells. Time-dependent material balances of holes and electrons and Poisson's equation described the semiconductor. The interface was included in terms of charge and flux boundary conditions. The model was limited by lack of convergence for electrode thicknesses greater than that of the space-charge region and did not treat explicitly the electrolyte and counterelectrode. Orazem and Newman[35-37] presented a numerical solution of the governing equations that included analysis of neutral, space-charge, and inversion regions in the semiconductor coupled with explicit treatment of the electrolyte and the counterelectrode. Interfacial reactions were treated explicitly; however, limitations to electrolytic mass transfer were not included in the analysis. Potential-dependent concentration variables were defined to reduce the numerical difficulties associated with concentrations that can vary up to 20 orders of magnitude in a short distance.[35,201] Errors associated with matching of solutions for various regions of the semiconductor were thereby avoided. Numerical methods for solving coupled ordinary differential equations are discussed by Newman and associates,[202,203] and a general method for treating boundary conditions is presented by White.[204]

A number of computer programs related to the liquid-junction photovoltaic cell have been developed. Leary et al.[205] for example, calculated carrier concentrations in polycrystalline films using a numerical solution of Poisson's equation coupled with overall charge neutrality within spherical grains. Their model was used for analysis of semiconductor gas sensors. Davis and colleagues[206-208] presented a computer program which uses simultaneous calculation of surface and solution equilibrium states to obtain the equilibrium condition of electrical double layers.

3. The Influence of Cell Design

The optimal design of liquid-junction photovoltaic cells shares constraints with solid-state photovoltaic cells.[25,209] Current collectors cast shadows and can reduce the amount of sunlight absorbed in the semiconductor. A constraint unique to the liquid-junction cell is the placement of the counterelectrode relative to the semiconductor–electrolyte interface. Shadows, which reduce efficiency and cause local currents in solid-state photovoltaic cells, may lead to localized corrosion in photoelectrochemical cells. Mass-transfer and kinetic limitations at the counterelectrode and resistance of the electrolyte can play important roles in the optimal design of the liquid-junction photovoltaic cell. These considerations are treated qualitatively by Parkinson.[210]

Ideally, modeling and optimization of photoelectrochemical cell configurations involves solution of the governing equations in three dimensions. A first approach toward the analysis of these devices involves coupling the solution of the one-dimensional equations (see the previous section) with primary resistance calculations for the specific cell configuration. The computational problem is still difficult but can be solved. This approach toward calculation of the influence of cell design is illustrated below and is based on the work of Orazem and Newman.[211] The system modeled was an n-type GaAs semiconducting anode in contact with an 0.8 M K_2Se, 0.1 M K_2Se_2, 1.0 M KOH electrolytic solution. The choice of this semiconducting electrode system was based upon the work of Heller and associates.[28] The semiconductor was assumed to be in the form of a thin film (see Mitchell for a review of thin-film photovoltaic technologies[29]). Interfacial reactions were included but were not limited by kinetics.[35–37] This approach requires calculation of the resistance to current flow associated with the two-dimensional systems. Some methods for calculation of this resistance were reviewed by Fleck et al.[212] Kasper,[213–217] Moulton,[218] Newman,[219] and Orazem and Newman[220] have presented primary resistance calculations for configurations of potential application to photoelectrochemical cells.

The design of a photoelectrochemical configuration is illustrated here for the slotted-semiconductor electrode presented in

Fig. 7a. A glass cover plate protects the cell. Sunlight passes through the cover plate and the electrolyte to illuminate the semiconductor surface. Electrical current passes between the semiconductor and the counterelectrode through slots cut in the semiconductor. Characteristic features of this configuration are that no shadows are cast upon the semiconductor and that reaction products could be separated if a membrane were placed between the semiconductor and the counterelectrode.

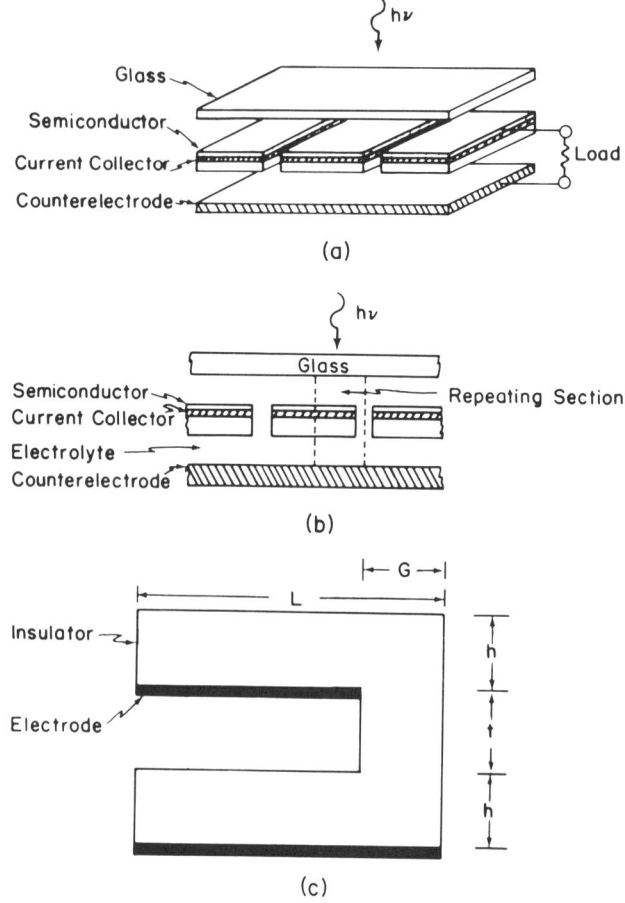

Figure 7. Slotted-electrode photoelectrochemical cell: (a) three-dimensional configuration, (b) two-dimensional representation, and (c) repeating section.

A two-dimensional representation of this cell is presented in Fig. 7b. The primary current distribution and the resistance of a cell containing a slotted electrode were calculated using numerical methods coupled with the Schwarz–Christoffel transformation.[220-222] The cell resistance is a function of three geometric ratios, chosen to be t/G, h/G, and L/h, where L is the half-length of the protruding electrode assembly, t is the thickness of the protruding electrode assembly, G is the half-gap between the electrode assemblies, and h is the separation between the electrode and the upper insulating wall. The separation between the counterelectrode and the lower edge of the semiconductor–electrode assembly is also given by h. These parameters are shown in Fig. 7c.

Four geometric parameters characterize this cell design. The distance between the counterelectrode and the semiconductor was chosen to be 0.5 cm, and the semiconductor assembly thickness was assumed to be 0.1 cm. The values chosen for this analysis were based on mechanical considerations. Smaller spacing could result in shorting of counterelectrode and semiconductor and/or trapping of gas bubbles. The influence of the counterelectrode could be reduced by increasing the flow rate or degree of mixing near the counterelectrode, thereby increasing the limiting current. Interfacial reactions were considered to be equilibrated. Kinetic limitations at the semiconductor–electrolyte interface may greatly reduce the performance of some semiconductor systems.

The primary resistance for this system is presented in Fig. 8 as a function of L/D with h/G as a parameter. The maximum power density is presented in Fig. 9 as a function of L/h with h/G as a parameter. The maximum power density for this system is obtained with a small gap. For $h/G = 0.5$ ($G = 1$ cm), the maximum power density was 47.8 W/m^2, and the maximum power efficiency was 5.4%. The current density under maximum power conditions was 15 mA/cm^2 delivered at 477.6 mV. For $h/G = 10$ ($G = 0.05$ cm), the maximum power density was 67.7 W/m^2, and the maximum power efficiency was 7.7%. At maximum power the current density was 15.2 mA/cm^2 delivered at 534.6 mV.

The hierarchy of photovoltaic cell efficiencies is presented in Table 1. Semiconductor effects, such as recombination, reduce the power efficiency of a GaAs-based device from a value of 37%, based solely upon band gap, to 15.3%. Reflection losses, with an

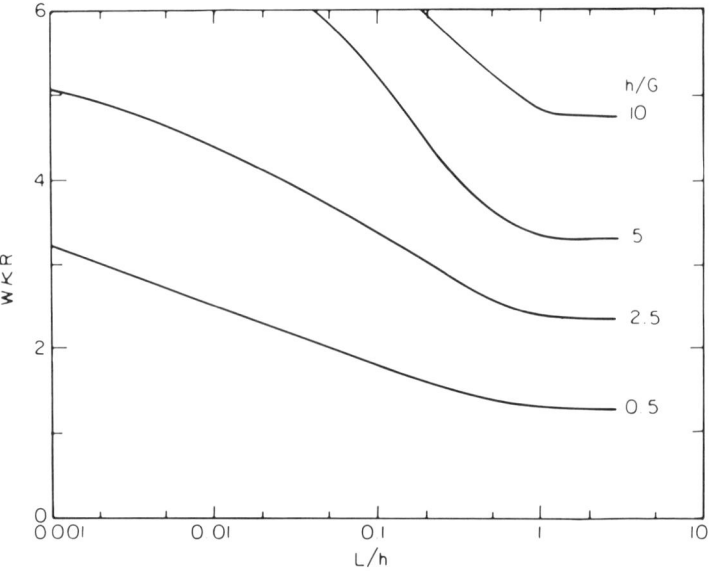

Figure 8. Primary resistance of a slotted-electrode cell as a function of L/h (Ref. 211). Reprinted by permission of the publisher, the Electrochemical Society, Inc.

arbitrarily chosen 80% efficiency of illumination, reduce this value to 12.2%. This value is consistent with the value measured in a bench-top experimental system for which the influence of counterelectrode limitations and electrolyte resistance can be minimized. This value can also be compared to the 12% efficiency obtained in the experimental work of Heller and Miller.[28,91,92] Accounting for the effect of cell design reduces the efficiency from 15.3% to 9.8%, and inclusion of illumination losses further reduces the cell efficiency to 7.7% for the slotted-electrode cell.

The maximum power efficiency is presented as a function of illumination intensity in Fig. 10 for the slotted-electrode cell. The cell was designed with the design parameters calculated to be optimal under peak AM-2 illumination. The power efficiency decreases with increasing illumination due to the influence of electrolyte resistance and kinetic and mass-transfer limitations at the counterelectrode. These phenomena become increasingly important as current densities increase, and mass-transfer limitations at the counterelectrode can result in an upper limit for cell currents.

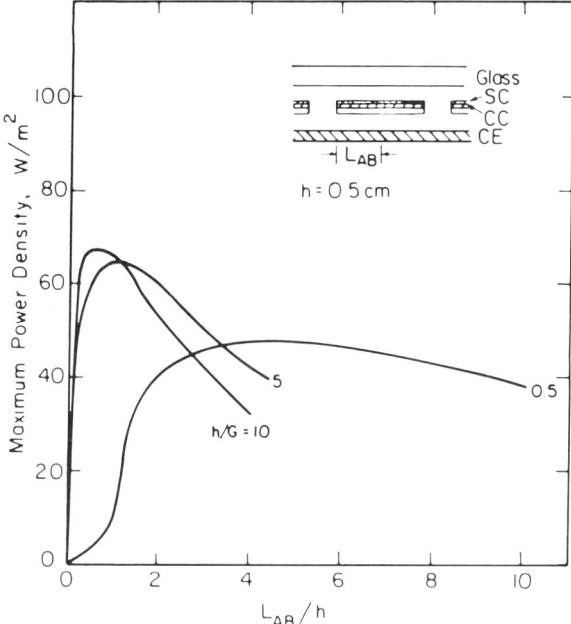

Figure 9. Maximum power density as a function of L/h for the slotted-electrode photoelectrochemical cell (Ref. 211). Reprinted by permission of the publisher, the Electrochemical Society, Inc.

The maximum power efficiency for the system without counterelectrode limitations is also presented in Fig. 10. These results are consistent with the use of a porous counterelectrode in the cell. The maximum cell current obtained under large magnitudes of illumination depends upon the ratio of the counterelectrode area to the semiconductor area. This ratio must be large for liquid-junction photovoltaic cells designed for large intensities of illumination, and for some configurations, a porous counterelectrode may be appropriate.[223] Inclusion of a cooling system in the cell design becomes important under these conditions.[224] The electrolyte itself can serve as a heat exchange in photoelectrochemical systems.

Similar calculations were performed for a cell with a wire-grid counterelectrode through which sunlight passes.[211] Current-potential curves are presented in Fig. 11 for these optimally designed cells as compared to the cell without interfacial kinetic limitations,

Table 1
Power Efficiency under Front Illumination

	No illumination losses	Illumination losses[a]	Experimental results[b]
Optimal band gap	45	36 (80%)	
GaAs band gap	37	30 (80%)	
Semiconductor-electrolyte junction	15.3	12.2 (80%)	12.0
Slotted-electrode cell design	9.8	7.7 (55.4%)	

[a] In some cases, the number in parentheses represents the fraction of AM-2 illumination (above the band gap) which actually enters the semiconductor, after accounting for reflection, shadowing, and absorption in intervening phases. In other cases, where detailed calculations were not made, it represents the ratio to column 1 because the nonlinear effect of illumination could not be assessed.
[b] References 28, 91, 92.

counterelectrode limitations, or electrolytic resistance. The cell with a slotted semiconductor has a larger power efficiency than the wire-grid counterelectrode cell and can be designed for separation of chemical products. The analysis of the system designed for separation of chemical products would need to include the electrical resistance of the membrane.

The allowable capital investment for a photovoltaic cell is given by

$$I = 8.76 P_{in} \eta \Delta c y_e \tag{48}$$

where P_{in} is the annual incident illumination intensity averaged over 24 h in W/m^2 (on this basis, the average insolation of the continental United States is 200–250 $W/m^{2\,9}$), η is the cell efficiency, Δc is the difference in selling price and operating cost in dollars per kW h, and y_e is the break-even point in years. Lenses or mirrors could be used to increase the amount of sunlight striking the semiconductor surface. Based upon a 7.7% power efficiency (averaged over 24 h), 250 W/m^2 incident illumination (averaged over 24 h), \$0.05/kW h profit, and a break-even period of 5 yr, an investment of \$42/m^2 is justified for the complete cell. Based upon a 15.3% power efficiency (this number is averaged over 24 h and

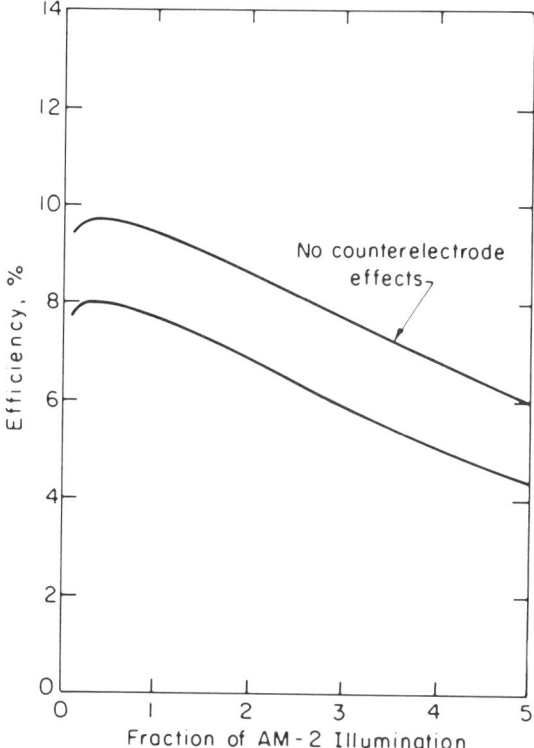

Figure 10. Maximum power efficiency as a function of the fraction of peak Am-2 illumination for a slotted-electrode photoelectrochemical cell with $h/G = 10$ and $L/h = 0.5$ (Ref. 211). Reprinted by permission of the publisher, the Electrochemical Society, Inc.

neglects the influence of cell design), an investment of $83/m^2$ is justified for the complete cell.

An increase of solar illumination by a factor of 5 while reducing the efficiency to 6% (see Fig. 10) yields an acceptable initial investment of $164/m^2$. If the mirrors and lenses needed to concentrate sunlight are cheaper than the semiconducting film, the cell may be most economical under high illumination. The values presented here can be compared to the cost estimate of $0.34 per peak watt presented by Weaver et al.[225] for a GaAs photoelectrochemical cell. Their estimate is based on materials cost (see also Refs. 226 and

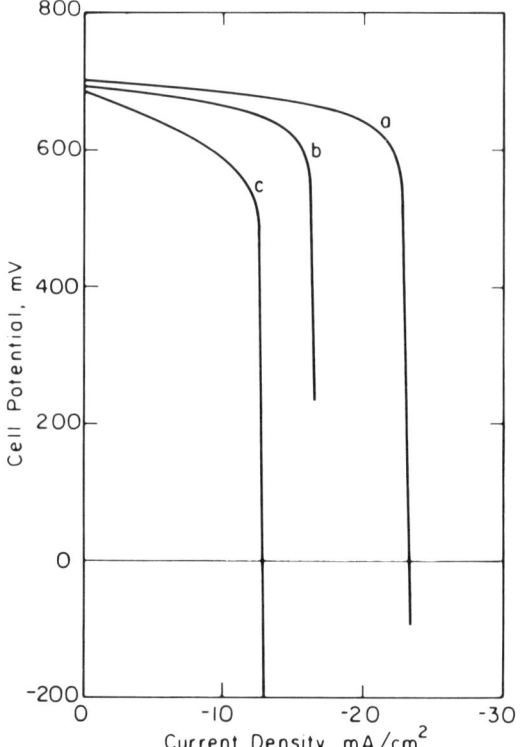

Figure 11. Cell potential as a function of current density for (a) a front-illuminated semiconductor without kinetic, electrolyte resistance, and counterelectrode limitations; (b) an optimally designed slotted-electrode cell with a porous counterelectrode; and (c) an optimally designed cell with a wire-grid counterelectrode of radius 0.5 cm (Ref. 211). Reprinted by permission of the publisher, the Electrochemical Society, Inc.

227) and assumes a cell efficiency of 13%. The influence of cell design was therefore neglected. Under AM-2 illumination, this value corresponds to $39/m^2.

IV. CONCLUSIONS

Development of a mathematical model of photoelectrochemical devices requires treatment of the diffuse double layer (or space charge region) in the semiconductor. The principles of electrochemical engineering can be readily applied to provide a chemical description of these devices in terms of potentials and concentrations of charged species. The macroscopic transport relations which govern the electrolyte and the semiconductor are coupled by a microscopic model of the interface. Analytic solution of the governing equations requires restrictive assumptions which can be avoided by use of numerical methods.

The optimization of photoelectrochemical devices for solar energy conversion depends on the choice of semiconductor, electrolyte, and cell design. The performance of the cell is strongly dependent upon the design, surface area, and placement of the counterelectrode and current collectors. This type of solar cell may be economical under concentrated illumination or in regions where electrical power has high value.

ACKNOWLEDGMENT

This work was supported by the Assistant Secretary for Conservation and Renewable Energy, Office of Energy Systems Research, Energy Storage Division of the U. S. Department of Energy, under contract No. DE-AC03-76SF00098.

NOTATION

1. Roman Characters

c_i molar concentration of species i (mol/cm^3)
D_i diffusivity of species i (cm^2/s)
E_i energy of species or site i (eV)
ΔE_i ionic adsorption energy (J/mol)
f_i molar activity coefficient of species i
F Faraday's constant (96,487 C/equiv)
G_{th} rate of thermal electron–hole pair generation (mol/s cm^3)
G_L rate of photo electron–hole pair generation (mol/s cm^3)
i current density (mA/cm^2)
i_0 exchange current density (mA/cm^2)
$k_{f,l}$ forward reaction rate constant for reaction l
$k_{b,l}$ backward reaction rate constant for reaction l

k_k rate constants for homogeneous reaction k
K_l equilibrium constant for reaction l
m solar absorption coefficient (1/cm)
M_i symbol for chemical formula of species i
n number of electrons involved in electrode reaction
n electron concentration (mol/cm^3)
n_i intrinsic electron concentration (mol/cm^3)
N total site concentration (mol/cm^3)
N_a total bulk electron-acceptor concentration (mol/cm^3)
N_d total bulk electron-donor concentration (mol/cm^3)
N_i flux of species i (mol/cm^2 s)
p hole concentration (mol/cm^3)
$p_{i,l}$ heterogeneous reaction order
$q_{i,l}$ heterogeneous reaction order
q_0 incident solar flux (mol/s cm^2)
r_l heterogeneous reaction rate (mol/s cm^2)
R universal gas constant (8.3143 J/mol K)
R_i net rate of production of species i (mol/s cm^3)
R_{rec} net rate of electron-hole recombination (mol/s cm^3)
s_i stoichiometric coefficient of species i in an electrode reaction
T absolute temperature (K)
u_i mobility of species i (cm^2 mol/J s)
V potential drop across depletion layer (V)
W depletion layer thickness (cm)
z_i charge number of species i

2. Greek Characters

β symmetry factor
Γ_k surface concentration of energy or species k (mol/cm^2)
δ_k distance between interfacial planes (gap denoted by k) (cm)
ε permittivity (C/V cm)
η photon efficiency
η_k total overpotential at interface k (V)
Θ fractional occupation of surface sites
κ conductivity (mho/cm)
λ Debye length (cm)
μ_i electrochemical potential of species i (J/mol)
Φ electrical potential (V)

3. Superscripts

0   equilibrium
Θ   secondary reference state at infinite dilution
*   secondary reference state in semiconductor

4. Subscripts

bulk   associated with the bulk
c associated with conduction band in semiconductor
CE   associated with the counterelectrode
cell   associated with the cell
e relating to electrons
h relating to holes
IHP   associated with inner Helmholtz plane
ISS   associated with inner surface states
l associated with reaction l
0   equilibrium value or initial value
OHP   associated with outer Helmholtz plane
OSS   associated with outer surface states
sc   associated with semiconductor
sol   associated with solution
v associated with valence band in semiconductor

REFERENCES

[1] A. J. Bard, Photoelectrochemistry and heterogeneous photocatalysis at semiconductors. *J. Photochem.* **10** (1979) 59-75.
[2] A. J. Bard, Photoelectrochemistry, *Science* **207** (1980) 139-144.
[3] A. Fujishima and K. Honda, Electrochemical photolysis of water as a semiconductor electrode, *Nature* **238** (1972) 37-38.
[4] J. Manassen, D. Cahen, and G. Hodes, Electrochemical, solid state, photochemical, and technological aspects of photoelectrochemical energy converters, *Nature* **263** (1976) 97-100.
[5] H. Gerischer and J. Gobrecht, On the power-characteristics of electrochemical solar cells, *Ber. Bunsenges. Phys. Chem.* **80** (1976) 327-330.
[6] H. Ehrenreich and J. H. Martin, Solar photovoltaic energy, *Phys. Today* **32**(9) (1979) 25-32.
[7] H. Gerischer, Heterogeneous electrochemical systems for solar energy conversion, *Pure Appl. Chem.* **52** (1980) 2649-2667.
[8] A. Heller, Electrochemical solar cells, *Solar Energy* **29** (1982) 153-162.

9. M. Green, Electrochemistry of the semiconductor-electrolyte interface, Chap. 5 in *Modern Aspects of Electrochemistry*, No. 2, Ed. by J. O'M. Bockris, Academic, New York, 1959.
10. H. Gerischer, Semiconductor electrode reactions, Chap. 4 in *Advances in Electrochemistry and Electrochemical Engineering*, Vol. 1, Ed. by P. Delahay, Interscience, New York, 1961.
11. M. D. Archer, Electrochemical aspects of solar energy conversion, *J. Appl. Electrochem.* **5** (1975) 17-38.
12. K. Rajeshwar, P. Singh, and J. DuBow, Energy conversion in photoelectrochemical systems: A review, *Electrochim. Acta* **23** (1975) 1117-1144.
13. L. A. Harris and R. H. Wilson, Semiconductors for photoelectrolysis, *Ann. Rev. Mater. Sci.* **8** (1978) 99-134.
14. A. J. Nozik, Photoelectrochemistry: Applications to solar energy conversion, *Ann. Rev. Phys. Chem.* **29** (1978) 189-222.
15. M. Tomkiewicz and H. Fay, Photoelectrolysis of water with semiconductors, *Appl. Phys.* **18** (1979) 1-28.
16. R. H. Wilson, Electron transfer processes at the semiconductor-electrolyte interface, *CRC Crit. Rev. Solid State Mater. Sci.* **10** (1980) 1-41.
17. R. Memming, Charge transfer processes at semiconductor electrodes, in *Electroanalytical Chemistry*, Ed. by Allen J. Bard, Marcel Dekker, New York, 1981, pp. 1-84.
18. S. R. Morrison, *Electrochemistry at Semiconductor and Oxidized Metal Electrodes*, Plenum Press, New York, 1980.
19. U. M. Khan and J. O'M Bockris, Photoelectrochemical kinetics and related devices, Chap. 3 in *Modern Aspects of Electrochemistry*, No. 14, Ed. by J. O'M. Bockris, B. E. Conway, and R. E. White, Plenum Press, New York, 1982.
20. A. S. Grove, *Physics and Technology of Semiconductor Devices*, Wiley, New York, 1967.
21. M. A. Omar, *Elementary Solid State Physics: Principles and Applications*, Addison-Wesley, Menlo Park, California, 1975.
22. S. M. Sze, *Physics of Semiconductor Devices*, Wiley-Interscience, New York, 1969.
23. N. B. Hannay, Semiconductor principles, Chap. 1 in *Semiconductors*, Ed. by N. B. Hannay, Reinhold, New York, 1959.
24. A. F. Ioffe, *Physics of Semiconductors*, Academic, New York, 1960.
25. J. Hovel, Solar cells, Vol. 11 of *Semiconductors and Semimetals*, Ed. by R. K. Willardson and Albert C. Beer, Academic, New York, 1975.
26. H. J. Hovel, Photovoltaic materials and devices for terrestrial solar energy applications, *Solar Energy Mater.* **2** (1980) 277-312.
27. S. Wagner, Physico-chemical problems in photovoltaic research, *Ber. Bunsenges. Phys. Chem.* **84** (1980) 991-995.
28. A. Heller, Conversion of sunlight into electrical power and photoassisted electrolysis of water in photoelectrochemical cells, *Acc. Chem. Res.* **14** (1981) 154-162.
29. K. W. Mitchell, Status of new thin-film photovoltaic technologies, *Ann. Rev. Mater. Sci.* **12** (1982) 401-415.
30. D. C. Grahame, The electrical double layer and the theory of electrocapillarity, *Chem. Rev.* **41** (1947) 441-501.
31. J. F. Dewald, Semiconductor electrodes, Chap. 17 in *Semiconductors*, Ed. by N. B. Hannay, Reinhold, New York, 1959.
32. M. J. Sparnaay, *The Electrical Double Layer*, Pergamon, New York, 1972.
33. J. Newman, *Electrochemical Systems*, Prentice-Hall, Englewood Cliffs, New Jersey, 1973.
34. S. J. Fonash, *Solar Cell Device Physics*, Academic, New York, 1981.

[35] M. E. Orazem, *Mathematical modeling and optimization of liquid-junction photovoltaic cells*, Ph.D. dissertation, Department of Chemical Engineering, University of California, Berkeley, June 1983, LBL-16131.

[36] M. E. Orazem and J. Newman, Mathematical modeling of liquid-junction photovoltaic cells: I. Governing equations, *J. Electrochem. Soc.* **131** (1984) 2569-2574.

[37] M. E. Orazem and J. Newman, Mathematical modeling of liquid-junction photovoltaic cells: II. Effect of system parameters on current-potential curves, *J. Electrochem. Soc.* **131** (1984) 2574-2582.

[38] E. A. Guggenheim, The conceptions of electrical potential difference between two phases and the individual activities of ions, *J. Phys. Chem.* **33** (1929) 842-849.

[39] R. Memming, Solar energy conversion by photoelectrochemical processes, *Electrochim. Acta* **25** (1980) 77-88.

[40] R. Parsons, Equilibrium properties of electrified interphases, Chap. 3 in *Modern Aspects of Electrochemistry*, No. 1, Ed. by J. O'M. Bockris, Academic, New York, 1954, pp. 103-179.

[41] A. J. Rosenberg, Activity coefficients of electrons and holes at high concentrations, *J. Chem. Phys.* **33** (1960) 655-667.

[42] W. W. Harvey, The relation between the chemical potential of electrons and energy band parameters of the band theory as applied to semiconductors, *J. Phys. Chem. Solids* **23** (1962) 1545-1548.

[43] P. T. Landsberg and A. G. Guy, Activity coefficient and the Einstein relation, *Phys. Rev. B: Condensed Matter* **28** (1983) 1187-1188.

[44] M. B. Panish and H. C. Casey, Jr., The solid solubility limits of zinc in GaAs at 1000°, *J. Phys. Chem. Solids* **28** (1967) 1673-1684.

[45] M. B. Panish and H. C. Casey, Jr., The 1040° solid solubility isotherm of zinc in GaP: the Fermi level as a function of hole concentration, *J. Phys. Chem. Solids* **28** (1968) 1719-1726.

[46] J. C. Hwang and J. R. Brews, Electron activity coefficients in heavily doped semiconductors with small effective mass, *J. Phys. Chem. Solids* **32** (1971) 837-845.

[47] N. G. Nilsson, Empirical approximation for the Fermi energy in a semiconductor with parabolic bands, *Appl. Phys. Lett.* **33** (1978) 653-654.

[48] J. S. Blakemore, Approximations for Fermi-Dirac integrals, especially of order 1/2 used to describe electron density in a semiconductor, *Solid State Electron.* **25** (1982) 1067-1076.

[49] X. Aymerich-Humet, F. Serra-Mestres, and J. Millan, An analytical approximation for the Fermi-Dirac integral $F_{1/2}(\eta)$, *Solid State Electron.* **24** (1981) 981-982.

[50] M. A. Shibib, Inclusion of degeneracy in the analysis of heavily doped regions in silicon solar cells and other semiconductor devices, *Solar Cells* **3** (1981) 81-85.

[51] H. P. D. Lanyon, The physics of heavily doped n^+-p junction solar cells, *Solar Cells* **3** (1981) 289-311.

[52] S. R. Dhariwal and V. N. Ojha, Band gap narrowing in heavily doped silicon, *Solid State Electron.* **25** (1982) 909-911.

[53] F. A. Lindholm and J. G. Fossum, Pictorial derivation of the influence of degeneracy and disorder on nondegenerate minority-carrier concentration and recombination current in heavily doped silicon, *IEEE Electron Device Lett.* **EDL-2** (1981) 230-234.

[54] R. A. Abram, G. J. Rees, and B. L. H. Wilson, Heavily doped semiconductor devices, *Adv. Phys.* **27** (1978) 799-892.

[55] G. D. Mahan, Energy gap in Si and Ge: Impurity dependence, *J. Appl. Phys.* **51** (1980) 2634-2646.

[56] D. S. Lee and J. G. Fossum, Energy band distortion in highly doped silicon, *IEEE Trans. Electron Devices* **ED-30** (1983) 626-634.

[57] S. Selberherr, *Analysis and Simulation of Semiconductor Devices*, Springer-Verlag, New York, 1984.

[58] M. E. Orazem and J. Newman, Activity coefficients of electrons and holes in semiconductors, *J. Electrochem. Soc.* **131** (1984) 2715-2717.

[59] D. B. Bonham and M. E. Orazem, Activity coefficients of electrons and holes in semiconductors with a parabolic density of states, *J. Electrochem. Soc.* (in press).

[60] A. G. Guy, Calculation of activity coefficients of conduction electrons in metals and semiconductors, *Solid State Electron.* **26** (1983) 433-436.

[61] P. T. Landsberg and S. A. Hope, Two formulations of semiconductor transport equations, *Solid State Electron.* **20** (1977) 421-429.

[62] M. S. Lundstrom, R. J. Schwartz, and J. L. Gray, Transport equations for the analysis of heavily doped semiconductor devices, *Solid State Electron.* **24** (1981) 195-202.

[63] A. H. Marshak and R. Shrivastava, Calculation of the electric field enhancement for a degenerate diffusion process, *Solid State Electron.* **25** 151-153.

[64] D. Kumar and S. K. Sharma, Theory of open circuit photo-voltage in degenerate abrupt $p-n$ junctions, *Solid State Electron.* **25** (1982) 1161-1164.

[65] P. T. Landsberg, *Thermodynamics and Statistical Mechanics*, Oxford University Press, New York, 1978, p. 327.

[66] R. B. Bird, W. E. Stewart, and E. N. Lightfoot, *Transport Phenomena*, Wiley, New York, 1960.

[67] R. A. Smith, *Semiconductors*, Cambridge University Press, London, 1959.

[68] R. G. Shulman, Recombination and trapping, Chap. 11 in *Semiconductors*, Ed. by N. B. Hannay, Reinhold, New York, 1959.

[69] J. L. Moll, *Physics of Semiconductors*, McGraw-Hill, New York, 1964.

[70] H. Gerischer, Semiconductor electrochemistry, Chap. 5 in *Physical Chemistry: An Advanced Treatise*, Vol. IX, Ed. by H. Eyring, Academic, New York, 1970.

[71] P. Delahay, *Double Layer and Electrode Kinetics*, Interscience, New York, 1965.

[72] H. Oman and J. W. Gelzer, Solar cells and arrays, in *Energy Technology Handbook*, Ed. by Douglas M. Considine, McGraw-Hill, New York, 1977, pp. 6.56-6.80.

[73] M. A. Russak, J. Reichman, H. Witzke, S. K. Deb, and S. N. Chen, Thin film CdSe photoanodes for electrochemical photovoltaic solar cells, *J. Electrochem. Soc.* **127** (1980) 725-733.

[74] M. S. Kazacos and B. Miller, Electrodeposition of CdSe films from selenosulfite solution, *J. Electrochem. Soc.* **127** (1980) 2378-2381.

[75] C.-H. J. Liu, J. Olsen, D. R. Saunders, and J. H. Wang, Photoactivation of CdSe films for photoelectrochemical cells, *J. Electrochem. Soc.* **128** (1981) 1224-1228.

[76] C.-H. J. Liu and J. H. Wang, Spray-pyrolyzed thin film CdSe photoelectrochemical cells, *J. Electrochem. Soc.* **129** (1982) 719-722.

[77] D. R. Pratt, M. E. Langmuir, R. A. Boudreau, and R. D. Rauh, Chemically deposited CdSe thin films for photoelectrochemical cells, *J. Electrochem. Soc.* **128** (1981) 1627-1629.

[78] K. Rajeshwar, L. Thompson, P. Singh, R. C. Kainthla, and K. L. Chopra, Photoelectrochemical characterization of CdSe thin film anodes, *J. Electrochem. Soc.* **128** (1981) 1744-1750.

[79] J. Reichman and M. A. Russak, Properties of CdSe thin films for photoelectrochemical cells, *J. Electrochem. Soc.* **128** (1981) 2025-2029.

[80] J. Reichman and M. A. Russak, Improved efficiency of n-CdSe thin-film photoelectrodes by zinc surface treatment, J. Appl. Phys. **53** (1982) 708-711.

[81] G. J. Houston, J. F. McCann, and D. Haneman, Optimizing the photoelectrochemical performance of electrodeposited CdSe semiconductor electrodes, J. Electroanal. Chem. **134** (1982) 37-47.

[82] B. Miller, A. Heller, M. Robbins, S. Menezes, K. C. Chang, and J. Thomson, Jr., Solar conversion efficiency of pressure-sintered cadmium selenide liquid-junction cells, J. Electrochem. Soc. **124** (1977) 1019-1021.

[83] A. Heller, K. C. Chang, and B. Miller, Spectral response and efficiency relations in semiconductor liquid junction solar cells, J. Electrochem. Soc. **124** (1977) 697-700.

[84] B. Miller and A. Heller, Semiconductor liquid junction solar cells based on anodic sulphide films, Nature **262** (1976) 680-681.

[85] L. M. Peter, The photoelectrochemical properties of anodic cadmium sulphide films, Electrochim. Acta **23** (1978) 1073-1080.

[86] C. C. Tsou and J. R. Cleveland, Polycrystalline thin-film CdSe liquid junction photovoltaic cell, J. Appl. Phys. **51** (1980) 455-458.

[87] R. A. L. Vanden Berghe, W. P. Gomes, and F. Cardon, Some aspects of the anodic behavior of CdS single crystals in indifferent electrolyte solutions, Ber. Bunsenges. Phys. Chem. **77** (1973) 289-293.

[88] A. Heller, B. Miller, S. S. Chu, and Y. T. Lee, 7.3% efficient thin-film, polycrystalline n-GaAs semiconductor liquid junction solar cell, J. Am. Chem. Soc. **101** (1979) 7633-7634.

[89] W. D. Johnston, Jr., H. J. Leamy, B. A. Parkinson, A. Heller, and B. Miller, Effect of ruthenium ions on grain boundaries in gallium arsenide thin film photovoltaic devices, J. Electrochem. Soc. **127** (1980) 90-95.

[90] R. Noufi and D. Tench, High efficiency GaAs photoanodes, J. Electrochem. Soc. **127** (1980) 188-190.

[91] A. Heller and B. Miller, Photoelectrochemical solar cells: Chemistry of the semiconductor-liquid junction, Chap. 12 in *Interfacial Photoprocesses: Energy Conversion and Synthesis*, Advances in Chemistry Series, 184, Ed. by M. S. Wrighton, American Chemical Society, Washington, D.C., 1980.

[92] B. A. Parkinson, A. Heller, and B. Miller, Enhanced photoeletrochemical solar-energy conversion by gallium-arsenide surface modification, Appl. Phys. Lett. **33** (1978) 521-523.

[93] J. Belloni, G. Van Amerongen, M. Herlem, J.-L. Sculfort, and R. Heindl, Photocurrents from semiconductor-liquid ammonia junctions, J. Phys. Chem. **84** (1980) 1269-1270.

[94] M. A. Butler, R. D. Nasby, and R. K. Quinn, Tungsten trioxide as an electrode for photoelectrolysis of water, Solid State Commun. **19** (1976) 1011-1014.

[95] M. P. Dare-Edwards, A. Hamnett, and J. B. Goodenough, The efficiency of photogeneration of hydrogen at p-type III/V semiconductors, J. Electroanal. Chem. **119** (1981) 109-123.

[96] A. B. Ellis, S. W. Kaiser, and M. S. Wrighton, Semiconducting potassium tantalate electrodes. Photoassistance agents for the efficient electrolysis of water, J. Phys. Chem. **80** (1976) 1325-1328.

[97] F.-R. F. Fan, H. S. White, R. Wheeler, and A. J. Bard, Semiconductor electrodes: XXIX. High efficiency photoelectrochemical solar cells with n-WSe$_2$ electrodes in an aqueous iodide medium, J. Electrochem. Soc. **127** (1980) 518-520.

[98] W. Gissler, P. L. Lensi, and S. Pizzini, Electrochemical investigation of an illuminated TiO_2 electrode, J. Appl. Electrochem. **6** (1976) 9-13.

[99] J. Gobrecht, R. Potter, R. Nottenburg, and S. Wagner, An n-CdSe/SnO$_2$/n-Si tandem electrochemical solar cell, *J. Electrochem. Soc.* **130** (1983) 2280–2283.

[100] K. Hirano and A. J. Bard, Semiconductor electrodes: XXVIII. Rotating ring-disk electrode studies of photo-oxidation of acetate and iodide at n-TiO$_2$, *J. Electrochem. Soc.* **127** (1980) 1056–1059.

[101] P. A. Kohl and A. J. Bard, Semiconductor electrodes: XVII. Electrochemical behavior of n- and p-type InP electrodes in acetonitrile solutions, *J. Electrochem. Soc.* **126** (1979) 598–608.

[102] H. H. Kung, H. S. Jarrett, A. W. Sleight, and A. Ferretti, Semiconducting oxide anodes in photoassisted electrolysis of water, *J. Appl. Phys.* **48** (1977) 2463–2469.

[103] Y. Nakato, N. Takamori, and H. Tsubomura, A composite semiconductor photoanode for water electrolysis, *Nature* **295** (1982) 312–313.

[104] K. Uosaki and H. Kita, Photoelectrochemical characteristics of semiconductor-metal/SPE/metal cells, *J. Electrochem. Soc.* **130** (1983) 2179–2184.

[105] D. Cahen, J. Manassen, and G. Hodes, Materials aspect of photoelectrochemical systems, *Solar Energy Mater.* **1** (1979) 343–355.

[106] A. W. Czanderna, Stability of interfaces in solar energy materials, *Solar Energy Mater.* **5** (1981) 349–377.

[107] L. E. Murr, Interfacial phenomena in solar materials, *Solar Energy Mater.* **5** (1981) 1–19.

[108] J. J. Loferski, Theoretical considerations governing the choice of the optimum semiconductor for photovoltaic solar energy conversion, *J. Appl. Phys.* **27** (1956) 777–784.

[109] J. J. Loferski, Recent research on photovoltaic solar energy converters, *Proc. IEEE* **51** (1963) 667–674.

[110] J. J. Wysocki, Photon spectrum outside the earth's atmosphere, *Solar Energy* **6** (1962) 104.

[111] M. Wolf, Limitations and possibilities for improvement of photovoltaic solar energy converters, Part I: Considerations for earth's surface operation, *Proc. IRE* **48** (1960) 1246–1263.

[112] W. Shockley and H. J. Queisser, Detailed balance limit of efficiency of p–n junction solar cells, *J. Appl. Phys.* **32** (1961) 510–519.

[113] K. C. Chang, A. Heller, B. Schwartz, S. Menezes, and B. Miller, Stable semiconductor liquid-junction cell with 9% solar to electrical conversion efficiency, *Science* **196** (1977) 1097–1098.

[114] A. B. Ellis, J. M. Bolts, S. W. Kaiser, and M. S. Wrighton, Study of n-type gallium arsenide- and gallium phosphide-based photoelectrochemical cells. Stabilization by kinetic control and conversion of optical energy to electricity, *J. Am. Chem. Soc.* **99** (1977) 2848–2854.

[115] P. Josseaux, A. Kirsch-De Mesmaeker, J. Riga, and J. Verbist, Improvements of CdS film photoanodic behavior by sulfur organic reducing agents, *J. Electrochem. Soc.* **130** (1983) 1067–1074.

[116] M. S. Wrighton, J. M. Bolts, A. B. Bocarsly, M. C. Palazzotto, and E. G. Walton, Stabilization of n-type semiconductors to photoanodic dissolution: II–VI and III–V compound semiconductors and recent results for n-type silicon, *J. Vac. Sci. Technol.* **15** (1978) 1429–1435.

[117] A. Heller, B. Miller, and F. A. Thiel, 11.5% Solar conversion efficiency in the photocathodically protected p-InP/V^{3+}-V^{2+}-HCl semiconductor liquid-junction cell, *Appl. Phys. Lett.* **38** (1981) 282–284.

[118] D. S. Ginley, R. J. Baughman, and M. A. Butler, BP-Stabilized n-Si and n-GaAs photoanodes, *J. Electrochem. Soc.* **130** (1983) 1999–2002.

[119] H. Gerischer, Photoelectrochemical solar cells, in *Proceedings of the 2nd Electrochemical Photovoltaic Solar Energy Conference*, Ed. by R. van Overstraeten and W. Palz, Reidel, Boston, 1979, pp. 408-431.

[120] S. Menezes, A. Heller, and B. Miller, Metal filmed-semiconductor photoelectrochemical cells, *J. Electrochem. Soc.* **127** (1980) 1268-1273.

[121] K. W. Frese, Jr., M. J. Madou, and S. R. Morrison, Investigation of photoelectrochemical corrosion of semiconductors: III. Effects of metal layers on the stability of GaAs, *J. Electrochem. Soc.* **128** (1981) 1939-1943.

[122] R. Noufi, D. Tench, and L. F. Warren, Protection of n-GaAs photoanodes with photoelectrochemically generated polypyrrole films, *J. Electrochem. Soc.* **127** (1980) 2310-2311.

[123] R. Noufi, D. Tench, and L. F. Warren, Protection of semiconductor photoanodes with photoelectrochemically generated polypyrrole films, *J. Electrochem. Soc.* **128** (1981) 2596-2599.

[124] T. Skotheim, I. Lundstrom, and J. Prejza, Stabilization of n-Si photoanodes to surface corrosion in aqueous electrolyte with a thin film of polypyrrole, *J. Electrochem. Soc.* **128** (1981) 1625-1626.

[125] T. Skotheim, L.-G. Petersson, O. Inganas, and I. Lundstrom, Photoelectrochemical behavior of n-Si electrodes protected with Pt-polypyrrole, *J. Electrochem. Soc.* **129** (1982) 1737-1741.

[126] F.-R. F. Fan, R. L. Wheeler, A. J. Bard, and R. N. Noufi, Semiconductor electrodes XXXIX. Techniques for stabilization of n-silicon electrodes in aqueous solution photoelectrochemical cells, *J. Electrochem. Soc.* **128** (1981) 2042-2045.

[127] R. A. Bull, F.-R. F. Fan and A. J. Bard, Polymer films on electrodes VII. Electrochemical behavior at polypyrrole-coated platinum and tantalum electrodes, *J. Electrochem. Soc.* **129** (1982) 1009-1015.

[128] K. Rajeshwar, M. Kaneko, and A. Yamada, Regenerative photoelectrochemical cells using polymer-coated n-GaAs photoanodes in contact with aqueous electrolytes, *J. Electrochem. Soc.* **130** (1983) 38-43.

[129] H. S. White, H. D. Abruna, and A. J. Bard, Semiconductor electrodes XLI. Improvement of performance of n-WSe_2 electrodes by electrochemical polymerization of o-phenylenediamine at surface imperfections, *J. Electrochem. Soc.* **129** (1982) 265-271.

[130] L. Fornarini, F. Stirpe, and B. Scrosati, Electrochemical solar cells with layer-type semiconductor anodes: Stabilization of the semiconductor electrode by selective polyindole electrodeposition, *J. Electrochem. Soc.* **130** (1983) 2184-2187.

[131] M. S. Wrighton, Photoelectrochemical conversion of optical energy to electricity and fuels, *Acc. Chem. Res.* **12** (1979) 303-310.

[132] H. Gerischer, On the stability of semiconductor electrodes against photodecomposition, *J. Electroanal. Chem.* **82** (1977) 133-143.

[133] G. L. Araujo, and E. Sanchez, Analytical expressions for the determination of the maximum power point and the fill factor of a solar cell, *Solar Cells* **5** (1982) 377-386.

[134] J. P. Charles, M. Zaghdoudi, P. Mialhe, and A. Marrakchi, An analytical model for solar cells, *Solar Cells* **3** (1981) 45-56.

[135] J. P. Charles, M. Abdelkrim, Y. H. Muoy, and P. Mialhe, A practical method of analysis of the current-voltage characteristics of solar cells, *Solar Cells* **4** (1981) 169-178.

[136] A. De Vos, The fill factor of a solar cell from a mathematical point of view, *Solar Cells* **8** (1983) 283-296.

[137] G. S. Kousik and J. G. Fossum, P^+-N-N^+ Solar cells with hole diffusion lengths comparable with the base width: A simple analytic model, *Solar Cells* **5** (1981-1982) 75-79.

[138] W. A. Miller and L. C. Olsen, Model calculations for silicon inversion layer solar cells, *Solar Cells* **8** (1983) 371-395.

[139] P. Panayotatos and H. C. Card, Recombination in the space-charge region of Schottky barrier solar cells, *Solid State Electron.* **23** (1980) 41-47.

[140] C. M. Singal, Analytical expressions for the series-resistance-dependent maximum power point and curve factor for solar cells, *Solar Cells* **3** (1981) 163-177.

[141] E. J. Soukup and D. R. Slocum, A model for the collection of minority carriers generated in the depletion region of a Schottky barrier solar cell, *Solar Cells* **7** (1982-1983) 297-310.

[142] G. P. Srivastava, P. K. Bhatnagar, and S. R. Dhariwal, Theory of metal-oxide-semiconductor solar cells, *Solid State Electron.* **22** (1979) 581-587.

[143] N. G. Tarr and D. L. Pulfrey, An investigation of dark current and photocurrent superposition in photovoltaic devices, *Solid State Electron.* **22** (1979) 265-270.

[144] R. Tenne, N. Muller, Y. Mirovsky, and D. Lando, The relation between performance and stability of Cd-chalcogenide/polysulfide photoelectrochemical cells: I. Model and the effect of photoetching, *J. Electrochem. Soc.* **130** (1983) 852-860.

[145] K. M. Van Vliet and A. H. Marshak, The Schottky-like equations for the carrier densities and the current flows in materials with a nonuniform composition, *Solid State Electron.* **23** (1980) 49-53.

[146] W. W. Gärtner, Depletion-layer photoeffects in semiconductors, *Phys. Rev.* **116** (1959) 84-87.

[147] M. A. Butler, Photoelectrolysis and physical properties of the semiconducting electrode WO_3, *J. Appl. Phys.* **48** (1977) 1914-1920.

[148] W. Kautek, H. Gerischer, and H. Tributsch, The role of carrier diffusion and indirect optical transitions in the photoelectrochemical behavior of layer type d-band semiconductors, *J. Electrochem. Soc.* **127** (1980) 2471-2478.

[149] L. McC. Williams, *Structural and photoelectrochemical properties of plasma deposited titanium dioxide*, Ph.D. thesis, University of California, Berkeley, December, 1982, LBL-14994.

[150] R. H. Wilson, A model for the current-voltage curve of photoexcited semiconductor electrodes, *J. Appl. Phys.* **48** (1977) 4292-4297.

[151] R. H. Wilson, A model for the current-voltage curve of photoexcited semiconductor electrodes, in *Semiconductor Liquid-Junction Solar Cells*, Ed. by A. Heller, The Electrochemical Society, Princeton, New Jersey, 1977.

[152] W. J. Albery, P. N. Bartlett, A. Hamnett, and M. P. Dare-Edwards, The transport and kinetics of minority carriers in illuminated semiconductor electrodes, *J. Electrochem. Soc.* **128** (1981) 1492-1501.

[153] W. J. Albery and P. N. Bartlett, The recombination of photogenerated minority carriers in the depletion layer of semiconductor electrodes, *J. Electrochem. Soc.* **130** (1983) 1699-1706.

[154] W. J. Albery and P. N. Bartlett, The transport and kinetics of photogenerated carriers in colloidal semiconductor electrode particles, *J. Electrochem. Soc.* **131** (1984) 315-325.

[155] J. Reichman, The current-voltage characteristics of semiconductor-electrolyte junction photovoltaic cells, *Appl. Phys. Lett.* **36** (1980) 574-577.

[156] J. Reichman, Collection efficiency of low-mobility solar cells, *Appl. Phys. Lett.* **38** (1981) 251-253.

[157] H. Reiss, Photocharacteristics for electrolyte-semiconductor junctions, *J. Electrochem. Soc.* **125** (1978) 937-949.

[158] W. L. Ahlgren, Analysis of the current-voltage characteristics of photoelectrolysis cells, *J. Electrochem. Soc.* **128** (1981) 2123-2128.

[159] J. F. McCann and D. Haneman, Recombination effects on current-voltage characteristics of illuminated surface barrier cells, *J. Electrochem. Soc.* **129** (1982) 1134-1145.

[160] J. F. McCann, S. Hinckley, and D. Haneman, An analysis of the current-voltage characteristics of thin-film front wall illuminated and back wall illuminated liquid junction and Schottky barrier solar cells, *J. Electrochem. Chem.* **137** (1982) 17-37.

[161] K. E. Heusler and M. Schulze, Electron-transfer reactions at semiconducting anodic niobium oxide films, *Electrochim. Acta* **20** (1975) 237-244.

[162] F. Williams and A. J. Nozik, Irreversibilities in the mechanism of photoelectrolysis, *Nature* **271** (1978) 137-139.

[163] R. H. Wilson, Observation and analysis of surface states on TiO electrodes in aqueous electrolytes, *J. Electrochem. Soc.* **127** (1980) 228-234.

[164] S. M. Ahmed and H. Gerischer, Influence of crystal surface orientation on redox reactions at semiconducting MoS_2, *Electrochim. Acta* **24** (1979) 705-711.

[165] A. Aruchamy and M. S. Wrighton, A comparison of the interface energetics for n-type cadmium sulfide- and cadmium telluride-nonaqueous electrolyte junctions, *J. Phys. Chem.* **84** (1980) 2848-2854.

[166] A. J. Bard, A. B. Bocarsly, F.-R. F. Fan, E. G. Walton, and M. S. Wrighton, The concept of Fermi level planning at semiconductor-liquid junctions. Consequences for energy conversion efficiency and selection of useful solution redox couples in solar devices, *J. Am. Chem. Soc.* **102** (1980) 3671-3677.

[167] F.-R. F. Fan and A. J. Bard, Semiconductor electrodes. 24. Behavior and photoelectrochemical cells based on p-type GaAs in aqueous solutions, *J. Am. Chem. Soc.* **102** (1980) 3677-3683.

[168] A. B. Bocarsly, D. C. Bookbinder, R. N. Dominey, N. S. Lewis, and M. S. Wrighton, Photoreduction at illuminated p-type semiconducting silicon photoelectrodes. Evidence for Fermi level pinning, *J. Am. Chem. Soc.* **102** (1980) 3683-3688.

[169] G. Nagasubramanian, B. L. Wheeler, and A. J. Bard, Semiconductor electrodes: XLIX. Evidence for Fermi level pinning and surface-state distributions from impedance measurements in acetonitrile solutions with various redox couples, *J. Electrochem. Soc.* **130** (1983) 1680-1688.

[170] D. C. Card and H. C. Card, Interfacial oxide layer mechanisms in the generation of electricity and hydrogen by solar photoelectrochemical cells, *Solar Energy* **28** (1982) 451-460.

[171] R. Haak and D. Tench, Electrochemical photocapacitance spectroscopy method for characterization of deep levels and interface states in semiconductor materials, *J. Electrochem. Soc.* **131** (1984) 275-283.

[172] W. Siripala and M. Tomkiewicz, Surface recombination at n-TiO_2 electrodes in photoelectrolytic solar cells, *J. Electrochem. Soc.* **130** (1983) 1062-1067.

[173] K. Uosaki and H. Kita, Effects of the Helmholtz layer capacitance on the potential distribution at semiconductor/electrolyte interface and the linearity of the Mott-Schottky plot, *J. Electrochem. Soc.* **130** (1983) 895-897.

[174] C. G. B. Garrett and W. H. Brattain, Physical theory of semiconductor surfaces, *Phys. Rev.* **99** (1955) 376-387.

[175] E. O. Johnson, Large-signal surface photovoltage studies with germanium, *Phys. Rev.* **111** (1958) 153-166.

[176] A. Many, Y. Goldstein and N. B. Grover, *Semiconductor Surfaces*, North-Holland, Amsterdam, 1965.
[177] S. R. Morrison, *The Chemical Physics of Surfaces*, Plenum Press, New York, 1977.
[178] J. O'M. Bockris and K. Uosaki, The theory of the light-induced evolution of hydrogen at semiconductor electrodes, *J. Electrochem. Soc.* **125** (1978) 223-227.
[179] R. Memming, The role of energy levels in semiconductor-electrolyte solar cells, *J. Electrochem. Soc.* **125** (1978) 117-123.
[180] K. W. Frese, Jr., A study of rearrangement energies of redox species, *J. Phys. Chem.* **85** (1981) 3911-3916.
[181] W. Kautek and H. Gerischer, A kinetic derivation of the photovoltage for electrochemical solar cells employing small-band gap semiconductors, *Electrochim. Acta* **27** (1982) 355-358.
[182] K. Rajeshwar, Charge transfer in photoelectrochemical devices via interface states: United model and comparison with experimental data, *J. Electrochem. Soc.* **129** (1982) 1003-1008.
[183] J.-N. Chazalviel, Electrochemical transfer via surface states: A new formulation for the semiconductor-electrolyte interface, *J. Electrochem. Soc.* **129** (1982) 963-969.
[184] P. Singh, K. Rajeshwar, R. Singh, and J. DuBow, Estimation of series resistance losses and ideal fill factors for photoelectrochemical cells, *J. Electrochem. Soc.* **128** (1981) 1396-1398.
[185] J. F. McCann, S. P. S. Badwal, and J. Pezy, The electrical analogue of an n-SnO_2-1 M NaOH-Pt cell, *J. Electroanal. Chem.* **118** (1981) 115-130.
[186] J. F. McCann and S. P. S. Badwal, Equivalent circuit analysis of the impedance response of semiconductor-electrolyte-counterelectrode cells, *J. Electrochem. Soc.* **129** (1982) 551-559.
[187] J. R. Macdonald, Accurate solution of an idealized one-carrier metal-semiconductor junction problem, *Solid-State Electron.* **5** (1962) 11-37.
[188] A. De Mari, An accurate numerical steady-state one-dimensional solution of the p-n junction, *Solid-State Electron.* **11** (1968) 33-58.
[189] A. De Mari, An accurate numerical steady-state one-dimensional solution of the p-n junction under arbitrary transient conditions, *Solid-State Electron.* **11** (1968) 1021-1053.
[190] S. C. Choo, Numerical analysis of a forward-biased step-junction p-i-n diode, *IEEE Trans. Electron Devices* **ED-18** (1971) 574-586.
[191] S. C. Choo, Theory of a forward-biased diffused-junction p-i-n rectifier—Part 1: Exact numerical solution, *IEEE Trans. Electron Devices* **ED-19** (1972) 954-966.
[192] J. E. Sutherland and J. R. Hauser, A computer analysis of heterojunction and graded composition solar cells, *IEEE Trans. Electron Devices* **ED-24** (1977) 363-372.
[193] M. Hack and M. Shur, Computer simulation of amorphous silicon based p-i-n solar cells, *IEEE Electron Device Lett.* **EDL-4** (1983) 140-143.
[194] M. F. Lamorte and D. H. Abbott, AlGaAs/GaAs cascade solar cell computer modeling under high solar concentration, *Solar Cells* **9** (1983) 311-326.
[195] M. F. Lamorte and D. H. Abbott, Influence of band gap on cascade solar cell efficiency, *Solar Cells* **10** (1983) 33-48.
[196] R. Radojcic, A. E. Hill and M. J. Hampshire, A numerical model of a graded band gap CdS_xTe_{1-x} solar cell, *Solar Cells* **4** (1980) 109-120.
[197] D. Laser and A. J. Bard, Semiconductor electrodes: VII. Digital simulation of charge injection and the establishment of the space charge region in the absence and presence of surface states, *J. Electrochem. Soc.* **123** (1976) 1828-1832.

[198] D. Laser and A. J. Bard, Semiconductor electrodes: VIII. Digital simulation of open-circuit photopotentials, *J. Electrochem. Soc.* **123** (1976) 1833-1837.

[199] D. Laser and A. J. Bard, Semiconductor electrodes: IX. Digital simulation of the relaxation of photogenerated free carriers and photocurrents, *J. Electrochem. Soc.* **123** (1976) 1837-1842.

[200] D. Laser, Modes of charge transfer at an illuminated semiconductor electrode: A digital simulation, *J. Electrochem. Soc.* **126** (1979) 1011-1014.

[201] N. M. Bogatov, L. I. Gromovoi, M. B. Zaks, and E. V. Lelyukh, A numerical solution of the one-dimensional boundary value problem of charge transfer in solar cells, *Geliotekhnika* **18** (1982) 7-14.

[202] J. Newman, Numerical solution of coupled, ordinary differential equations, *Ind. Eng. Chem. Fund.* **7** (1968) 514-517.

[203] R. White, C. M. Mohr, Jr., P. Fedkiw, and J. Newman, The fluid motion generated by a rotating disk: A comparison of solution techniques, *Lawrence Berkeley Laboratory Report*, LBL-3910, November 1975.

[204] R. E. White, *Simultaneous reactions on a rotating-disk electrode*, Ph.D. thesis, University of California, Berkeley, March 1977, LBL-6094.

[205] D. J. Leary, J. O. Barnes, and A. G. Jordan, Calculation of carrier concentration in polycrystalline films as a function of surface acceptor state density: Application for ZnO gas sensors, *J. Electrochem. Soc.* **129** (1982) 1382-1386.

[206] J. A. Davis, R. O. James, and J. O. Leckie, Surface ionization and complexation at the oxide-water interface: I. Computation of electrical double layer properties in simple electrolytes, *J. Colloid Interface Sci.* **63** (1978) 480-499.

[207] J. A. Davis and J. O. Leckie, Surface ionization and complexation at the oxide-water interface: II. Surface properties of amorphous iron oxyhydroxide and adsorption of metal ions, *J. Colloid Interface Sci.* **67** (1978) 91-107.

[208] J. A. Davis and J. O. Leckie, Surface ionization and complexation at the oxide-water interface: 3. Adsorption of ions, *J. Colloid Interface Sci.* **74** (1980) 32-43.

[209] P. A. Iles, Evolution of silicon solar cell design, *9th IEEE Photovoltaic Specialists Conference*, Silver Springs, Maryland, 1972, pp. 1-5.

[210] B. Parkinson, An evaluation of various configurations for photoelectrochemical photovoltaic solar cells, *Solar Cells* **6** (1982) 177-189.

[211] M. E. Orazem and J. Newman, Mathematical modeling of liquid-junction photovoltaic cells: III. Optimization of cell configuration, *J. Electrochem. Soc.* **131** (1984) 2582-2589.

[212] R. N. Fleck, D. N. Hanson and C. W. Tobias, Numerical evaluation of current distribution in electrochemical systems, *Lawrence Berkeley Laboratory Report*, UCRL-11612, 1964.

[213] C. Kasper, The theory of the potential and the technical practice of electrodeposition: I. The general problem and the cases of uniform flow, *Trans. Electrochem. Soc.* **77** (1940) 353-363.

[214] C. Kasper, The theory of the potential and the technical practice of electrodeposition: II. Point-plane and line-plane systems, *Trans. Electrochem. Soc.* **77** (1940) 365-384.

[215] C. Kasper, The theory of the potential and the technical practice of electrodeposition: III. Linear polarization on some line-plane systems, *Trans. Electrochem. Soc.* **78** (1940) 131-146.

[216] C. Kasper, The theory of the potential and the technical practice of electrodeposition: IV. The flow between and to circular cylinders, *Trans. Electrochem. Soc.* **78** (1940) 147-161.

[217] C. Kasper, The theory of the potential and the technical practice of electrodeposition: V. The two-dimensional rectangular enclosures, *Trans. Electrochem. Soc.* **82** (1942) 153-185.

[218] H. F. Moulton, Current flow in rectangular conductors, *Proc. London Math. Soc. (Ser. 2)* **3** (1905) 104-110.

[219] J. Newman, The fundamental principles of current distribution in electrochemical cells, in *Electroanalytical Chemistry*, Ed. by A. J. Bard, Vol. 6 (1973), pp. 187-352.

[220] M. E. Orazem and J. Newman, Primary current distribution and resistance of a slotted-electrode cell, *J. Electrochem. Soc.* **131** (1984) 2857-2861.

[221] F. Bowman, *Introduction to Elliptic Functions with Applications*, Wiley, New York, 1953.

[222] M. Abramowitz and I. A. Stegun, *Handbook of Mathematical Functions*, Dover, New York, 1964.

[223] D. Canfield and S. R. Morrison, Electrochemical storage cell based on polycrystalline silicon, *Lawrence Berkeley Laboratory Report*, LBL-14639, 1982.

[224] N. Muller and D. Cahen, Photoelectrochemical solar cells: Temperature control by cell design and its effects on the performance of cadmium chalcogenide-polysulfide systems, *Solar Cells* **9** (1983) 229-245.

[225] N. L. Weaver, R. Singh, K. Rajeschwar, P. Singh, and J. DuBow, Economic analysis of photoelectrochemical cells, *Solar Cells* **3** (1981) 221-232.

[226] P. Singh and J. D Leslie, Economic requirements for new materials for solar photovoltaic cells, *Solar Energy* **24** (1980) 589-592.

[227] F. M. Kamel and A. Muhlbauer, Impact of solar cell efficiency on the design of photovoltaic systems, *Solar Cells* **11** (1984) 269-280.

3

Electron Transfer Effects and the Mechanism of the Membrane Potential

L. I. Boguslavsky

A. N. Frumkin Institute of Electrochemistry, Academy of Sciences of the USSR, 117071 Moscow, USSR

I. INTRODUCTION

Classical electrochemistry on metal electrodes, or as Bockris[1] defined it, electrodics, operates, as a rule, with the electrode potential, determined in the simplest case by a single reaction, called the potential-determining reaction.

The purpose of measurements of the electrode potential is, above all, the desire to obtain an energy scale that would serve as a reference for other parameters of the process under investigation. These parameters can be the current passing through the electrode, the reaction product yield, changes in adsorption, etc. The use of the electrode potential as an energy scale only makes sense if the process takes place on an equipotential surface, which is always the case for metal electrodes.

Membrane electrodes present a much more complex object for investigation than metal electrodes. The redistribution of charges in the course of the reaction between the components in the solution around the membrane and the components inside the membrane can lead to generation of a potential jump between electrolyte and membrane. However, for the formation of a transmembrane potential it is essential that the membrane should be permeable to certain ions, as for instance, hydrogen ions. Consequently, although redox reactions[2,3] can occur in the membrane/electrolyte system, just as on a metal electrode, they cannot always be detected by measuring the membrane potential alone.[4] On the contrary, potential jumps at both interfaces between membrane and aqueous electrolyte solutions can be measured.[5-7]

When attempting to investigate the potentials of membranes of organisms, we come up against a new difficulty, namely, that the object being examined is a device functioning in a complex manner not at all suited for observing the potential or any other electric parameter. As a rule, only disintegration of the investigated object reveals changes in the electric potentials, which can then be compared with those in other quantities to be controlled. The approach, based on the presumption that dismembering of a complex object does not change the functions of its separate elements, has become traditional in biochemistry and has led to amazing successes, thereby fully justifying itself. Nevertheless, it is not absolutely faultless, since the possibility always exists that the observed phenomenon is common to the destructed objects and not to the initial ones.

Moreover, since numerous reactions take place simultaneously in biological membrane systems, the membrane potential can very often not be the result of a single process, but rather a steady-state potential reflecting the joint functioning of several systems maintaining the homeostasis of the organism. For example, the potential observed in model systems when the enzymes of the respiratory chain of mitochondrian changes upon addition of "food" such as NADH, ascorbic acid, and a number of other reductants. Exhaustion of the energy source or addition of inhibitors of the redox reaction leads to disappearance of the potential. It is convenient, although rather arbitrary, to divide all systems participating in the formation of the membrane potential into systems of active and passive ion transport. When divided in this way, the mechanisms of energy supply of both animals and plants prove similar to other systems of active transport, since the transport of electrons and protons is controlled by the functioning of these energy-transforming systems.

The present chapter is limited chiefly to examination of redox processes that could serve as the source of potential on bilayer lipid membranes—the simplest model of a biological membrane.

II. MODELING NONENZYMATIC SYSTEMS OF ELECTRON TRANSFER IN THE INITIAL PART OF THE RESPIRATORY CHAIN OF MITOCHONDRIA

1. Respiratory Chain

The two main processes of providing the organisms with energy—photosynthesis and respiration—are localized in the membranes of chloroplasts, mitochondria, and bacteria.[8-10] The energy coming from the environment can be accumulated in organic substances liable to oxidation—substrates—or could be delivered by light quanta.

The process of the oxidation of substrates is a multistage one and the energy is used by organisms during transfer of electrons along the so-called electron transport chain—a complex of closely bound enzymes inserted in the membrane (Fig. 1).

When investigating the electron transport processes in mitochondria and chloroplasts and bacteria, it is simplest to assume that the interaction of the chain components follows the mass action law.[11-14] That would mean that the free movement of individual elements of the chain in the membrane is possible and also transfer of the charge by accidental collisions. However, transfer of electrons during both respiration and photosynthesis passes along the electron transfer chain organized into definite structural complexes. Consequently a molecule possessing an electron can donate it to a

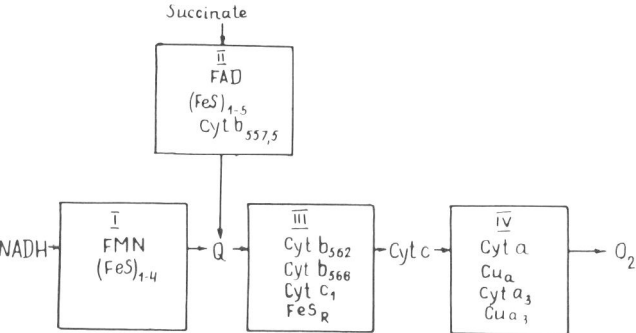

Figure 1. Multienzyme complexes forming part of the respiratory chain of mitochondria.[10]

molecule only in its immediate vicinity, and only if there is a vacancy for the electron, i.e., if the molecule is oxidized. Moreover, it is necessary to take into account the possible dependence of the rate constant of electron transfer between two neighboring elements of the chain on the state (for example, the degree of oxidation) of all the other elements. This property of cooperativeness is the most important feature of transport along the electron transport chain.[15-17]

A significant difference between the free collisions model and the structural model lies in the fact that the latter predicts a higher reaction order with respect to donor or acceptor concentration under certain conditions. The degree of reduction (or oxidation) of the carrier depends on how far it is located from the beginning of the chain. Using the structural approach, it is possible to find the place of each element in the transport chain by determining the dependence of the rate of the process on the degree of reduction of this element.[14,18]

Another important feature of the electron transport chain is that the macromolecular protein complexes responsible for electron transfer along the electron transport chain have a dynamic structure. That was clearly shown, for example, in the investigation of electron transfer in the reaction centers of purple bacteria Rhodospirileum rubrum in oxidizing conditions.[6]

When a parameter, for example, humidity, temperature, charge, affecting the structure of the protein complex attained a certain value after treatment with preparations inhibiting the intramolecular mobility, it was found that the structure of the complex in the oxidized and reduced states is different.[16,19-22] Earlier, a similar conclusion was drawn from an analysis of experiments using spin labels which are the indicators of intramolecular mobility.[23]

2. Potentials of the Respiratory Chain Elements

Calculation of the change in the free energy of redox reactions taking place in complexes and its comparison with that accepted for reactions in solutions indicates that in the presence of cooperativeness in the electron transfer or in the absence of an equilibrium with the medium, a conventional thermodynamics analysis based on the use of the concentrations of oxidized and reduced forms should be replaced by methods of nonequilibrium thermodynamics.[10,24,26]

Although a thermodynamic analysis of the functioning of the respiratory chain should be conducted using methods of nonequilibrium thermodynamics, in practice it is the so-called midpoints potentials determined by the method of potentiometric titration with the use of mediators[13] that are used for characterizing the redox properties of the components of the respiratory chain. The method devised for this purpose is founded on the supposition that the mediator is oxidized or reduced only by the chain components of interest to us. The correctness of this supposition is usually controlled by independent methods, as for instance, the spectroscopic or ESR method.

A table of the redox potentials of the components of the respiratory chain of mitochondria obtained this way is given below.[13]

Each result was obtained on a number of appropriate mediators, not just one, and therefore does not depend on the nature of the mediator and its concentration (See Table 1). Table 1 shows that all the components can be divided into four groups according to their potentials: with potentials of -300, 0, 230, and

Table 1
The Half-Reduction Potentials of the Components of the Mitochondrial Respiratory Chain (mV), pH = 7.2 [a]

Component	n value	φ_1 (uncoupled)	φ_2 (+ATP)	$\varphi_1 - \varphi_2$
Cytochrome a_3	1.0	365 ± 10 mV	135 ± 10 mV	-230 mV
Cytochrome c_1	1.0	215 ± 10 mV	175 ± 10 mV	-40 mV
Cytochrome a	1.0	210 ± 10 mV	260 ± 10 mV	50 mV
Ubiquinone	2.0	45 ± 10 mV	45 ± 15 mV	0 mV
Cytochrome b_K	1.0	30 ± 10 mV	30 ± 10 mV	0 mV
Fe-S center 5	1.0	40 ± 15 mV	40 ± 15 mV	0 mV
Fe-S SDH	1.0	30 ± 10 mV	30 ± 10 mV	0 mV
Cytochrome b_T	1.0	-30 ± 10 mV	245 ± 15 mV	275 mV
Fe-S center (3 + 4)	1.0	-245 ± 15 mV	-245 ± 15 mV	0 mV

[a] Reference 13.

400 mV. Within each of these groups the standard redox potentials are similar. Consequently the change in the free energy ΔG for electron transfer within each group should be close to zero. The change in ΔG takes place when an electron passes from one potential group to another and can be used for the synthesis of ATP. Since the table data refer to suspensions of mitochondria where individual enzymes form chains, it should be assumed that the surface of the mitochondrial membrane is not an equipotential and at each given potential only certain parts of its containing enzyme with the appropriate potential are capable of an electron-exchange reaction with the components in the solution. Actually, the distribution of the respiratory chain elements according to potentials indicates the sequence of the movements of the electron from the substrate. It should, however, be pointed out that for a number of chain components their redox potential shifts depending on whether the ATP complex is included or excluded (uncoupled mitochondria or +ATP) (see Table 1). This signifies that the redox potentials obtained cannot be used without a special examination in each case as to what extent the conditions in which they were obtained correspond to the conditions of the given experiment in which they are to be used.

3. Ubiquinones in the Respiratory Chain

The participation of lipophylic quionones, possessing an isoprenoid side chain of different length, is a distinctive feature of electron transport both for animals and the plant kingdom. Plastoquinones are to be found in the membranes of chloroplasts,[27,28] while mitochondria contain ubiquinones.[29-31]

It is well known that the extraction of the coenzyme Q by isooctane abolishes the activity of the electron transport chain. On the contrary introduction of the coenzyme Q into lipid bilayers makes it possible to model the electron transport through biomembranes.[32-36]

There exist several hypotheses as to how the ubiquinone functions in the membrane. The most common one is that coenzyme Q floats freely in the membrane[37] or forms a mobile Q pool[27] or Q cycle[38,39] Ubiquinones differ in the length of the side isoprenoid chain, consisting usually of 6-10 carbon atoms.

Investigations of various analogs of coenzyme Q^{40} have shown that the isoprenoid chain does not take part in charge transfer, but it regulates the lipophylic properties of the molecule as a whole, which ensures the optimal contact of the quinoid ring with its respective partner in the electron exchange reaction. Owing to carbonyl groups, the quinoid ring interacts strongly with metal atoms. Coenzyme Q exchanges electrons with Fe-S centers of dehydrogenase, which is an essential condition for electron transport from complexes I and II of the respiratory chain. In the process of redox transformation ubiquinones, like ordinary quinones, form radicals fairly easily. In the system investigated they can be generated by the interaction of nicotinamide with coenzyme Q at the interface. Using the ESR method it is shown for submitochondrial particles that electron transfer from NADH to coenzyme Q does indeed lead to generation of ubisemiquinones.[41,42] They form from 0.2% to 1.5% of the total quantity of ubiquinone. The formation of semiubiquinone radicals was also observed when investigating succinate-Q-reductase.[42] the rate of the process of generation of $Q_6^{\cdot}H$ and $Q_6^{\cdot -}$ radicals depends on the aqueous solution pH because of the existence of an acid–base balance

$$\cdot Q_6^- + H_2O \rightleftharpoons Q_6H^{\cdot} + OH^- \qquad (1)$$

whose pH equals 5.9.

It should be noted that, apparently, only the complex of coenzyme Q with protein is a really functioning system in native coupling membranes. Although an equilibrium does exist between the free and the bound coenzyme Q, nevertheless the rate of redox transformations of free Q is considerably less than that of other redox systems in the electron transport chain. Possibly, the protein in the Q–protein complex stabilizes the radical $Q_6H^{\cdot}$, as a result of which the latter can readily take part in the electron-exchange reaction.[31]

4. Participation of Membrane Lipids in the Functioning of the Respiratory Chain

The enzymatic redox reactions in native systems are coupled with the destruction of lipids. This means that interaction occurs between the two redox systems, one of which includes the enzymes of the respiratory chain, while the other includes the lipids of the membrane.[43,44] For example, in the microsomes isolated from the liver cells the oxidation of NADPH catalyzed by enzymes is accompanied by peroxidation of lipids.[45]

The oxidation of lipids is due primarily to the nonsaturated fatty acid residues. For instance, in lipids extracted from the heart muscle for one atom of phospholipid phosphorus there are on the average 3.2 double bonds. The functioning of mitochondria depends on the nonsaturated state of lipids. When these are replaced by compounds with saturated hydrocarbon chains the electron transport in the respiratory chain is disturbed.[8] The oxidation of phospholipids eventually leads to destruction of biomembranes. It should be noted at this point that the strongly interacting chain components with different redox potentials (Table 1) inevitably lead to heterogeneity of the membrane surface, and, as a result, to corrosion of its material, i.e., to oxidation of phospholipids. The lipid layer of the membrane could roughly be likened in this case to a thin film of insulator.[46] As a result of the oxidation of lipids, a considerable quantity of oxidation products accumulate in the membrane and they are physiologically active. It is with the accumulation of these products in the organism that the disturbance of its various physiological functions is linked, including such diseases as cardiac coronary diseases, cancer,[47] as well as aging.[48] The products of oxidation of lipids react with amino acids, proteins, and DNA.[49] In order to compensate for the process of membrane destruction a considerable part of the energy generated in mitochondria is used up for synthesis of lipids to replace the oxidized lipids. It should be noted that the oxidation process of lipids in membranes is not merely an annoying and undesirable chain of side reactions causing destruction of the membrane, but plays a certain physiological role. Comparatively recently[50] investigators have found in the mitochondria of plants that apart from the path of the electron along the electron transport chain via cytochromes (Fig. 1), there exists a second path—a nonenzymatic one which is not inhibited by cyanides. Both these paths include as an essential element ubiquinone, which is an electron carrier in membranes.[51] Ubiquinone reduces the products of peroxidation of lipids (Fig. 2). The oxidation of ubiquinone can proceed along the nonenzymatic path or with the participation of enzymes, for instance, it can be

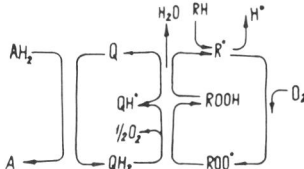

Figure 2. Nonenzymatic route of electron transport AH_2-Q-RH-O_2.[50]

activated by peroxides formed during oxidation of the substrates in the mitochondrial respiratory chain. One of the reaction products in the interaction of ubiquinone with peroxides being the semiquinone radical ˙QH, this path of the electron is inhibited by trappers of free radicals propyl galate, such as substituted phenols specifically. The electron transport along this path is, apparently, regulated by changes in the concentration of the oxides which are generated during the action of dehydrogenases.[52] A higher level of the reduction of the respiratory chain at which it functioned more intensively inhibits the nonenzymatic path. Consequently, the nonenzymatic chain of electron transport based on peroxidation of membrane lipids, which is insensitive to inhibition by cyanides, can be regarded as an emergency path for electron transport in living systems.

Extensive investigations of the auto-oxidation of lipids in biological membranes, both of native objects[34,35] and model systems,[43] lead to the conclusion that the functioning of enzymes of the respiratory chain and the oxidation of lipids are coupled processes. However, for a long time it remained unclear whether oxidation of lipids was in any way related to generation of the membrane potential. Comprehensive investigation of the nature of membrane potentials arising during redox reactions in mitochondria and in the photosynthesis appartus of plants would be possible only when details of the molecular structure of electron and proton carriers in the membrane are known. Studies are still being made, not only on their molecular structure, but also on their exact position in the membranes of organelles. It is this circumstance, as well as the difficulty of inserting proteins into an artificial membrane, that prompts investigators to study the potentials and conductivity of simpler objects—bilayer phospholipid membranes, which contain certain coenzymes of the respiratory chain. A reconstruction presupposing only the building of the nonenzymatic chain is, of course, an oversimplification of the native system, but certain features of such a system can be accounted for. With this in mind, an attempt was made[34,35,55-57] to build, using a lipid bilayer, a part of the respiratory chain which would include the following components:

$$NADH \to FMN \to Q_6 \to K_3Fe(CN)_6 \qquad (2)$$

The results of the investigations of transmembrane potentials in this system under various conditions are given below in Sections 5-9. However, first it would be more reasonable to consider a simpler chain.

5. Transmembrane Potentials in the Chain NADH–Coenzyme Q–O$_2$

Investigating membranes containing coenzyme Q or model compounds—various naphthaquinones soluble in lipids—Ismailov et al.[34,54] found that when a reducing agent such as NaBH$_4$ or NADH is added at one side, a transmembrane potential is generated. Initially the observed potential was attributed to electron transport across the bilayer. However, as was first suggested by Yaguzhinsky,[34,35,55] the results obtained could also be attributed to changes in pH of the layer adjacent to the membrane (Fig. 3). Confirmation was provided by the results obtained when investigating the reaction of oxidation of NADH of oxygen, which takes place in the vicinity of the border of the bilayer lipid membrane and is accompanied by consumption of hydrogen ions[34,35,55-57]:

$$NADH + Q_6 \rightleftharpoons NAD^+ + Q_6H^-$$
$$Q_6H^- + \tfrac{1}{2}O_2 + H^+ \to Q_6 + H_2O \qquad (3)$$

Earlier, the unstirred layer adjacent to the bilayer membrane was investigated by Le Blanc.[58] The effect of the unstirred layer on the transport of ions was observed in the individual cells,[59] thylakoids,[60] and other objects.[61-67]

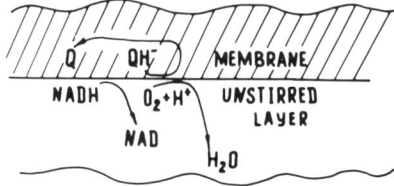

Figure 3. Alkalization of the layer adjacent to the membrane as a result of oxidation of NADP by air oxygen. The reaction takes place at the interface between electrolyte and the membrane containing coenzyme Q_6.[35]

Figure 4. Value of transmembrane potential on bilayer from phosphatidylethanolamine with ubiquinone Q_6 as a function of NADP concentration in aqueous solution.[56] When NADP is introduced: pH does not change (1), pH shifts in the alkaline direction (2), if the potential shift were determined only by pH gradient when NADP is added (3). Initial value of pH 8.8. Buffer 5 mM citric acid, 5 mM Tris, 50 mM NaCl, 6.10^{-6} M EDTA, 10^{-6} M TTFB.

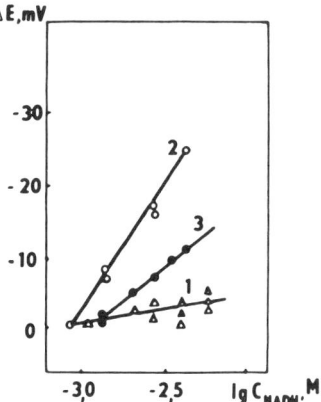

When investigating the transmembrane potential in the systems†

$$pH_1, \quad NADH, \quad O_2 \quad \begin{vmatrix} BLM, \\ TTPB \end{vmatrix} \quad O_2, \quad pH_2 \qquad (4)$$

$$pH_1, \quad NADH, \quad O_2 \quad \begin{vmatrix} Q_6, \\ BLM, \\ TTPB \end{vmatrix} \quad O_2, \quad pH_2 \qquad (5)$$

Sokolov and Shipunov[56] found that for reaction (3) to take place it is necessary for pH to change at the very moment NADH is added to the aqueous solution. It turned out that the value of the potential in the systems investigated depended on the pH value at the very moment NADH is added. If a NADH solution was introduced into the cell on one side of the membrane (chain 4) whose pH equaled the initial pH to an accuracy of 0.01, then the transmembrane potential changed so negligibly that within the accuracy of the experiment and its reproducibility the change could be disregarded (Fig. 4, curve 1). But if at first the pH in the cell on one side of the membrane was changed, and then NADH was added without changing the pH, the potential after the addition of NADH also remained practically unchanged. In this case the potential was determined only by the initial pH gradient in keeping with the

† Bilayer lipid membrane (BLM) containing weak acid tetrachlorotrifluorinemethylbenzimidazole (TTFB) is protonselective.

Nernst law (Fig. 4, curve 3). However, when the pH of the NADH solution differed from the pH of the solution in the cell, then the pH in the cell changed at the very moment NADH was added. A transmembrane potential was generated, the value of which exceeded the one that could be expected from the pH gradient on both sides of the membrane. A comparison of the curves 2 and 3 reveals that the maximum value of the observed potential is 2-2.5 times greater than that determined by the Nernst law. How much greater it was depended on NADH concentration.

The connection between the observed effect and the reaction of oxidation of NADH by oxygen can be established by investigating the dependence of this effect on the composition of solution and membrane, the buffer capacity of the solutions, and the rate of stirring.[63] At a high buffer capacity the dependence of potential on the changes in pH tends to a value corresponding to the Nernst law. Intensive stirring on the side to which NADH was added leads, as in the case of stirring on both sides, to an irreversible drop in the potential. The potential drop can be explained by a decrease of the difference in the concentration of protons in the layer adjacent to the membrane and in the bulk of the solution. In the absence of ubiquinone in the membrane (chain 5) the value of the generated potential proves less than when it is introduced into the membrane, which points doubtless to the participation of ubiquinone in the potential generation process.

It was also established that the nature of the lipids which make up the bilayer substantially affects the value and even the sign of the generated potential. Alongside of the reaction that is accompanied by alkalization of the layer adjacent to the membrane there also occurs the reaction of peroxidation of lipids accompanied by the ejection of protons into the layer adjacent to the membrane (Fig. 5).

The formation of protons was detected experimentally during the peroxidation of lipids of bilayer lipid membranes.[68,69] It was shown by direct measurements of the aqueous solution pH that X-ray irradiation of the aqueous solution on one side of the bilayer leads to ejection of protons into the solution. As a result, a transmembrane potential difference is generated on BLM. Since after X-ray irradiation peroxide anions are formed in the solution, peroxidation of lipids can be initiated by the anion radical O_2^-.[70]

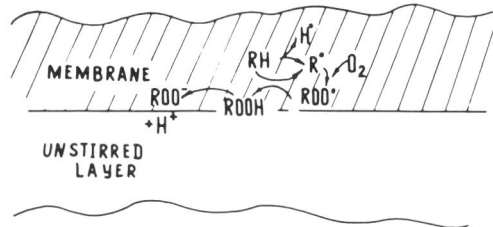
UNSTIRRED
LAYER

Figure 5. Acidification of the layer adjacent to the membrane as a result of peroxidation of membrane lipids by air oxygen. This reaction is usually coupled with the main reaction of electron transport along the electron transport chain. See also Fig. 2.

The participation of lipids in the generation of potential on the membrane becomes most evident from a comparison of the results of experiments with bilayers formed from phospholipids containing nonsaturated or saturated fatty acid residues (chain 6) in air atmosphere or in an inert atmosphere:

$$pH_1, \quad NADH, \quad O_2 \quad \left| \begin{array}{c} BLM, \\ \text{ubiquinone } Q_8, \\ TTPB \end{array} \right| \quad O_2, \quad pH_1 \quad (6)$$

In system (6) peroxidation of membrane lipids can be initiated not only by ubiquinone, but also by nicotinamides. This can be explained if one bears in mind that, as can be seen from reaction (3), redox equilibrium between the oxidized and reduced forms of coenzymes is accomplished through intermediate stages at which radicals are formed. For instance, NADH is not oxidised immediately to NAD^+ but passes through the stages of formation of $NAD^{\cdot}$ or $NADH^+$. Since nicotinamide does not penetrate into the membrane phase, it itself cannot cause oxidation of hydrocarbon residues of fatty acids, it could be supposed that NADH initiated peroxidation through peroxide anions $\cdot O_2^-$, which can be formed according to the following reaction:

$$NAD^{\cdot} + O_2 \rightarrow NAD^+ + \cdot O_2^- \quad (7)$$

at a sufficiently high rate.

Figure 6 shows the dependence of $\Delta\varphi$ in chain (6) on the value of pH changes, beginning with the initial pH value 8.8 after equal

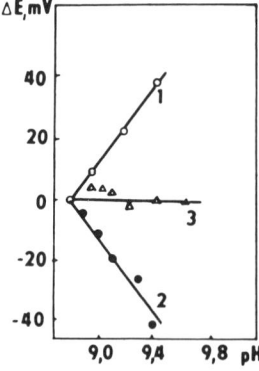

Figure 6. Dependence of transmembrane potential on abrupt changes of pH caused by equal additions of concentrated NaOH solution on both sides of a bilayer made from phosphatidylethanolamine (1, 3), containing residues of unsaturated fatty acids, dipalmitoyllecithin (2) containing saturated hydrocarbon chains.[56] Curve (3) was obtained in an argon atmosphere. In all cases NADP in the concentration of 2.7×10^{-3} M was present on one side of the membrane. The experiment was conducted in the presence of 10^{-6} M TTFB.

additions of a concentrated alkali solution on both sides of the bilayer. In this case pH on both sides of the membrane was the same. A straight line 1 was obtained on BLM formed from phosphatidylethanolamine containing residues of nonsaturated fatty acids. A potential with a minus sign arises on the membrane on the side of NADH. In this case the predominating reactions are accompanied by generation of protons (Fig. 5).

If the bilayers are formed on dipalmitoyllecithin, the molecule of which contains saturated hydrocarbon chains, then the potential jump has a minus sign on the side to which nicotinamide was added (curve 2). After removal of air oxygen by blowing argon through the cell no transmembrane potentials arise when all the reagents are introduced (curve 3). None of the reactions with the participation of oxygen (Figs. 3 and 5) is possible. It can be supposed that the difference in the results obtained with different lipids could be attributed to different stability of phospholipids against oxidation. The residues of saturated fatty acids in dipalmitoyllecithin (curve 2, Fig. 6) have a lower reactivity and are less susceptible to oxidative struction than nonsaturated hydrocarbon chains. Because of this, it is the reaction of oxidation of NADH by air oxygen that predominates in this case. Depending on the nature of the hydrocarbon tail, lipids can remain either inert, or involved in the chain of redox reactions between NADH, coenzyme Q, and oxygen.

It could be supposed that in the presence of nicotinamide and ubiquinone at least two competitive processes take place on a bilayer membrane. One is due to the reaction between nicotinamide, coenzyme, and oxygen (Fig. 3), and the other is the result of peroxidation of lipids (Fig. 5). If the former proceeds with the absorption of protons the latter involves their liberation. Consequently, the potentials in Fig. 6 reflect the changes in pH in the layer adjacent to the membrane. In the absence of oxygen pH in this layer remains unchanged, and according to the schemes on Figs. 3 and 5 none of the reactions is possible. If oxygen is present in the system, but the lipid is not oxidized then the layer adjacent to the membrane is depleted in protons as a result of the reactions occurring according to the scheme shown in Fig. 3, the more, the greater the change in pH. In experiments with a membrane formed of oxizable lipids enrichment of the layer adjacent to the membrane in protons is observed. Since the membrane is selective in regard to hydrogen ions as a result of the addition of TTPB, it registers the ΔpH in the layer adjacent to the membrane on both sides of the bilayer.

The functioning of the mitochondrial respiratory chain can be accompanied by the liberation of protons on one side of the membrane and their absorption on the other side. According to Mitchell's theory,[71,72] in this case a so-called proton pump is functioning, i.e., there is active transport of protons across the membrane, which makes possible generation of a transmembrane potential and synthesis of ATP. However, as was shown earlier when investigating the nonenzymatic electron transport chain on a bilayer, liberation or absorption of protons on one interface membrane and between the layer adjacent to the membrane could prove sufficient for the formation of a transmembrane proton gradient. The possibility of creating such a pseudopump of protons was demonstrated by Sokolov *et al.*[73] Possibly, many of the effects observed in the functioning of membrane enzyme systems could be attributed to the processes of peroxidation of lipids, which investigators often overlook.[74]

6. Participation of FMN in the Oxidation of Membrane Lipids

Flavine mononucleotide (FMN) forms a part of the first complex of the respiratory chain of mitochondria, and its effect on the transmembrane potential value during redox reactions in the system $NADH-Q_6-O_2$ was shown by Ismailov *et al.* taking as an example bilayer lipid membranes.[54,55,57,75]

But before passing on to the complete system (2), it is necessary to understand how flavine mononucleotide affects the transmembrane potential and the conductivity of the bilayer.

Figure 7 shows the dependence of the transmembrane potentials on the FMN content in the system

$$\text{FMN, } O_2 \text{ |BLM| } O_2 \qquad (8)$$

When FMN is added to a nonmodified membrane without Q_6 (curve 1), a small potential with a minus sign is observed on that

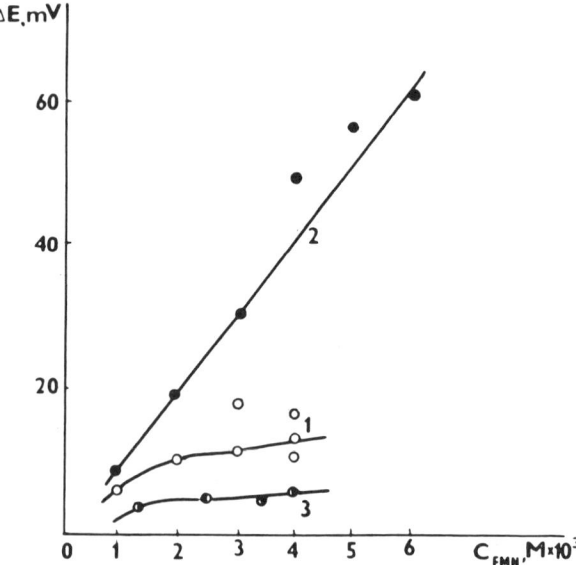

Figure 7. Transmembrane potential value on bilayer lipid membrane (made from egg lecithin + cholesterol) depending on FMN concentration in the system

FMN, O_2 |BLM| O_2, buffer	(curve 1)
FMN, O_2, buffer $\begin{vmatrix} \text{BLM} \\ \text{TTFB} \end{vmatrix}$ O_2, buffer		(curve 2)
FMN, O_2, buffer $\begin{vmatrix} \text{BLM} \\ \text{TTFB} \\ \text{tocopheryl} \\ \text{acetate} \end{vmatrix}$ O_2, buffer		(curve 3)

Buffer 10 mM Tris + 10 mM NaCl, pH 7.5, TTFB 10^{-5} M.[57]

side of the membrane on which FMN was introduced. In order to elucidate the nature of the processes taking place in that case system (8) was modified by the addition of TTPB, which made the membrane proton selective (chain 9):

$$\text{FMN, } O_2 \quad \begin{vmatrix} \text{BLM,} \\ \text{TTFB} \end{vmatrix} \quad O_2 \qquad (9)$$

After this addition the membrane became an electrode, sensitive to differences in proton concentrations in the layers adjacent to the membrane on either side of BLM. What is more, a considerable increase in the transmembrane potential (curve 2) was observed. If an inhibitor of peroxidation (tocopheryl acetate) was added to the membrane in the volume ratio of 1:3 (chain 10):

$$\text{FMN, } O_2 \quad \begin{vmatrix} \text{BLM,} \\ \text{tocopheryl} \\ \text{acetate, TTFB} \end{vmatrix} \quad O_2 \qquad (10)$$

then the addition of the FMN to the membrane did not lead to generation of any considerable potential (curve 3). The effect of Q_6, which is soluble only in a lipid, on the value of the generated potential (chain 11) is that

$$\text{FMN, } O_2 \quad \begin{vmatrix} \text{BLM,} \\ \text{TTFB, } Q_6 \end{vmatrix} \quad O_2 \qquad (11)$$

like tocopheryl acetate, Q_6 which is soluble only in a lipid, reduces transmembrane potential only slightly.

Since in the systems (8)–(11) FMN is introduced in the oxidized form, what most probably happens is that it catalyzes the reaction of oxidation by oxygen air of phospholipids which make up the bilayer membrane. The addition of TTFB (see chain 9) causes a sharp increase in the value of the potential (curve 2). The sign of the potential corresponds to the transport of a positively charged particle to the side opposite to that on which FMN was added. This can be explained by the fact that when lipids are oxidized on one side of the membrane, the layer adjacent to the membrane is enriched in protons, for instance, according to the following formal scheme:

$$ROOH + FMN \longrightarrow FMN^{\cdot-} + H^+ + ROO^{\cdot} \quad (12)$$

$$FMN^{\cdot-} \xrightarrow{O_2} FMN + O_2^{\cdot-} \quad (13)$$

$$RH + ROO^{\cdot} \longrightarrow ROOH + R^{\cdot} \quad (14)$$

When flavines are oxidized by air oxygen, one could expect the peroxide anion to be formed, but if the reaction proceeded in that way we would not have observed the effect of the decrease of pH in the layer adjacent to the membrane.

The correctness of the potential generation mechanism with the participation of peroxides in these conditions is substantiated by experiments on the action of tocopheryl acetate on the system being investigated (10) (curve 3). Tocopheryl acetate is hydrolyzed to form tocopherol, which is an antioxidant. The reaction of inhibiting the oxidation of lipids in membranes according to the relay model[76] could be represented summarily by the process

$$R^{\cdot} + InH \rightarrow RH + In^{\cdot} \quad (15)$$

according to which the antioxidant InH reacts with the free radical $R^{\cdot}$ before it has time to interact with oxygen. As a result, the chain of reactions represented by the reactions (12)-(14) slows down, which sharply reduces pH in the layers adjacent to the membrane and the value of the potential observed in the presence of TTFB. So the action of coenzyme Q_6 as an antioxidant is much weaker in system (11). The kinetics of potential generation in systems (8)-(11) has a clearly defined induction period, which most probably points to the chain mechanism of the reaction of formation of particles determining the transmembrane potential.

It should be noted that at present along with the molecular mechanism of peroxidation of lipids based on the theory of the liquid phase oxidation of hydrocarbons,[77] there exists the concept of a relay model of peroxidation of lipids of biological membrane,[76] which is more like the description of the oxidation processes of solid polymers.

7. Participation of FMN in the Transmembrane Transport of Protons

As was already shown in the previous section, at pH 7.5 FMN catalyzes the reaction of the formation of lipid peroxides, which can stimulate the conductance of the bilayer.[78] However, in order to detect the transmembrane transport of protons in the form of the transmembrane potential generation it is necessary that there should be absorption or liberation of hydrogen ions during the reactions at the interface. It follows from relation (12) that the reaction with the participation of FMN can be accompanied by proton liberation, which is confirmed by curve 2 in Fig. 7. The transmembrane potential (curve 1, Fig. 7) is also observed if there is no modifier in the membrane which makes it proton selective (chain 8). The problem of increasing of the conductance of both natural membranes and bilayers has been investigated many times.[79-81] As was shown by Sokolov *et al.*, addition of lipid hydroperoxides to bilayer membranes effectively increases their conductance only if hydroperoxide is added to both solutions flowing around the bilayer. That seems to indicate that the channel formation mechanism is more likely than the carrier mechanism.[82]

Figure 8 shows the changes in the value of the transmembrane potential generated in the system

$$\text{NADH, FMN, } O_2 \left| \begin{array}{c} \text{BLM,} \\ Q_6 \end{array} \right| O_2 \qquad (16)$$

at pH 7.5 as a function of the FMN concentration. The first point to the left corresponds to the potential generated in system (17) which differs from system (16) by the absence of FMN:

$$\text{NADH, } O_2 \left| \begin{array}{c} \text{BLM,} \\ Q_6 \end{array} \right| O_2 \qquad (17)$$

When the FMN content in system (16) is increased the sign of the transmembrane potential is reversed. In the presence of TTFB:

$$\text{NADH, FMN, } O_2 \left| \begin{array}{c} \text{BLM,} \\ Q_6, \text{TTFB} \end{array} \right| O_2 \qquad (18)$$

no substantial change takes place in the transmembrane potential. Figure 8c shows the change in the conductance of BLM in this system.

Since NADH decreases [reaction (3)] and FMN increases [reaction (12)] the proton concentration in the layer adjacent to

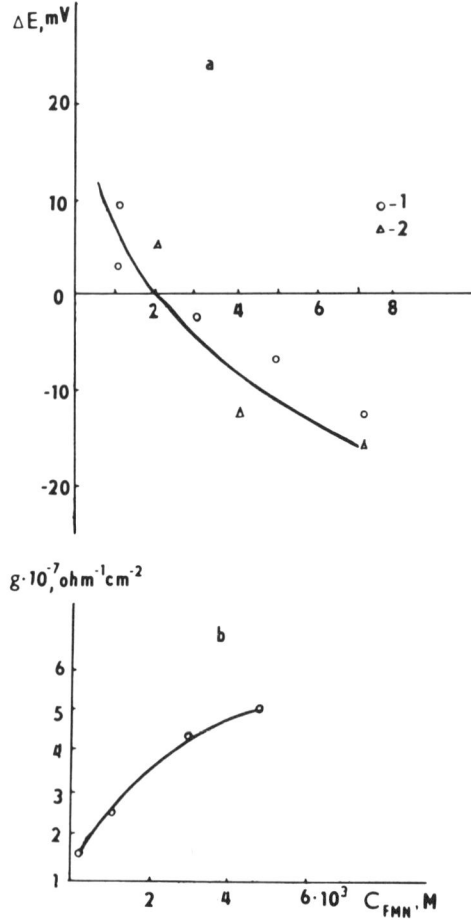

Figure 8. Dependence of transmembrane potential value (a) and BLM conductance (b) at pH 7.5 on FMN concentration in the systems

$$\text{NADP, FMN, O}_2 \quad \begin{vmatrix} \text{BLM,} \\ \text{TTFB} \end{vmatrix} \quad \text{O}_2 \quad \quad (1)$$

$$\text{NADP, FMN, O}_2 \quad |\text{BLM}| \quad \text{O}_2 \quad \quad (2)^{57}$$

the membrane, the value and sign of the potential are determined by the ratio of the reagents (Fig. 8a).

When pH equals 5.8 FMN does not catalyze the reaction of peroxidation of lipids, but effectively catalyzes the reduction of Q_6 in system (16). Indeed, in these conditions addition of FMN and NADH to the system without Q_6 does not cause E.M.F. to arise on the membrane. In the presence of Q_6 comparatively low FMN concentrations increase the conductance and the transmembrane potential on BLM when NADH is added. The addition of FMN into the functioning system NADH-Q_6-O_2 in the presence of an uncoupler does not change the value of the potential generated on BLM.

At pH 7.5 in the presence of FMN addition of TTFB affects the value of the potential only very slightly, which points to sufficiently high proton conductance of BLM in these conditions. Apparently, this is due to the fact that addition of FMN, just as in the case of pH equal to 5.8, increases the rate of formation of the reduced Q_6 form:

$$Q_6 + NADH \xrightarrow{FMN} Q_6H^- + NAD^+ \qquad (19)$$

The increase of the conductance of BLM when FMN is added to the functioning system (Fig. 8b) might also be connected with increase in the proton concentration in the layer adjacent to the membrane, which should lead to increase to the form Q_6H_2 which is inactive in electron transport but should possess the well-known property of phenol, namely, it should increase the proton conductance of BLM. Accumulation of Q_6H_2 in BLM is due to the fact that in solutions with a high buffer capacity the depletion in proton in the layer adjacent to the membrane is eliminated as a result of which the reaction consuming hydrogen ions

$$Q_6H^- + H^+ \rightleftharpoons Q_6H_2 \qquad (20)$$

shifts to the right. When reaction (20) takes place at the BLM/solution interface the reduction product of Q_6-Q_6H_2 becomes a natural proton carrier in BLM (Fig. 9). This is also substantiated by the data pointing to the existence of proton conductance of BLM modified by coenzyme Q_6 under conditions of a high buffer capacity at acid pH. Apart from this, as was shown above, in these conditions

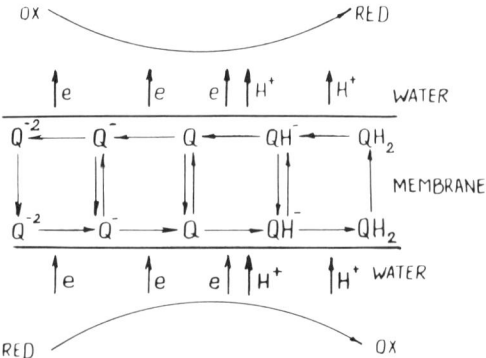

Figure 9. Coenzyme Q as a natural proton-exchanger. Proton transport reaction is coupled with electron-exchange reaction. The right-hand side of the scheme corresponds to processes at lower pH of aqueous phase than the left-hand side.

lipid peroxides (ROOH) accumulate (see Fig. 8b) and their presence can increase the conductance.

8. Participation of FMN in the Transmembrane Transport of Electrons

In order to obtain a potential as a result of transmembrane electron transport, ferricyanide was introduced into the system on the side of the membrane opposite to NADH.

The dependence of the transmembrane potential on the FMN concentration at different oxidizer $(K_3Fe(CN)_6)$ concentrations in the system

$$\text{NADH, FMN, O}_2 \left| \begin{matrix} \text{BLM,} \\ Q_6 \end{matrix} \right| K_3Fe(CN)_6, O_2 \quad (21)$$

is shown in Fig. 10.

The family of curves demonstrates that increase in the $K_3Fe(CN)_6$ concentration leads to increase in the transmembrane potential, and its sign corresponds to the transmembrane electron transport from NADH to ferricyanide. In these conditions the potential is removed by addition of TTFB, which acts as a proton shunt, unlike the case when there is no ferricyanide in the system (compare with Fig. 8a).

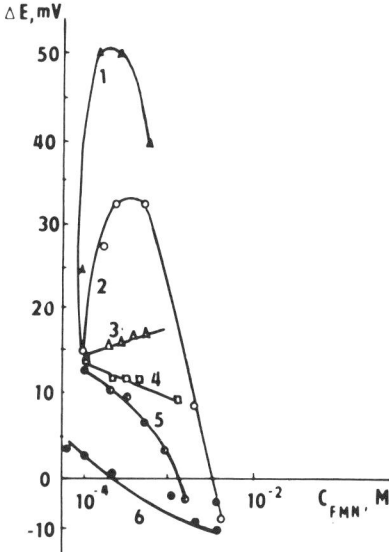

Figure 10. Dependence of transmembrane potential value on FMN concentration in the chain: NADH, FMN, O_2 |BLM, Q_6| $K_3Fe(CN)_6$ at content of $K_3Fe(CN)_6$ in the concentration 3.5×10^{-2} M (1), 2.5×10^{-2} M (2), 2×10^{-2} M (3), 1.5×10^{-2} M (4), 0 M (5). (6) The same system without Q_6. Content of $K_3Fe(CN)_6$ as for (3). NADP concentration in all systems 5×10^{-3} M, buffer 5 mM Tris, 100 mM NaCl, pH 3–7.5.[57]

At high FMN concentrations the potential corresponding to the electron transport to ferricyanide decreases because, in addition to the transmembrane electron transport, there also takes place peroxidation of lipid, which leads to the enrichment in protons of the layer adjacent to the membrane. Simultaneously, lipid peroxides formed in BLM and also a reduced form of Q_6, which makes the bilayer proton selective. The presence of such compounds in BLM decreases the potential connected with electron transport and transforms the pH gradient into a potential of the opposite sign.

9. Interaction of FMN with Other Chain Components

Thus, FMN can perform the following functions. FMN accelerates the reaction of oxidation of lipids by air oxygen, and it increases the rate of the heterogeneous reaction of reduction of coenzyme Q_6 by pyridine nucleotide. In discussing the biological significance of the first process, let us examine the electron transport reaction taking place in the membranes of mitochondria. It is known that electron transport in the initial part of the respiratory chain from

NADH to ferricyanide, which proceeds effectively in aeorbic conditions, stops completely in anaerobic conditions. It could be supposed that in anaerobic conditions no peroxides necessary for electron transport are formed. Thus, it could be assumed that one of FMN's functions is that it maintains in mitochondrial membrane a certain optimal concentration of peroxide radicals, which are needed as catalysts of the electron transport reaction during respiration. It was shown that the formation of peroxides with the participation of FMN disappears in acid media. On the other hand, when peroxides are formed the layer adjacent to the membrane becomes enriched in protons. Hence, it follows that the peroxide formation process might possibly be a self-regulating one, the rate of which cannot rise endlessly. This circumstance once more substantiates the supposition that this process might play a very important functional role in membrane redox reactions.

The experiments described above showed that in a system where NADH oxidation by oxygen proceeds chiefly on one side of the membrane, addition of the oxidizer $K_3Fe(CN)_6$ on the other side of BLM (system 21) leads to a qualitative change in the process: the reaction in the layer adjacent to the membrane is replaced by transmembrane electron transport (see Fig. 11).

Consequently, it can be seen that depending on the ratio of the redox potentials in the solutions on either side of the membrane the examined electron transport chain can be located either on one side of the membrane, or directed across the membrane.[57]

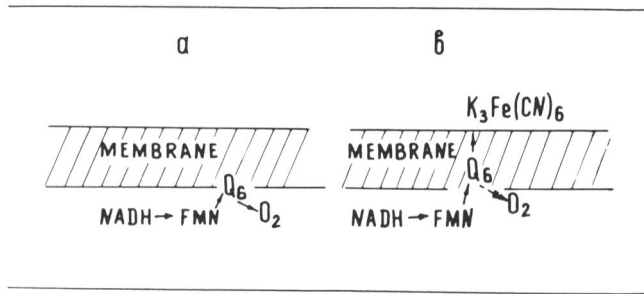

Figure 11. Two paths of electron transport in chain NADP-FMN-Q_6-O_2 ($K_3Fe(CN)_6$).

III. POTENTIAL GENERATION ON BILAYER MEMBRANES CONTAINING CHLOROPHYLL

Chlorophyll, which is responsible for photosensitisation during photosynthesis, itself undergoes redox transformation in the course of this process.[83-85]

The energetics of the electron transport reactions with the participation of chlorophyll is described in detail in the review by Silly.[86] Briefly, the role of light in photosynthesis is that the electron transport reaction from the donor (water, in the case of higher plants) to the reaction center where chlorophyll is located and then to the acceptor becomes possible when the chlorophyll molecule undergoes excitation. This process yields products with a high reduction potential. What animals receive in a ready-made form, in plants is synthesized under the action of light. These transformations have been sufficiently well studied for true chlorophyll solutions in photochemical oxidation and reduction reactions.[85] However, in natural conditions chlorophyll in the reaction center of the photosynthetic apparatus is inserted in the membrane of thylakoids, and, therefore, enters into photochemical reactions at the interface.

1. Chlorophyll at the Membrane/Electrolyte Interface

The lipophilic tails of chlorophyll molecules are inserted in the hydrocarbon part of the membrane, while the porphyrin rings belong to its polar part. The chlorophyll inserted in a bilayer or located at the water/hydrocarbon interface differs substantially from the chlorophyll at the air/water interface in that in its case the porphyrin rings are located not in the same manner as at the air/water interface. They are located at an angle of about 60° to the oil/water interface.[87,88]

According to the data obtained from measurements of the dichroism of monolayers of lecithin with chlorophyll, the angle of inclination of chlorophyll to the interface in this case equals about 35°,[89,90] the porphyrin rings lying in the region of the polar heads of the lipid. Possibly, this mutual arrangement of the porphyrin rings is the optimal one, overlapping the electron orbitals of the neighboring molecules. At any rate it should be noted that the redox reactions involving electron transport in bilayer lipid membranes and at the interface between immiscible liquids do not take place in the water/air system.[88]

2. Redox Potentials of Chlorophyll

In the presence of an oxidizer of corresponding potential, chlorophyll inserted in a bilayer lipid membrane undergoes redox transformations at the interface. It should be remembered that chlorophyll aggregation and its interaction with the water and protein molecules surrounding it substantially change its redox properties. The standard redox potential of CHL/CHL^+ depends on the size of the aggregate $(CHL \cdot 2H_2O)_n$. In the monomer state $E_{1/2}$ equals 0.78 V, while in the state of aggregation it already amounts to 0.93 V.[91] Actually, however, it is necessary to take into account not only the degree of chlorophyll aggregation, but also the protein surroundings and other elements forming part of the reaction center, as well as its functional state. How great the contribution of these factors can be is evident from a comparison of the ground and excited states of chlorophyll and its electrochemical potentials in the solution and in the reaction center. In the transition from the solution to the reaction center P_{680},[85,86] the standard potential φ_0 shifts by about -0.30 V, while the difference between the standard potentials of P_{680} and P_{700} amounts to 0.5-0.8 V. Just as it was noted in the analysis of Table 1, the shift of the redox potential is due to the fact that the investigated molecules are actually part of the structure all the elements of which are closely bound.

3. Reactions of Chlorophyll Inserted in the Membrane with the Redox Components in an Aqueous Solution under Illumination

The chain of redox processes taking place under illumination at the interfaces between a bilayer lipid membrane and electrolyte causes a photopotential to be set up.[92-95] The stationary values of the photopotential depend on the redox potentials of the components in aqueous solutions. The spectral dependence of the photopotential corresponds approximately to the absorption spectrum of the pigment, while the photopotential depends on the solution pH and the applied external voltage. The most comprehensive information about the processes taking place in the membrane can be obtained using light pulses generating photopotential, in combination with a potentiostatic device allowing the transmembrane potential to be maintained at a given level. The products of the reactions are registered spectroscopically, as well as by ESR and NMR methods.

The data on various reaction products appearing and disappearing as electric processes take place in the membrane provide the most complete information. When this method was applied to investigate the processes on bilayers containing di-n-amyl ether magnesium mesoporphyrin IX, redox reactions were registered at both interfaces, as well as diffusion of uncharged and charged forms of pigments through the bulk of the bilayer.[96] The general scheme of the reactions of the pigments inserted in the bilayer with the redox components of the solution is shown in Fig. 12. Until illumination is switched on, the redox reaction usually taking place is the one characterized by cycle I. Illumination causes additional appearance of P^+ particles over and above the quantity produced in cycle I. However, the greater part of P^+ particles react with the redox system on the same side of the bilayer. Actually, this results in an

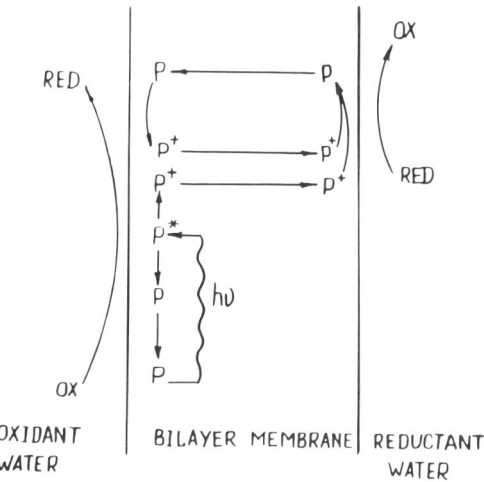

Figure 12. Scheme of phototransformations of porphyrin in bilayer lipid membrane in contact with redox components in aqueous solution.

increase in the exchange current of the reaction between the redox components in the solution and those in the membrane.[97]

The most common way of investigating redox reactions is to use the methods enabling one to follow the fate of the products of redox reactions taking place in the membranes of liposomes and microemulsions.[98-101] The use of these systems provides the possibility of actively controlling the potential at the interface. However, the redox potentials of the process components are chosen on the basis of the table data.

When the membranes which contain chlorophyll and are in contact with redox systems are illuminated, a photopotential connected with the formation of the cation CHL^+ is generated in aqueous solutions. The first stage of the process is the appearance of an excited chlorophyll molecule:

$$CHL + h\nu \to CHL^* \qquad (22)$$

The further fate of the excited molecule depends on the concrete conditions of the experiment. As an example we can take the processes occurring in liposome membranes[101] containing sodium ascorbate on the inside and Cu^{2+} on the outside. After an excited CHL molecule appears at the interface between the membrane and the water containing an oxidizer, the following reaction takes place:

$$CHL^* + Cu^{2+} \to CHL^+ + Cu^{1+} \qquad (23)$$

The greater part of the CHL^+ cations immediately react with the reaction product Cu^{1+}:

$$CHL^+ + Cu^{1+} \to CHL + Cu^{2+} \qquad (24)$$

As a result of the back reaction (24), on the opposite side of the bilayer the CHL^+ concentration is three orders of magnitude lower than that of the Cu^{1+} which initially resulted from reaction (23).

Attention should be drawn at this point to a distinctive feature important for phospholipid vesicles. It was found that the electron transport reaction goes with different efficiency on the outside and inside of the bilayer, which is due to different lateral mobility of chlorophyll in the external and internal monolayers and different possibilities of recombination of the charges arising under illumination.[101] On the opposite side, inside the liposome the reducer is present and so reaction (25) becomes possible:

$$CHL^+ + Asc.(red) \to CHL + Asc.(ox) \qquad (25)$$

The cation CHL^+ generated in reactions (24) and (25) enters into further reactions with the reducer at the same interface where it originated or on the opposite side of the membrane.

However, such a simple process scheme is possible only when other active electron acceptors—air oxygen or lipids capable of oxidation—are excluded from the system. Experimental data obtained by the ESR method indicate that under illumination O_2^- and $OH^{\cdot}$ are formed in chlorophyll-containing systems in the presence of air oxygen.[102]

4. Transmembrane Potentials in the Chain OX–CHL–RED

Using the same approach as in analyzing chain (2), it would be of interest to trace how common is the coupling of the main redox reaction with the oxidation of membrane lipids. To this end, an investigation was made of the dark transmembrane potentials on bilayers formed from egg lecithin with cholesterol, modified by chlorophyll in the presence of such redox systems as ferri-ferrocyanide, NADP-ferrocyanide, NAD-ferrocyanide.[103]

Figure 13 shows the concentration dependence of the transmembrane potential on bilayer formed from lecithin in the

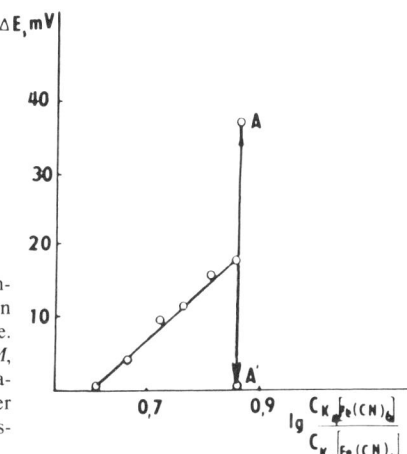

Figure 13. Dependence of transmembrane potential on the concentration ratio of ferrocyanide-ferricyanide. Supporting electrolyte: KCl, $10^{-3}\,M$, at point A TTFB added in concentration $2 \times 10^{-6}\,M$, at point A' buffer added in concentration $100\,mM$ Tris-HCl.[103]

ferricyanide–ferrocyanide system. The ratio of the concentrations of the reduced and oxidized forms of the redox pair are plotted on the abscissa. The investigated electrochemical chain can be represented in the following form:

$$K_4Fe(CN)_6/K_3Fe(CN)_6, \quad O_2 \quad \left| \begin{array}{c} BLM, \\ CHL \end{array} \right|$$

$$K_4Fe(CN)_6/K_3Fe(CN)_6, \quad O_2 \tag{26}$$

The experiment was arranged in such a way that the membrane was formed in symmetrical conditions, when the ratios of the oxidized and reduced forms of the redox pair on both sides of the membrane were identical, and a reducing or an oxidizing agent was introduced into one of the cell compartments.

Although the slope of the curve in Fig. 13 is 58 mV for a tenfold change in the reducer:oxidizer ratio, a more careful examination reveals that the bilayer lipid membrane with inserted chlorophyll cannot be simply likened to a platinum electrode. Indeed, if using the weakly liposoluble acid TTFB the membrane is made proton selective, i.e., sensitive to a proton gradient, then an additional potential jump is observed (see Fig. 13, point A), which indicates the existence of proton gradient in the layers adjacent to the membrane† directed to the side opposite to electron transport. It could be supposed that in the presence of a redox system and chlorophyll the membrane itself should possess proton-exchanger properties like a glass electrode. And indeed it follows from Fig. 14 that the membrane is sensitive to a proton gradient, but only in the presence of redox system and chlorophyll.

So, the difference in the redox potentials of aqueous solutions of redox systems on either side of the membrane leads to a pH gradient in the layers adjacent to the membrane as a result of the redox reactions.

It was interesting to use $NADP^+$ as an acceptor instead of ferricyanide, as $NADP^+$ forms part of the electron transport chain in photosynthesis. If a reducer, ferrocyanide, was added on the

---

† The concentration in question is that of the protons in the layer, since in the bulk of the solution on either side of the membrane pH of the solutions are identical. The thickness of the unstirred layer depends on conditions and reaches a value of some micrometers.[58,63,64,67]

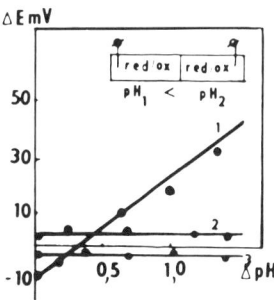

Figure 14. Dependence of transmembrane potential on proton concentration gradient in ferrocyanide–ferricyanide system (1). Potential sign is positive in part of chain with pH 7, buffer: citrate-phosphate. The same dependence on membrane modified by chlorophyll in absence of redox system (2). Same dependence in presence of redox-system on nonmodified membrane.[103]

other side (chain 27):

$$K_4Fe(CN)_6, \; O_2 \; \left| \begin{array}{c} BLM, \\ CHL \end{array} \right| \; NADP, \; O_2 \qquad (27)$$

then the observed transmembrane potential (curve 1, Fig. 15) was at least double that observed in the absence of ferrocyanide (curve 2, Fig. 15). That means that the process is apparently much more intensive if both an oxidizer and a reducer are present together. An increase in the transmembrane potential can also be observed when TTFB in the concentration $2 \times 10^{-6}\,M$ is added to the last two chains (points A and B on curves 1 and 2).

If instead of $10^{-3}\,M$ KCl citrate-phosphate buffer solutions ($2 \times 10^{-3}\,M$) or Tris-buffer ($2 \times 10^{-3}\,M$) was used as supporting electrolyte no generation of a transmembrane potential was observed (points A_1 and B_1, Fig. 15).

Using the same tests as for the redox system ferrocyanide–ferricyanide we found that when redox reactions take place on both sides of the membrane a pH gradient is formed in the layer adjacent

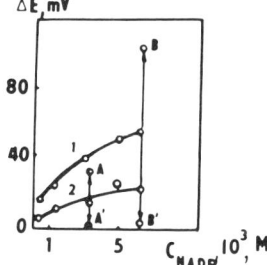

Figure 15. Dependence of transmembrane potential on NADP concentration in presence of $K_4Fe(CN)_6$, $7.5 \times 10^{-3}\,M$ (1), 0.0 M (2). Supporting electrolyte, KCl, $10^{-3}\,M$, TTFB concentration $2 \times 10^{-6}\,M$.[103]

to the membrane which can be removed by increasing the buffer capacity of the solution.

Summing up the data obtained from the investigation of the transmembrane potentials on bilayers, as well as the results of the studies of oxidation processes in the membranes of mitochondria and chloroplasts, we can say that the redox reactions taking place in biological membranes, as well as at the interface between bilayer lipid membranes and electrolyte, lead to coupling of the main redox reaction with the reaction of peroxidation of lipids. And it makes no difference whether the oxidation of the lipids was caused by redox processes of the coenzymes of the respiratory chain of mitochondria or by participation of chlorophyll, porphyrins, or ubiquinones. Certain common features can be noted in each of these cases. First, both in the electron transport chain and during the oxidation of lipids, redox processes take place with participation of protons. Second, these redox processes lead to the formation at the membrane/electrolyte interface of an unstirred layer enriched or depleted in protons, which is not in equilibrium with the bulk concentration of protons in the solution. Third, membranes prepared from lipids capable of being oxidized (egg lecithin and phosphatidylethanolamine) even without additions of substances that could act as proton exchangers as ubiquinones become proton selective due to the accumulation of oxidation products.† Fourth, the transmembrane potential in proton selective systems is determined primarily by the proton gradient in the layers adjacent to the membrane, but by no means by their concentration in the bulk of the solutions washing the membranes. In a more complex case, when simultaneously with proton transport there also occurs electron transport, the transmembrane potential depends on the ratio of the rates of two processes.

IV. POSSIBLE MECHANISMS OF THE MOTION OF ELECTRONS AND PROTONS IN THE MEMBRANE

The processes of photosynthesis and respiration proceed as a sequence of elementary acts of electron transport along the electron

---

† It should be added that mitochondrial lipids are oxidized even more easily than egg lecithin or phosphatidylethanolamine.

transport chain. The transition of electrons from one component to the next in the electron transport chain is accompanied by consecutive redox transformations in it. The physical mechanism of the motion of electrons in the electron transport chain is so far an open question. Even less is known about the mechanism of the motion of protons. It is quite possible that the mechanism could be different in various parts of the respiratory chain. Moreover, it cannot be ruled out that there exists some kind of interrelation is based on the concepts[104-106] according to which the jumps of the electron from one redox component to another can increase the effective diffusion coefficient of ions participating in the electron exchange reaction. This point of view was substantiated during an investigation of a number of polymers containing redox groups capable of exchanging electrons.[106,107] However, the similarity between the electron transport chains and polymers matrix has a high conformation mobility, which, apparently, should be taken into account when developing any theory of charge motion in protein structures.[16,108-110]

1. Hypotheses on the Mechanism of Electron Motion in Biological Membranes

The electron in the electron transport chain is not free like in a metal wire. Therefore the electron motion in each act involves surmounting an energy barrier. As was shown in Refs. 16 and 108–110, a substantial role in this process is played by the conformations of the macromolecular components of the electron transport chain. Nevertheless, the simplest model systems of electron transport realized on bilayer lipid membranes were virtually based on the concept of a membrane as a thin liquid hydrocarbon in which a substance capable of redox transformations is dissolved, the products of this reaction being able to diffuse inside the bilayer. The electron transport from the aqueous phase containing a reducer amounts to injection of charges into the nonaqueous phase if it contains an electron acceptor:

$$^{w}Red_1 + {}^{m}A \rightarrow {}^{w}Ox_1 + {}^{m}D \qquad (28)$$

At the same time, on the other side of the membrane, if there occurs a transmembrane charge transfer, the reduced form ^{m}D

reacts with the electron acceptor in the water:

$$^mD + Ox_2 \rightarrow {}^mA + Red_2 \qquad (29)$$

Consequently, the electron transport across the membrane proves possible if there exists a redox pair dissolved in the membrane which acts as a charge carrier:

$$\begin{array}{c|c|c} Red_1 & {}^mA & Ox_2 \\ \bigg) & \bigg(\bigg) & \bigg(\\ Ox_1 & {}^mD & Red_2 \\ & Membrane & \end{array} \qquad (30)$$

This process was carried out experimentally by Liberman *et al.*[111] on a flat bilayer membrane, as well as on liposome membranes containing ferrocene, phenazinemetasulfate, and benzoquinone.[112] A phenomenological description of the mechanism of carriers is given in Ref. 113. When photo processes on bilayers in the presence of chlorophyll were investigated, it was also presumed that the cation CHL^+ diffuses through the bilayer.[101] However, it remains unclear whether this process could really be effective since the chlorophyll molecule has a very long phylot tail. It therefore remains an open question whether the carrier mechanism does indeed function in respect to redox processes in biological membranes.

When bilayers are used, and specifically when their photo-properties are investigated, investigators often resort to the tunnel mechanism of electron transport through the bilayer. That is equivalent to assuming that the membrane is approximated as a thin insulator or semiconductor film. A charge injected in the membrane is localized either on the molecules which make up the membrane or on the additions having a greater affinity for the electron than the molecules of which the membrane is made up:

$$\begin{array}{c|c|c} Red_1 & & Ox_2 \\ \bigg) & \xrightarrow{membrane} & \bigg(\\ Ox_1 & & Red_2 \end{array} \qquad (31)$$

Processes of type (31) take place in thin single crystals of aromatic hydrocarbons or polymer films. The investigations of insulating films dividing two redox systems with different redox potentials made it possible to elucidate many of the regularities of electron transport across the interface between an insulating membrane and electrolyte solution.[46,114] For instance, it was discovered when investigating insulating polymer films in contact with various redox systems that the redox system O_2/H_2O is capable of hole injection into the insulating film.[46]

According to the model proposed by Ilani and Berns, in a bilayer lipid membrane containing chlorophyll and carotenoids a charge arising at the interface under illumination [see reactions (24)-(27)] is transported along the system conjugated bonds of carotenoid.[93] This supposition has not yet been confirmed, but it does not contradict the views that carotenoids could be located not only along the bilayer, as believed in the past, but also across it. The factor determining the type of location of carotenoids could be chemical nature of the lipid from which the membrane is formed.

It should, however, be noted that the transmembrane electron transport through the entire thickness of the phospholipid bilayer is exceedingly doubtful, although some authors uphold this view.[92,93] In natural membranes there exist sites, small in extent, where electron transfer between hemes occurs as a result of the overlapping of the orbitals of neighboring molecules, and the probability of such a transfer increases with the degree of overlapping of the orbitals. In cytochrome c_3 there are four hemes mutually perpendicular and the rate of electron exchange between them is low. In complexes with a parallel arrangement of the hemes, transfer of electrons is rapid.[115] The distance between the hemes and their mutual orientation are an important factor in regulating the rate of electron transport.

Electron transfer during photosynthesis, according to the tunnel mechanism, was examined in detail in Refs. 116-121. The most interesting of all for the modeling of electron transport in biological membranes is the case when the redox reaction at the membrane/electrolyte interface leads to ion permeability, and not only to electron permeability. Let us assume that ion B^- is insoluble in the membrane, and, therefore the membrane is impermeable to it. However, if at the interface the ion undergoes redox transformations:

$$^wB^- \to {^wB^0} + e \qquad (32)$$

as a result of which a neutral form arises capable of dissolving in the membrane, then

$$^wB^0 \to {}^mB^0 \qquad (33)$$

and one can obtain a transport scheme in which ion permeability is actually controlled by electrochemical injection of charges through the membrane, and by the value of the permeability of the membrane to the neutral product B

$$^wB^- \to \left| \begin{array}{c} {}^mB^0 \longrightarrow {}^mB^0 \\ e \xrightarrow{\text{membrane}} e \end{array} \right| \to {}^wB^- \qquad (34)$$

Ion permeability of this type was discovered when investigating the permeability of bilayer phospholipid membranes in the iodine/iodide system,[3,4] as well as the photopotentials of membranes containing chlorophyll and plastoquinone.[122]

2. Ion Permeability of Bilayer Membranes in the Iodine/Iodide System Controlled by Redox Reactions at the Interface

Iodine is one of the most effective substances in the presence of which the conductance of phospholipid bilayers increases. Various interpretations of the results of measurements of the conductance of bilayers in the presences of the I^0/I^- systems have been the subject of discussion. A number of authors[123,124] believed that in the presence of iodine electron conductance is possible in the membrane as a result of the formation of iodine complexes with phospholipids. In contrast to this view, it was claimed[125] that only ionic conductance is possible. The impedance of bilayers in the presence of iodine was first measured by Louger et al.[2] By measuring the impedance of bilayers it is possible to detect the motion of ions inside the membrane, during the measurement of the impedance at a sufficiently high frequency which would not give ions to pass from the membrane into the solution.[3,4,126] the impedance of a bilayer membrane formed from egg lecithin with cholesterol in the presence of 10^{-3} M I_2 in an aqueous solution, measured by Lebedev et al.[3,4] is given in Figs. 16a and 16b as a function of the concentration of KI added to the solution.

Before discussing a possible model of the process, let us first examine some reactions at the interface between the membrane

Electron Transfer Effects and Mechanism of Membrane Potential 149

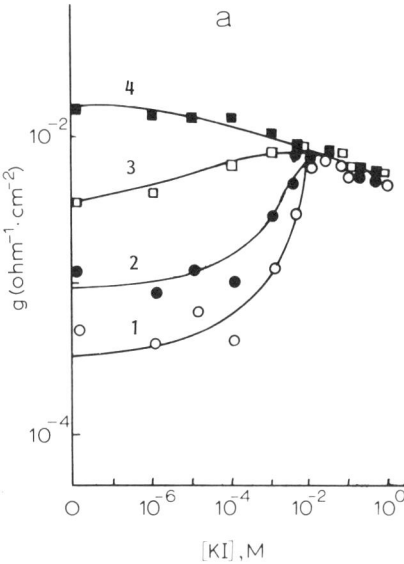

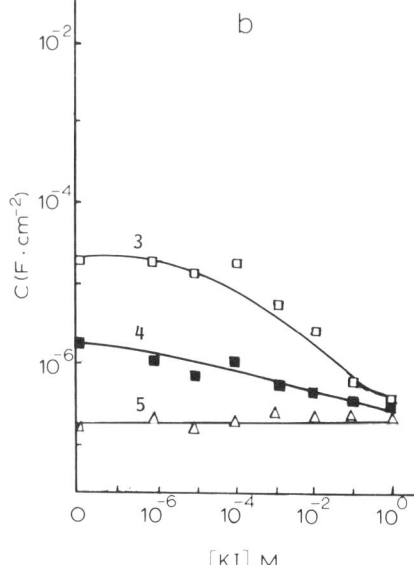

Figure 16. Frequency characteristics of conductance (a) and capacitance (b) of BLM in presence of 10^{-3} M I_2^0 depending on KI content in solution. 1, Direct current; 2, 0.1 cps; 3, 20 cps; 4, 500 cps; 5, 5000 cps.[3]

modified by iodine and the solution of the iodine/iodide system. Let us assume that at the lipid/aqueous iodine/iodide solution interface the following reactions take place:

$$^wI \to {^wI^-} + p \tag{35}$$

$$^wI \to {^mI} \tag{36}$$

Next we shall try to explain some of the experimental results (Figs. 16a, 16b) obtained when measuring the impedance. If the concentration ratio $I^-/I_2^0 < 1$, then an extremely low conductance of the membrane for a direct current is observed because of the small number of ions capable of discharging on the BLM surface. However, measurements on a high frequency have shown that in the membrane itself there are a sufficient number of current carriers and its conductance is two orders of magnitude higher than the overall conductance. The resistance at the interface is, in this case, of most importance, and the limiting step at the interface manifests itself in an extremely high capacitance of the membrane (Fig. 16b).

When the iodide concentration is increased the conductance also increases up to a certain limit, until the molecular iodine concentration changes noticeably as a result of the reaction

$$^wI^- + {^wI_2} \rightleftharpoons {^wI_3^-} \tag{37}$$

which competes with the electron exchange reaction (35). Starting from a certain KI concentration, the charge transfer step inside the membrane becomes the limiting step (right-hand side of Fig. 16a). Increased capacitance disappears. So does the dispersions of conductance with frequency. If the impedance of the system has no frequency dispersion, then the conductance and capacitance are determined only by the bulk properties.

The full scheme is as follows:

$$\begin{array}{c} I_3^- \to \\ \text{Aqueous phase} \end{array} \left| \begin{array}{c} I_n \longrightarrow I_n \\ \nearrow \qquad\qquad \searrow \\ \nwarrow p \longleftarrow_{\text{membrane}} p \swarrow \end{array} \right| \begin{array}{c} \to I_3^- \\ \text{Aqueous phase} \end{array} \tag{38}$$

In the scheme proposed above the carriers–holes–p appear in the membrane as a result of the injection during the redox reaction

(35). The flux of holes combined with the neutral iodine flow across the membrane signifies that the membrane is thermodynamically permeable to iodine ion. The mechanism of electron transport through bilayers containing iodine was never made a subject of special investigation. It should, however, be borne in mind that in a nonaqueous solution iodine forms consisting of iodine atom aggregates (I_n) capable of accepting and donating electrons.

3. Electron and Proton Transport in Bilayers Containing Chlorophyll and Quinones under Illumination

Quinones are essential components of the electron transport chain of both bacteria and higher plants.[127] Since quinones can act as primary acceptors, investigations were made of the charge transfer processes in models in which porphyrin and quinone are bound by covalent bonds at different distances from one another.[128] An investigation of the mechanism of formation and disappearance of radicals during the interaction of chlorophyll and quinone under laser illumination showed that the most optimal conditions for the charge separation process exist at the interface, where the recombination of charges[129] in different phases is difficult.

Plastoquinones perform the same role in the photosynthetic apparatus of plants as ubiquinone does in the mitochondrial membrane. Masters and Mauzerall[122] used a bilayer lipid membrane made of egg lecithin containing chlorophyll "a" in order to elucidate the effect of plastoquinones PQ_5 and PQ_9 on photopotential. The addition of plastoquinones increases the dark and photoconductivities about 20 times.

While membranes with chlorophyll not containing quinones are insensitive to the pH gradient, membranes containing plastoquinone are strongly affected by it. The scheme of the coupling of the two processes is shown in Fig. 17 and formally resembles the model proposed for the iodine/iodide system. It should be noted that electron and proton transport follows the same direction. The authors believe that the charge is carried by diffusion of charged and uncharged forms of plastoquinones and chlorophyll. Other possible mechanisms of membrane conductance in the presence of quinones will be examined below.

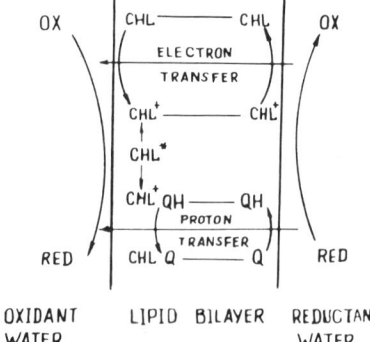

Figure 17. Scheme of coupling of phototransfer of protons and electrons in bilayer membrane containing chlorophyll and quinone in presence of redox components in aqueous solution.

4. Possible Conductance Mechanisms in Bilayers Containing Ubiquinone

The existence of lipophilic carriers of protons is typical for energy transducing membranes. These membranes are proton and electron exchangers simultaneously. Since in redox transformations ubiquinone participates both as an electron and a proton,[3] then, formally speaking, the conductance of membranes containing ubiquinone belongs to the same type as system (34).

Let us first consider the concepts of the mechanism by which coenzyme Q inserted in the membrane could transfer the charge (proton or electron). According to Robertson and Boardman,[130] coenzyme Q traps protons as they appear in the course of redox reactions and, therefore, is a proton carrier. A significant feature of this hypothetical mechanism is the ability of lateral movement in the hydrocarbon region, in the middle of the membrane. In the same part of the membrane where CH_2 groups adjoin the polar heads and are as if in a solid state, diffusion is made difficult. The same protons which could have come from the aqueous phase lose their solvate shell and join the polar groups of quinones. This line of argumentation is close to the interpretation of the experiments carried out with the octane/water interface.[131] When ubiquinone in the oxidized form (quinone form) adds two electrons from nonheme iron and sulfohydryl groups of proteins participating in the electron transport chain, then the strongly negative charge appearing at that point promotes the movement of the molecule in the direction of the aqueous phase towards a more polar region. Then a reaction with two protons takes place at the interface, protons being trapped by ubiquinone and losing their solvate shell. The ubiquinone molecule again becomes less polar and moves to the middle of the bilayer.

It should be noted that the proper motion of lipophilic electron and proton carriers in the membrane is not at all obligatory. The presence in the ubiquinone molecule of a hydrocarbon tail with conjugated bonds indicates the possibility of the motion of protons along the ubiquinone molecule, since a proton in a nonaqueous phase can be trapped by a conjugated bond forming a sort of Lewis acid.

So it is the ubiquinone molecule in a mitochondrial membrane or a similar molecule in a thylakoid membrane that carries the proton across the interface and, therefore, according to this hypothesis, is an essential participant of the process of resolvation.

This view could be augmented by the latest investigations on the effect of the length of the side isoprenoid chain on the fluidity of bilayers. Experiments with polarization have shown that fluorescence of perylene in bilayer vesicles containing ubiquinones with side chains of different length, the short-tail quinones ($n = 2.3$) reduce the fluidity of membranes. This effect is more pronounced in reduced quinones. When quinones with long tails ($n = 10$) are introduced in the bilayer, the fluidity of membranes rises with increase in their content in the membrane. Greater fluidity does not depend on the oxidation state of quinones.[132] This result contradicts the concepts that quinones with long tails penetrate deep into the bilayer and can form there aggregates in which charge transfer follows a different mechanism.[133,134]

5. Hypothesis on the Mechanism of Proton Transport in Biological Membranes

Proton transport across the membrane is a feature common to the bioenergic processes of a large variety of different organellas. The redox transformations of substrates are accompanied by the movement of proton in the transhydrogenase,[135] cytochromoxydase.[136,137] Proton passes through the ATP-synthetase complex. According to the chemiosmotic theory, both in its original form[71] and in the modernized version,[38,72] proton moves through the membrane. The local proton theory[138,139] presupposes effective motion of proton. The molecular mechanism of proton transport across a membrane containing protein structure was proposed by Nagle *et al.*[140,141] and is based on a model according to which the chains of hydrogen bonds formed by the side groups of proteins serve as a sort of wire along which proton conductance is possible by a mechanism similar to that of conductance in ice.[142,143] This hypothesis is supported by the NMR study of a number of compounds, as, for example, polyhistidine or polyglutamic acid, which shows how easily the hydrogen bond is polarized.[144,145]

Apart from the protein matrix, where it is possible for protons to move effectively in keeping with a mechanism somewhat similar to that proposed for the motion of protons in ice there is another path for protons through hydrophobic barrier of which the membrane is an example. This transition is based on the observations of phase transitions in a bilayer by X-ray diffraction methods. In this mechanism developed for mitochondria the main role falls to cardiolipin, which accounts for 33% of the total amount of lipids in the mitochondrial membrane.[146] The protonation of the head groups of cardiolipin contained in the membranes of mitochondria caused a phase transition of the bilayer into an inverted hexagonal phase.[147] This process is promoted by calcium ions whose reactions with the head groups of lipids favor neutralization of the membrane charge and effective dehydration of the polar heads.

Investigation of intact mitochondria, conducted by the ^{31}P-NMR method, indicate that 95% of the membranes are organized in the form of bilayers. Consequently, nonbilayer structures, which are less of a barrier to proton transport, if they exist at all, are fairly rare, or extremely unstable states.[148]

As for the physical properties of the medium which is a conductor of H^+, there is still nothing known about it. The information there is is of considerable interest in the experiments[149] with films obtained by deposition of mitochondria with unruptured membranes, in which the mobility of charges resulting from the injection of electrons was studied by the low-energy pulse method. No movement of charges through the film was observed in these studies. When the membrane was ruptured by freezing and subsequent defreezing an ohmic conductance appeared. However, movement of carriers through the specimens was observed when the specimens contained about 1% moisture. When the film was absolutely dry, the carriers shifted within 30 Å. If the film contained moisture, the mobility of positive carriers amounted to 2.10^{-2} cm^2/V sec. The proton mobility in aqueous solutions at infinite dilution equaled 3.6×10^{-3} cm^2/V sec. The mobility of negative charges obtained in these experiments amounted 5×10^{-2} cm^2/V sec. The mobility of OH$^-$ groups equaled 1.8×10^{-3} cm^2/V sec. Although nothing can be said about the nature of the carriers of these charges, it should, however, be noted that they could have been not only electrons and holes, but also protons whose mobility could be very great in an ordered structure. The activation energy of the mobility proved very small, 0.15 eV.

That this unexpectedly high mobility value is indeed connected with the membrane structure was indicated by the results of experiments in which after the specimens had been heated to 125°C for 15 min, the signal indicating the movement of positive charges disappeared and subsequent holding of the films in water vapor did not restore the former mobility value that existed in the native specimen. According to rough estimates, the shift of the charges in a denatured film does not exceed 10–30 Å.

6. Potentials of Coupling Membranes

The problem of coupling of redox reactions with the reaction of ATP synthesis taking place on coupling membranes has certain points in common with the questions discussed for transmembrane potentials in simple model systems. Initially it was believed that phosphorylation in chloroplasts and mitochondria, in principle, follows the same pattern as substrate phosphorylation, that is, they are coupled homogenous reactions. That point of view was reflected in the so-called chemical hypothesis of coupling, which chronologically was the first.[150] From this theory it followed that there exist concrete intermediate macroergic compounds. A macroerg is, first and foremost, the energy of an individual chemical bond formed via concrete intermediate compounds which, at least in principle, can be identified. Strictly speaking, within the framework of this model neither the membrane nor the transport of ions through it has any relation to coupling.

These concepts were later replaced by several theories in which the membrane of the organelle was believed to play a certain role in the coupling process. We shall consider Mitchell's chemiosmotic theory, which has been developed in the greatest detail. According to Mitchell's theory, in all coupled bioenergetic reactions which take place in the membrane of a living cell the transmembrane electrochemical potential of protons acts as a mediator. The general idea is beautiful, simple, and very close to the engineering activities of human beings: one kind of reaction, the energy-producing one, charges the membrane. For this purpose there exist primary proton pumps which use the light energy or that of redox reactions to induce the motion of protons through the membrane. Apart from energy-producing processes there exist energy-consuming processes that use the energy of the primary pumps.

According to Papa,[137] the mechanism which links the redox enzymes with the ATPase complex is the proton flow. Each of these complexes is linked with the other complexes through the proton flow. In terms of thermodynamics the functioning of the primary proton pump results in an electrochemical potential gradient across the membrane, which includes the concentration and electric terms:

$$\Delta \bar{\mu}_{H^+} = F \Delta \varphi - 2.3 RT \Delta pH \qquad (39)$$

Since both terms on the right-hand side are vectors, $\Delta \bar{\mu}_{H^+}$ too is a vector motive force. In Mitchell's theory it is the membrane which is the essential condition for the vector nature of the flow. ΔpH refers to the bulk of the aqueous phases in contact with the membrane.

Another remarkable property of primary proton pumps is that all the redox reactions take place with the participation of the compounds in which both proton and electron participate in the reaction simultaneously. The simplest and best known type of such a redox system is quinone/hydroquinone. So all the redox systems, despite their different potentials, have a common ingredient—protons.

The next important postulate of Mitchell's theory concerns the consumers of the energy produced by primary pumps and presupposes the presence in the organella membranes of secondary proton pumps which use the transmembrane proton flow for ATP synthesis and a number of other processes. Essential to this theory is the statement that the membranes containing inserted pumps must be permeable to protons, since that is necessary for the formation of loops. Finally, membranes must contain carriers for a number of other anions and cations which, as experiments have shown, are transported through the membrane of the energy-coupling organelles in the process of functioning. The coupling by means of a membrane of the energy-producing and energy-consuming reactions with proton as common ingredient could be described within the framework of classical thermodynamics of nonequilibrium processes.[151]

What helped to make Mitchell's theory such a success was above all its simplicity and the ease with which the model could be visualized. From this theory it was clear what parameters had to be verified. Since it was presumed that the proton pump throws protons through the hydrophobic barrier of the membrane, the first task was to reconstruct in the simplest form an energy-coupling membrane, insert the corresponding enzymes into the bilayer, and observe the generation of the transmembrane potential. A flat bilayer membrane proved unsuitable for the insertion of such high molecular objects. However, a number of authors who used liposomes succeeded in inserting primary proton pumps into the membrane of liposomes.[152-154] Although the transmembrane potential cannot be measured on a liposome membrane, nevertheless by introducing certain tracers, as for instance ANS, it proved possible to demonstrate that the functioning of proton pumps changes the fluorescence of the dye.[155-158]

Drachev et al.[159] using a filter impregnated with lipids, were able to observe generation of a potential in liposomes with an inserted cytochromoxidase. But the absolute value of the transmembrane potential is calculated from an extremely complicated equivalent scheme.

Another method of measuring the potential is based on the shift of the absorption spectrum of a dye molecule in strong electric field (electrochromism). This method was used by Witt to determine the changes in the potential in chloroplasts.[160,161]

The results of such measurements cannot be interpreted unambiguously. As an example we could cite the widely used method of fluorescent probes, and in particular the experiments conducted with ANS. ANS is widely applied as a test for studying the generation of the transmembrane potential of mitochondria and in a number of other membrane systems.[155] However, there exist two contradictory ways of measuring ANS fluorescence during changes in the functional state of the membranes of mitochondria and chloroplasts. According to one model the ANS$^-$ anion diffuses through the membrane when the transmembrane potential changes. The new equilibrium leads to a change in fluorescence since the concentration of ANS in the matrix mitochondria changes.[156] And from the change in fluorescence it is possible to determine the change in the transmembrane potential. The permeability of the membrane to ANS is a point in favor of this concept. The alternative model presumes that ANS adsorption is sensitive to the surface potential of the membrane.[157,158] However, the latest data obtained by Robertson and Rottenberg[162] indicate that the change in ANS fluorescence is of a complex nature and both effects influence the change in the magnitude of fluorescence. Consequently, neither the change in the surface potential, nor that in the transmembrane potential can be determined quantitatively from this kind of experiment.

What proved a triumph for Mitchell's hypothesis was the experiments involving simultaneous insertion of ATP synthtases and of one of the primary pumps.[163] The synthesis of ATP was accomplished on liposomes. Addition of substances increasing the conductance of the membrane disturbed the coupling both in liposomes and in mitochondria.

The success of the chemiosmotic theory signified, first and foremost, a new level of understanding of biochemical processes. It was shown that the membrane is a structural element in which a vector biochemical reaction is possible, and thanks to Mitchell this definition has become a reliable characteristic of biochemical process. Alongside of the proofs in favor of the chemiosmotic theory there existed certain facts (their number increasing with time) which were extremely hard to explain on the basis of the above-mentioned postulates.

The chemiosmotic theory, as was pointed out correctly in Ref. 72, has nothing to offer by way of explanation of the nature of the aqueous phase on either side of the membrane. It is a homogenous medium into which protons are liberated and from which they are absorbed; in other words, the concentration of ingredients, including protons, remains constant right up to the interface. The interpretation of pH measurements during the investigation of the action of proton pumps in membranes fails to take into account the existence of the layer adjacent to the membrane. However, as was shown in previous sections, even on bilayer membranes an unstirred layer is formed, whose pH could differ substantially from that in the bulk of the solution. Much the same picture apparently exists in natural systems.[60,66,164] In mitochondrial and thylakoid membranes the picture is complicated by the presence of ubiquinones, COO^- groups of proteins, and anions of acid lipids with proton-exchange properties. This makes the polar part of coupling membranes and the layer adjacent to the membrane of a sort of proton buffer. Generations of protons and their consumers can communicate through such a buffer. An argument in favor of this model is provided by the investigations of the process of photophosphorylation in thylakoid chloroplasts under flash illumination. Dilli and Shreiber[165] indeed showed that between the source of protons—a protolytic redox reaction—and their consumer—ATP-synthetase—there exists something like a buffer pool, which is not in equilibrium with the bulk. If this pool is depleted in protons in advance, then despite the light flash no process of photophosphorylation takes place. Judging from the value of $pK_a = 7.8$ of these buffer groups, the main acting group is the NH_2 group. According to Mitchell's theory, the concentration of protons is presumed to be averaged also along the membrane plane, which means that when a substance creating a proton shunt is introduced in the bilayer, no matter how far from the proton pump a channel will be formed, in all cases this will lead to disappearance of the transmembrane potential. Apart from the fact that all concentrations are averaged along the membrane, its surface is assumed to be equipotential. However, it is sufficient to glance at Fig. 18,[10] where a diagram of the functioning of proton pumps in the membrane is shown with due regard for real scales, in order to understand that the membrane surface cannot be equipotential. Investigations have shown that pH in the layer adjacent to the membrane changes along the bilayer from point to point.[166] The existence of such a heterogeneity along the membrane means that the concepts of an isotropic medium adjacent to the membrane are much too crude and therefore lead to incorrect conclusions about the physical nature of the coupling membrane.

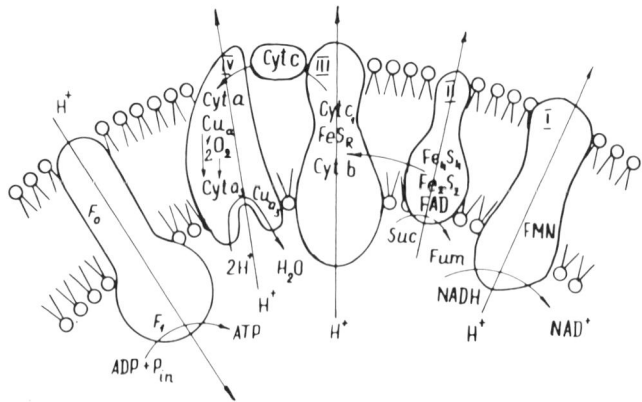

Figure 18. Coupling membrane of mitochondria.[10]

The actually operating forces should be not macroscopic, but microscopic, and their scale of operation, if such a criterion is introduced, should not exceed the average distance between the coupled enzymes.

A heterogeneous natural membrane is probably not a suitable object to which the concept of membrane potential as an energy scale could always be applied fruitfully. This has apparently been confirmed by the latest experiments conducted on membranes of mitochondria.[167] However, when experiments are set up in such a way that the close interaction of proton generator and proton consumers is disturbed, the concentrations near the membrane are equalized with the bulk concentrations and formula (39) is fully observed. But like any formula in thermodynamics, it does not presuppose any concrete mechanism. The chemiosmotic model introduced into this formula the idea of a sort of fuel cell, in which the role of the load is played by ATP synthetase, carrying out ATP synthesis. All the subsequent refinement and detailing of the theory were in actual fact merely an adaptation of the initial model to the new facts about the structure and peculiarities of the functioning of coupling membranes, namely, the "perfecting" of the fuel cell model.

Mitchell's theory was always a phenomenological theory. He himself tried to outline some of the basic features of the mechanism,[168] but his attempt could hardly be described as successful.

Yet in the very same year that Mitchell's theory appeared, Williams formulated the concept of a localized proton processing with the main emphasis on local interactions.[138]

A question arises whether the coupling of the reactions of the enzymes inserted in a coupling membrane takes place in such a way that the motive force of ATP synthesis is $\Delta\bar{\mu}_{H^+}$ accumulated in the bulk of the aqueous phases that wash the membrane, or whether there exists direct interaction between the membrane enzymes when, according to Williams' concept, the microsurroundings and local conditions are determining factors. It may, at first glance, seem immaterial to find the answer to this question, but in actual fact it represents two fundamentally different approaches to the understanding of the functioning of energy-coupling membranes.

Another point of view, which, if interpreted broadly, could be seen as an attempt to provide a physical substantiation of Williams' concept, appeared comparatively recently. It presumes that the reactions, including the redox reactions, take place in a medium formed by the enzyme itself. In this case, in keeping with Fine's theory,[169] it could be imagined that the vibration of the lattice approximating the medium in which the reaction takes place does not relax into heat quickly enough for thermodynamic equilibrium to be established. On the contrary, what is likely to occur is a process in the course of which the perturbation causes coherent vibrations. In examining the enzyme-substrate complexes, such as coupling photosynthetic or mitochondrial membranes,[116,118] the vibrations of the nuclei of the lattice in the subsystems are related by means of the electron-phonon interaction to the electron subsystem. This interaction can no longer be regarded as small. This question was discussed in a number of works by Blumenfeld,[19,26,116] and we shall not consider it here.

V. CONCLUSIONS

We have examined certain approaches to the investigation of redox reactions on membranes. Just as in the case of any electrode redox reaction, a potential arises on the membrane. In the simplest case the potential follows the Nernst equation. In other, more complex cases, when simultaneously ionic permeability is possible or when one of the reaction components can pass through the membrane the value of the potential begins to depend on the kinetic parameters of the system and the potential itself loses it initial simple meaning of an energy scale. That is the case even in the simplest example of bilayer membrane during redox reactions when products appear ejected into the reaction zone and an unstirred layer is formed, which has long been known in electrochemistry as an obstacle to the investigation of the true kinetics of an electrochemical process on the electrode. Here the stirred layer begins to play the key role in the value and sign of the transmembrane potential. For reactions having a common intermediate, such as proton, and when the membrane is proton permeable the transmembrane potential provides evidence of a proton gradient in the unstirred layers. However, more complex cases are also possible where the physical meaning of the observed potential can be elucidated only after special investigations in which changes in the potential are compared with those or other values measured independently.

The appearance of an unstirred layer could also be considered from another more general point of view, according to which the unstirred layer adjacent to the membrane is the result of the nonequilibrium state of the system in keeping with Prigozhin's concept leading to the appearance in it of a new structural element. This fact is worthy of attention, since the mitochondrial membrane, chloroplast membranes, and other biological membranes are packed with effectively functioning enzyme systems. Actually, this leads to Williams' localized proton processing hypothesis as the main coupling element. In actively functioning systems the forces operating in membranes are determined by local concentrations, which are in no way comparable to these in the solution bulk, and therefore the transmembrane potential, which by definition is measured at the points distant from the interface, cannot reflect the local processes taking place in these conditions.

However, if it is assumed that processes for which the pumps are responsible proceed so ineffectively that all the concentration gradients have time to become equalized, then in that case the membrane will be in equilibrium with the bulk of the solutions washing it, which fits the concept proposed by Mitchell.

Probably the entire structure of the coupling membrane is adapted to the conditions in which it functions effectively (that is, there exist corresponding proton and ion routes without their having to be in equilibrium with the bulk of the membrane), and in this sense Mitchell's coupling route is a reserve route which is used when the membrane is functioning ineffectively or is, perhaps, damaged. But does that mean that the transmembrane potential does not operate in native conditions? Most probably not. When the parameters of the system deviate from the optimal ones, shifts occur in the concentrations of organellas and it is likely that the transmembrane potential acts as a regulator—but only as a highly inertial regulator of the second stage. The first stage is local forces acting on the particles or molecules of the enzyme at a given point. The question could be formulated more drastically: is the membrane necessary for the coupling process in organisms, or does there exist a simpler structural element—the membrane/electrolyte interface— where this process could take place? If the simplest structural element essential for the coupling process were not the coupling membrane, but the coupling membrane/electrolyte interface, then within the framework of such a structural element only the adsorption potential and distribution potential could be considered, but no such concept as transmembrane potential exists for such a system. However, only future investigations will reveal the extent to which the ideas outlined here reflect the true picture.

REFERENCES

[1] J. O'M. Bockris and A. K. N. Reddy, *Modern Electrochemistry*, Vol. 1, p. 31, Plenum Press, New York, 1970.
[2] P. Läuger, J. Richter, and W. Lesslauer, *Ber. Bunsenges. Phys. Chem.* **71** (1967) 906.
[3] L. I. Boguslavsky, F. I. Bogolepova, and A. V. Lebedev, *Chem. Phys. Lipids* **6** (1971) 296.
[4] A. V. Lebedev and L. I. Boguslavsky, *Dokl. Acad. Nauk SSSR* **189** (1969) 1122.
[5] R. C. MacDonald and A. D. Bangham, *J. Membrane Biol.* **7** (1972) 29.
[6] O. Alvares and R. Latorre, *Biophys. J.* **21** (1978) 1.
[7] V. S. Sokolov, V. V. Chernii, and I. G. Abidor, *Dokl. Acad. Nauk SSSR* **251** (1980) 236.
[8] A. L. Lehninger, *The Mitochondrion Molecular Basis of Structure and Function*, Benjamin, New York, 1964.
[9] E. Racker, *A New Look at Mechanisms in Bioenergetics*, Academic, New York, 1976.
[10] A. B. Rubin and V. P. Shipkarev, *Electron Transport in Biological Systems*, Nauka, Moscow, 1984.
[11] B. Chance, G. Williams, W. Homes and J. Higgins, *J. Biol. Chem.* **217** (1955) 439.

[12] A. B. Rubin and A. S. Focht, *Biofizika* **10** (1965) 236.
[13] D. F. Wilson, M. Erecinska, T. Ohnishi, and P. L. Dutton, *Bioelectrochem. Bioenergetics* **1** (1974) 3.
[14] S. D. Varfolomeev, A. Naki, and I. V. Berezin, *Mol. Biol.* **11** (1977) 1100.
[15] S. Malkin, *Biochim. Biophys. Acta* **126** (1966) 433.
[16] P. P. Noks, E. P. Lukashev, A. A. Konoenko, P. S. Venediktov, and A. B. Rubin, *Molek. Biol.* **11** (1977) 1090.
[17] T. L. Hill and B. Chance, *J. Theor. Biol.* **72** (1978) 17.
[18] V. A. Saks, V. V. Kurijanov, and V. N. Lyzikov, *Biochim. Biophys. Acta* **213** (1972) 42.
[19] L. Blumenfeld, *J. Theor. Biol.* **58** (1976) 269.
[20] B. Cartling and A. E. Ehrenberg, *Biophys. J.* **23** (1978) 451.
[21] T. Graan and D. R. Ort, *J. Biol. Chem.* **258** (1983) 2831.
[22] G. R. Moore, *FEBS Lett.* **161** (1983) 171.
[23] J. D. Mc Elroy, D. C. Mauzerall, and G. Feher, *Biochim. Biophys. Acta* **333** (1974) 261.
[24] H. Rottenberg, *Biochim. Biophys. Acta* **549** (1979) 225.
[25] H. Arata and M. Nishimura, *Biophys. J.* **32** (1980) 791.
[26] L. A. Blumenfeld, *Physics of Bioenergetic Processes*, Springer, Berlin, 1983.
[27] M. Hirano, K. Satoh, and S. Katoh, *Photosynthesis Res.* **1** (1980) 149.
[28] M. Aoki and S. Katoh, *Biochim. Biophys. Acta* **682** (1982) 307.
[29] K. Folkers and Y. Yamamura, *Biochemical and Chemical Aspects of Coenzyme Q*, Elsevier, Amsterdam, 1977.
[30] B. L. Trumpower, Ed., *Function of Quinones in Energy Conserving Systems*, Academic, New York, 1982.
[31] C.-A. Yu and L. Yu, *Biochim. Biophys. Acta* **639** (1981) 99.
[32] P. Hinkle, *Biophys. Res. Commun.* **41** (1970) 1375.
[33] P. Hinkle, J. Kim, and E. Racker, *J. Biol. Chem.* **247** (1972) 1338.
[34] A. D. Ismailov, L. I. Boguslavsky, L. S. Yaguzhinsky, and V. P. Skulachev, *Dokl. Akad. Nauk SSSR* **210** (1973) 709.
[35] L. S. Yaguzhinsky, L. I. Boguslavsky, and A. D. Ismailov, *Biochim. Biophys. Acta* **368** (1974) 22.
[36] C. Ragan and P. Hinkle, *J. Biol. Chem.* **250** (1978) 8475.
[37] H. Schneider, J. J. Iemasters, N. Höchil, and C. R. Hackenbrock, *J. Biol. Chem.* **255** (1980) 3748.
[38] P. Mitchell, *J. Theor. Biol.* **62** (1976) 327.
[39] D. G. Grooth, R. Van Grondelle, J. C. Romijn, and M. P. J. Pulles, *Biochim. Biophys. Acta* **503** (1978) 480.
[40] C. A. Boicelli, C. Ramponi, E. Casali, and L. Masotti, *Membrane Biochem.* **4** (1980) 105.
[41] D. Bäckstroöm, B. Norling, A. Ehrenberg, and L. Ernster, *Biochim. Biophys. Acta* **197** (1970) 108.
[42] C. A. Yu, S. Nagaoka, L. Yu, and T. E. King, *Arch. Biochem. Biophys.* **204** (1980) 59.
[43] R. M. Kaschnitz and Y. Hatefi, *Arch. Biochem. Biophys.* **171** (1975) 292.
[44] A. Kröger and M. Klingenberg, *Eur. J. Biochem.* **34** (1973) 358.
[45] V. E. Kagan, S. V. Kotelcvcev, and Yu. P. Kozlov, *Dokl. Akad. Nauk SSSR* **217** (1974) 213.
[46] L. I. Boguslavsky, in *Comprehensive Treatise of Electrochemistry*, Vol. 1, Ed. by J. O'M. Bockris, B. E. Conway, and E. Yeager, Plenum Press, New York, 1980, p. 329.

[47] R. J. Shamberger, in *Food and Biological Systems*, Plenum Press, New York, 1980, p. 639.
[48] H. Shimasaki, N. Heta, and O. S. Privett, *Lipids* **15** (1980) 236.
[49] K. Fusimoto, W. W. Neff, and E. N. Frankel, *Biochim. Biophys. Acta* **795** (1984) 100.
[50] P. Rustin, J. Dupont, and C. Lance, *Arch. Biochem. Biophys.* **225** (1983) 630.
[51] B. T. Storey, in *The Biochemistry of Plants*, Ed. by D. D. Davies, Vol. 2, Academic, New York, p. 125, 1980.
[52] H. J. Forman and A. Baveris, in *Free Radicals in Biology*, Vol. 15, Ed. by W. A. Pryor, Academic, New York, 1982, p. 65.
[53] M. Logani and R. E. Davis, *Lipids* **15** (1980) 485.
[54] A. D. Ismailov, L. I. Boguslavsky, and L. S. Yaguzhinsky, in *Biophysics of Membranes*, Kaunas, 1973, p. 208.
[55] A. D. Ismailov, L. I. Boguslavsky, and L. S. Yaguzhinsky, *Dokl. Akad. Nauk SSSR* **216** (1974) 674.
[56] Yu. A. Shipunov, V. S. Sokolov, L. S. Yaguzhinsky, and L. I. Boguslavsky, *Biofizika* **21** (1976) 280.
[57] L. I. Boguslavsky, L. S. Yaguzhinsky, and A. D. Ismailov, *Bioelectrochem. Bioenergetics* **4** (1977) 155.
[58] O. Le Blanc, *Biochim. Biophys. Acta* **193** (1969) 350.
[59] D. M. Miller, *Biochim. Biophys. Acta* **266** (1972) 85.
[60] J. H. M. Suverijn, L. A. Huisman, J. Rosing, and A. Kemp, *Biochim. Biophys. Acta* **305** (1973) 185.
[61] T. E. Andreoli and S. L. Troutman, *J. General Physiol.* **57** (1971) 464.
[62] A. B. R. Thoinson, *J. Membr. Biol.* **47** (1979) 39.
[63] S. B. Hladky, *Biochim. Biophys. Acta* **30** (1973) 261.
[64] F. A. Wilson and J. M. Dietschy, *Biochim. Biophys. Acta* **363** (1974) 112.
[65] J. Gutknecht and D. C. Tosteson, *Science* **182** (1973) 1258.
[66] J. Dainty and C. R. House, *J. Physiol.* **182** (1966) 66.
[67] R. L. Preston, *Biochim. Biophys. Acta* **688** (1982) 422.
[68] A. Petkau, *Can. J. Chem.* **49** (1971) 1187.
[69] Yu. A. Vladimirov and A. I. Archakov, *Lipid Peroxidation in Biological Membranes*, Nauka, Moscow, 1972.
[70] A. Masa, I. Takahasi and K. Asada, *Arch. Biochem. Biophys.* **226** (1983) 558.
[71] P. Mitchell, *Nature (London)* **191** (1961) 144.
[72] H. V. Westerhoff, B. A. Melandri, G. F. Azzone and D. B. Kell, *FEBS Lett.* **165** (1984) 1.
[73] V. S. Sokolov, Yu. A. Shipunov, L. S. Yaguzhinsky, and L. I. Boguslavsky, in *Biophysic of Membranes*, Nauka, Moscow, 1981, p. 59.
[74] E. S. Hyman, *J. Membrane Biol.* **37** (1977) 263.
[75] L. I. Boguslavsky and L. S. Yaguzhinsky, in *Bioelectrochemistry and Electrosynthesis of Organic Compounds*, Nauka, Moscow, 1975, p. 305.
[76] I. I. Ivanov, *Molek. Biologiya* **8** (1984) 512.
[77] N. M. Emanuel, E. T. Denisov, and Z. K. Maizus, *Chain Reactions of Hydrocarbon Oxidation in the Liquid phase*, Nauka, Moscow, 1975.
[78] H. van Zutphen and D. G. Cornwell, *J. Membrane Biol.* **13** (1973) 79.
[79] Yu. P. Kozlov, V. B. Ritov, and B. E. Kagan, *Dokl. Akad. Nauk SSSR* **212** (1973) 1239.
[80] V. F. Antonov, A. S. Ivanov, E. A. Kornepanova, N. S. Osin, V. V. Petrov and Z. I. Truchmanova, in *Normal and Pathological Free Radical Oxidation of Lipids*, Nauka, Moscow, 1976.

[81] R. M. Locke, E. Rial, I. D. Scott, and O. G. Nicholls, *J. Biochem.* **129** (1982) 373.
[82] V. S. Sokolov, T. D. Churakova, B. G. Bulgakov, B. E. Kagan, M. V. Bilenko, and L. I. Boguslavsky, *Biofizika* **26** (1981) 147.
[83] A. A. Krasnovsky, in *Biochemistry and Biophysics of Photosynthesis*, Ed. by A. A. Krasnovsky, Nauka, Moscow, 1965.
[84] D. I. Arnon, *Proc. Natl. Acad. Sci. USA* **68** (1971) 2883.
[85] V. A. Shuvalov and A. A. Krasnovsky, *Biofizika* **26** (1981) 544.
[86] G. R. Sccly, *Photochem. Photobiol.* **27** (1978) 639.
[87] D. Mauzerall and F. T. Hong, in *Porphyrins and Metalloporphyrins*, Ed. by M. Smith, Elsevier, Amsterdam, 1975, p. 702.
[88] A. G. Volkov, M. A. Bibikova, A. F. Mironov, and L. I. Boguslavsky, *Bioelectrochem. Bioenergetics* **10** (1983) 477.
[89] A. J. Hoff, *Photochem. Photobiol.* **13** (1974) 51.
[90] A. Steinmann, G. Stark, and P. Läuger, *J. Membr. Biol.* **9** (1972) 177.
[91] I. G. Kadochnikova, D. A. Zakrzhevsky, and Yu. E. Kalashnikov, *Biofizika* **28** (1983) 917.
[92] H. Ti Tien, *Nature (London)* **219** (1968) 272.
[93] A. Ilani and D. S. Berns, *Biophysik* **9** (1973) 209.
[94] H. Ti Tien and S. P. Verma, *Nature (London)* **227** (1970) 1238.
[95] F. Hong and D. Mauzerall, *Biochim. Biophys. Acta* **266** (1972) 584.
[96] F. Hong and D. Mauzerall, *Nature New Biol.* **240** (1972) 154.
[97] R. R. Dogonadze and A. M. Kuznecov, *Elektrochimiya* **1** (1965) 1008.
[98] S. Papa, M. Lorusso, D. Boffoli, and E. Bellomo, *Eur. J. Biochem.* **137** (1983) 405.
[99] W. E. Ford, J. W. Otvos, and M. Calvin, *Proc. Natl. Acad. Sci. USA* **76** (1979) 3590.
[100] W. E. Ford and G. Tollin, *Photochem. Photobiol.* **36** (1982) 647.
[101] K. Kurihara, M. Sukigara, and Y. Toyoshima, *Biochim. Biophys. Acta* **547** (1979) 117.
[102] G. van Ginkel and J. K. Raison, *Photochem. Photobiol.* **32** (1980) 793.
[103] L. I. Boguslavsky, B. T. Lozhkin, and B. A. Kiselev, *Dokl. Akad. Nauk SSSR* **222** (1975) 73.
[104] H. Dahmas, *J. Phys. Chem.* **72** (1968) 362.
[105] F. B. Kaufman and E. M. Engler, *J. Am. Chem. Soc.* **101** (1979) 547.
[106] D. A. Buttry and F. C. Anson, *J. Electroanal. Chem.* **130** (1981) 333.
[107] G. Inzelt, Q. Chamberg, I. F. Krustle and R. W. Day, *J. Am. Chem. Soc.* **106** (1984) 3336.
[108] F. W. Cope, *Ann. N.Y. Acad. Sci.* **204** (1973) 416.
[109] E. E. Petrov, *Int. J. Quant. Chem.* **16** (1979) 133.
[110] B. J. Hales, *Biophys. J.* **16** (1976) 471.
[111] E. A. Liberman and V. P. Topali, in *Biophysics of Membranes*, Kaunas, 1971, p. 548.
[112] D. W. Deamer, R. C. Price and A. R. Croffts, *Biochim. Biophys. Acta* **274** (1972) 323.
[113] V. C. Markin, L. I. Krishtalik, E. A. Liberman, and V. P. Topali, *Biofizika* **10** (1969) 256.
[114] L. I. Boguslavsky, A. A. Chatisshvili, A. I. Gubkin, and V. A. Ogloblin, *Elektrokhimiya* **16** (1978) 648.
[115] M. V. Makinen, S. A. Schichman, S. C. Hill, and H. B. Gray, *Science* **222** (1983) 929.
[116] L. A. Blumenfeld and D. S. Chernavskii, *J. Theor. Biol.* **39** (1973) 1.
[117] J. J. Hopfield, *Proc. Natl. Acad. Sci. USA* **71** (1974) 3640.
[118] N. M. Chernavskaya and D. S. Chernavskii, *Electron Tunnelling in Photosynthesis*, Moscow State University, Moscow, 1977, p. 17.

[119] V. I. Goldansky, *Nature (London)* **279** (1979) 109.
[120] E. Buhkg and J. Jortner, *FEBS Lett.* **109** (1980) 117.
[121] R. F. Goldstein and A. Bearden, *Proc. Natl. Acad. Sci. USA* **81** (1984) 135.
[122] B. R. Masters and D. Mauzerall, *J. Membr. Biol.* **41** (1978) 377.
[123] B. Bhowmik, G. L. Jendrasiak, and B. Roseberg, *Nature* **215** (1967) 842.
[124] G. L. Jendrasiak, *Chem. Phys. Lipids* **3** (1969) 68.
[125] E. A. Liberman, V. P. Topali, L. M. Zofina, and A. M. Shkrob, *Biofizika* **14** (1965) 56.
[126] Yu. A. Chizmadzhev, V. S. Markin, and R. N. Kuklin, *Biofizika* **16** (1971) 437.
[127] C. A. Wraight, *Photochem. Photobiol.* **30** (1979) 767.
[128] Te-Fu Ho, A. R. McIntosh, and J. R. Bolton, *Nature* **5770** (1980) 254.
[129] J. K. Hurley, F. Castelli, and Gr. Tollin, *Photochem. Photobiol.* **34** (1981) 623.
[130] R. N. Robertson and N. K. Boardman, *FEBS Lett.* **60** (1975) 1.
[131] L. S. Yaguzhinsky, L. I. Boguslavsky, A. G. Volkov, and A. B. Rachmaninova, *Dokl. Akad. Nauk SSSR* **221** (1975) 1465.
[132] A. Spisni, G. Sarter, and L. Masotti, *Membr. Biochem.* **4** (1980) 149.
[133] A. Futami, E. Murt, and G. Hauska, *Biochim. Biophys. Acta* **547** (1979) 573.
[134] H. Katsikas and P. J. Quinn, *FEBS Lett.* **133** (1981) 230.
[135] R. M. Pennington and R. R. Fisher, *J. Biol. Chem.* **256** (1981) 8963.
[136] A. A. Konstantinov, *Dokl. Akad. Nauk SSSR* **237** (1977) 713.
[137] S. Papa, *J. Bioenerg. Biomembr.* **14** (1982) 69.
[138] R. J. Williams, *J. Theor. Biol.* **1** (1961) 1.
[139] R. J. P. Williams, *Biochim. Biophys. Acta* **505** (1978) 1.
[140] J. F. Nagle and H. J. Morowitz, *Proc. Natl. Acad. Sci. USA* **75** (1978) 298.
[141] J. F. Nagle and S. Tristram-Nagle, *J. Membr. Biol.* **74** (1983) 1.
[142] B. E. Conway, in *Modern Aspects of Electrochemistry*, No. 3, Ed. by J. O'M. Bockris and B. E. Conway, Butterworths, London, 1964.
[143] J. Bruinink, *J. Appl. Electrochem.* **2** (1972) 239.
[144] R. Lindemann and G. Zundel, *Biopolymers* **17** (1978) 1285.
[145] P. P. Rastogi, W. Kristof, and G. Zundel, *Biochem. Biophys. Res. Commun.* **95** (1980) 902.
[146] P. V. Ionnu and B. T. Colding, *Prog. Lipid. Res.* **17** (1979) 279.
[147] J. M. Seddon, R. D. Kaye, and D. Marsh, *Biochim. Biophys. Acta* **734** (1983) 347.
[148] P. R. Cullis, B. De Kruijff, M. J. Hope, R. Nayar, and S. L. Schmid, *Can. Biochem. J.* **58** (1980) 1091.
[149] A. V. Vannikov and L. I. Boguslavsky, *Biofizika* **14** (1969) 421.
[150] V. P. Skulachev, *Transformation of Energy in Biomembranes*, Nauka, Moscow, 1972, p. 6.
[151] H. Rottenberg, *Biochim. Biophys. Acta* **549** (1979) 225.
[152] V. Kagawa and E. Racker, *J. Biol. Chem.* **246** (1971) 5477.
[153] K. H. Leung and P. S. Hinkle, *J. Biol. Chem.* **250** (1975) 8467.
[154] E. Racker, *Energy Transducing Mechanisms*, Butterworths, London, 1974.
[155] A. Azzi, *Q. Rev. Biophys.* **8** (1975) 236.
[156] A. A. Jasaites, L. V. Chu, and V. P. Skulachev, *FEBS Lett.* **31** (1973) 241.
[157] A. Azzi, P. Ghererdini, and M. Santato, *J. Biochem.* **246** (1971) 2035.
[158] S. McLaughlin and H. Harary, *Biochemistry* **15** (1976) 1941.
[159] L. A. Drachev, A. A. Jasaitits, A. D. Kaulen, A. A. Kondrashin La Van Chu, A. Yu. Semenov, I. I. Severina and V. P. Skulachev, *J. Biol. Chem.* **251** (1976) 7072.
[160] H. T. Witt and A. Zicker, *FEBS Lett.* **37** (1973) 307.
[161] R. Tieman, G. Reuger, P. Gräber, and H. T. Witt, *Biochim. Biophys. Acta* **546** (1979) 498.
[162] D. E. Robertson and H. Rottenberg, *J. Biol. Chem.* **258** (1983) 11039.

[163] E. Racker and W. Stoeckenius, *J. Biol. Chem.* **249** (1974) 662.
[164] A. B. Thomson and J. M. Dietschy, *J. Theor. Biol.* **64** (1977) 277.
[165] R. A. Dilley and U. Shreiber, *J. Bioenerg. Biomembr.* **16** (1984) 173.
[166] Y. de Kouchkovsky and F. Haraux, *Biochem. Biophys. Res. Commun.* **99** (1981) 205.
[167] P. Walz, *Biochim. Biophys. Acta* **505** (1979) 279.
[168] P. Mitchell, *FEBS Lett.* **78** (1977) 1.
[169] V. M. Fain, *J. Chem. Phys.* **65** (1976) 1854.

4

Electrochemistry of Hydrous Oxide Films

Laurence D. Burke

Chemistry Department, University College Cork, Cork, Ireland

Michael E. G. Lyons

Chemistry Department, Trinity College, Dublin, Ireland

I. INTRODUCTION

For the purpose of this review it is convenient to classify oxides into two groups: (1) *compact, anhydrous oxides* such as rutile, perovskite, spinel, and ilmenite in which oxygen is present only as a bridging species between two metal cations and ideal crystals constitute tightly packed giant molecules; and (2) *dispersed, hydrous oxides* where oxygen is present not just as a bridging species between metal ions, but also as O^-, OH, and OH_2 species, i.e., in coordinated terminal group form. In many cases the latter materials when in contact with aqueous media contain considerable quantities of loosely bound and trapped water, plus, occasionally, electrolyte species. Indeed, with highly dispersed material (dispersion here refers to the molecular level, i.e., microdispersion, not to the finely divided state where the oxide microparticles may still be compact in character, i.e., macrodispersion), the boundary between the solid and aqueous phases may be somewhat nebulous as the two phases virtually intermingle. While compact oxides are usually prepared by thermal techniques, e.g., direct combination of the elements, decomposition of an unstable salt, or dehydration of a hydrous oxide, the dispersed oxides are almost invariably prepared in an aqueous environment using, for instance, base precipitation or electrochemical techniques. Very often the materials obtained in such reactions are deposited in the kinetically most accessible, rather than the thermodynamically most stable, form; thus they are often amorphous or only poorly crystalline and prone to rearrangement in a manner that is strongly influenced by factors such as temperature, pH, and ionic strength. Well-known examples of this technologically important group of compounds include certain varieties of aluminum hydroxide,[1] nickel hydroxide and oxyhydroxide,[2,3] and iridium oxide.[4] As will be discussed more fully later, microdispersion is usually due to the presence of strand, layer, tunnel, or cage structures which allow not just small ions but also in many cases solvent molecules to permeate the oxide or hydroxide phase.

The problem of classifying a particular sample of oxide as being hydrous or anhydrous is frequently not simple; a useful example in this area is the case of iridium dioxide, which, apart from apparent densities[5,6] of 11.68 and 2.0 g cm^{-3}—corresponding to the anhydrous and highly hydrated material, respectively—can also be deposited with a variety of intermediate density values.[7] It is worth noting that the surface of an anhydrous oxide such as thermally prepared RuO_2 hydrates (or hydroxylates) when the material is brought into contact with aqueous media. An appreciation of the effect of such hydration is essential for the interpretation of many aspects of the behavior of such surfaces.[8] The more open structure of hydrous materials leads to greater reactivity for various reactions, e.g., charge storage and electrochromic behavior, electrocatalysis, and dissolution. One process which may be somewhat inhibited by the disperse nature of the hydrous material is electron transfer.[9]

Hydrous oxides are of major interest in many areas of technology, e.g., corrosion and passivation of metals, formation of decorative, protective, and insulating films, aqueous battery systems, catalysis and electrocatalysis, electrochromic display systems, pH monitoring devices, soil science, colloid chemistry, and various branches of material science. Detailed accounts of some of the nonnoble hydrous metal oxide systems, especially aluminum,[1] have appeared recently. In the case of the noble metals such as platinum or gold most of the electrochemical work to date has been concerned with compact monolayer, and submonolayer, oxide growth.

However, interest in this area has broadened considerably in recent years with the development by Beer[10] of RuO_2-activated titanium anodes for use in the chlor-alkali and related industries. This has provided a major impetus for the study of noble metal—and indeed that of some of the more active nonnoble metal—oxide electrochemistry. The slightly more recent development of the potential cycling technique for producing hydrous oxide layers[11,12]—or at least an appreciation of the processes involved and the scope for development (the basic effect, i.e., the development of enhanced charge storage on cycling, was known in the case of iridium at a much earlier stage[13])—has fortunately enabled work on the anhydrous and hydrous oxides to progress, with obvious advantages, side-by-side.

The present review is concerned mainly with the electrochemical formation and redox behavior of the hydrous oxides of those transition metals centered within and around Group VIII of the periodic table. There have been a number of recent reviews[14-18] of monolayer oxide growth on these metals so that this area will not be treated here in an exhaustive manner. Structural data for many of the systems (especially direct evidence obtained by investigation of hydrous films themselves) are very sparse at the present time. However, some idea of the type of material involved can be obtained from structural studies of oxide battery materials[2,3,19]; a useful introduction to the structural complexities in this area in general is Alwitt's account[1] of the aluminium oxide system. An important feature of hydrous oxides, not normally as evident with their anhydrous analogs, is their acid-base behavior and in particular the influence of the latter on the redox properties of the hydrous material. Because of its central role in many oxide (especially hydrous oxide) processes, and its relative neglect in the electrochemistry of these systems until quite recently, this acid-base character of oxide systems will be reviewed here in some detail.

II. FORMATION OF HYDROUS OXIDES

Hydrous oxide preparation is a routine procedure in gravimetric analysis, the transition metal ions usually being precipitated in hydrous form by addition of base. While oxide suspensions prepared in this manner may be used to examine double layer phenomena, e.g., measurement of the point of zero charge and studies of ion adsorption behavior, the investigation of the redox and electrocatalytic properties requires that the hydrous oxides, ideally present in a uniform, homogeneous layer, be attached to some type of inert electronic conductor. There are various ways of preparing such coated electrodes, although in many cases the resulting system may not be as uniform as desired. One of the simplest methods is direct anodization of the metal in a suitable electrolyte using galvanostatic or potentiostatic techniques. This is frequently regarded as a dissolution–precipitation (i.e., hydrolysis) procedure—although as Alwitt[1] has pointed out in the case of aluminum there is no evidence for the direct participation of dissolved metal ion species. This technique is effective also for some of the noble metals, e.g., platinum[20-22] and gold[23]; it seems to be ineffective in at least one case, iridium, as the hydrous film dissolves[11,24] under the type of highly anodic dc conditions that are successful in the case of platinum.

Cathodic electroprecipitation is a technique used commercially to prepare nickel hydroxide deposits in the battery field.[25] In this case a nickel salt is present in solution at low pH (ca. 3.0) and hydrogen gas evolution around the cathode causes a local increase in pH, resulting in the precipitation of an adherent layer of nickel hydroxide at the metal surface. Similarly, anodic electroprecipitation is used commercially[26] to produce layers of another highly active battery material, γ-MnO_2.

One of the most versatile and convenient techniques used in recent times to generate hydrous oxides in a form suitable for examining their redox behavior is that of potential cycling.[11,12,27] In this case the potential of an electrode of the parent metal, which may be noble or nonnoble, is cycled repetitively between suitable limits in an aqueous solution of appropriate pH. The type of oxide growth cycle used—sinusoidal, square, or triangular wave—apparently makes little difference; the triangular wave is most convenient as changes in the current/voltage response, i.e., cyclic voltammetry, can, if necessary, be employed during the oxide growth reaction to monitor changes in redox behavior associated with the latter.

The mechanism of hydrous oxide growth on repetitive cycling is now reasonably well understood, at least at a qualitative level.

For most metals, but especially gold, platinum, iridium, and rhodium, extension of oxide growth beyond the monolayer level under conventional galvanostatic or potentiostatic conditions is usually quite slow—obviously due to the presence of the initial, compact oxide, product layer which acts as a barrier to further growth. Under potential cycling conditions the anodic limit plays quite a crucial role. There is probably a combination of thermodynamic and kinetic factors involved, but evidently the upper limit must be sufficiently anodic that compact oxide formation exceeds significantly the single monolayer level so that on subsequent reduction a disturbed, highly disordered layer of metal atoms is produced at the electrode surface. Thus with platinum[27] and gold,[28] two metals where oxide monolayer behavior is well defined, the optimum lower limit lies at a potential value at, or below, the maximum of the monolayer oxide reduction peak. On subsequent reoxidation the disturbed layer of metal atoms is evidently converted to hydrated, or partially hydrated, oxide—complete hydration under these conditions may involve several redox cycles (see Fig. 1)—with a fresh inner compact layer being regenerated at the metal surface on each anodic sweep. On repetitive cycling the porous outer layer increases in thickness at the expense of the underlying metal. Lack of stirring dependence in such oxide growth reactions[29,30] suggests that solution species, i.e., a dissolution/hydrolysis mechanism, are not involved.

One further promising procedure, which has been used to date mainly to prepare iridium oxide films, involves the use of sputtering techniques. These have been described recently in some detail by Kang and Shay.[7] The films were deposited onto a cooled substrate in an argon/oxygen (80/20) atmosphere at a net pressure of ca. 33 μm of mercury. To obtain a film of highly dispersed material by this technique traces of water in the vacuum chamber are apparently essential—although high levels were found to have a detrimental effect.

III. ACID–BASE PROPERTIES OF OXIDES

An acid is a substance which donates a proton to, or accepts a pair of electrons from, another substrate which is classified as being a

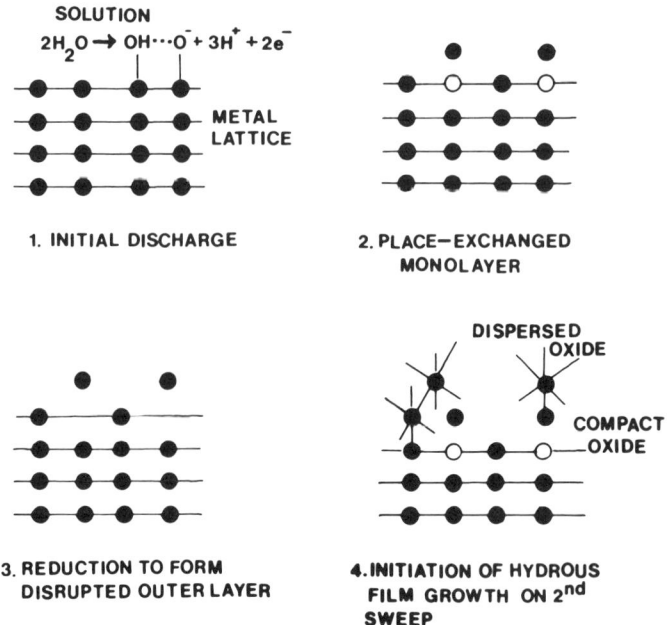

Figure 1. Schematic outline of the processes involved in hydrous oxide growth under potential cycling conditions.

base. Although it has received surprisingly little attention in electrochemistry, apart from adsorption[31] and double layer[32] studies, the behavior of solid oxides as acids and bases is well recognized in the study of heterogeneous catalysis.[33] The relationship between typical acid and base sites at an oxide surface may be illustrated as follows:

$$\begin{array}{ccc} \text{Lewis} & \text{Lewis} & \text{Bronsted} \quad \text{Bronsted} \\ \text{acid} & \text{base} & \text{acid} \quad\quad \text{base} \end{array} \tag{1}$$

When an oxide is in contact with aqueous media the distinction between oxide, hydrated oxide, and hydroxide (especially when the material is present in microdispersed form) tends to be rather vague as the following equilibria are usually assumed to exist:

$$(O^{2-})_s + H_2O = (O^{2-} \cdot H_2O)_s = 2(OH^-)_s \quad (2)$$

There is the additional complication also (as outlined in more detail later) that many oxides are amphotheric, i.e., they tend to function as bases (adsorbing protons or loosing hydroxide ions and thereby acquiring positive charge) at low pH and as acids (adsorbing hydroxide ions or loosing protons and thereby becoming negatively charged) at high pH. The confusion and uncertainty that can arise in this area is highlighted by the recent controversy[34,35] as to which of these two processes, proton or hydroxide insertion/withdrawal reactions, is involved in the charge neutralization stage of the redox reactions of hydrous iridium oxide layers.

In most discussions of the redox properties of metal and metal oxides to date oxide species (apart possibly from reaction intermediates) are regarded as being formally uncharged—in some surface oxidation work[36] attention has of course been drawn to the fact that metal-oxygen bonds are polar. However, there are some well-known acidic metal hydroxide species, and one that is highly relevant here, since it involves a noble metal and its structure has been established,[37] is platinic acid, $H_2Pt(OH)_6$. This compound, usually prepared by hydrolysis of $PtCl_6^{2-}$, might also be considered as the hydrated dioxide $PtO_2 \cdot 4H_2O$—although structurally it is an octahedrally coordinated hexahydroxoplatinum (IV) compound. The precipitate obtained in the hydrolysis reaction is clearly a solid acid; it consists of an agglomerate of *anionic* octahedrally coordinated hydroxy species with an equivalent number of protons present—the latter probably being rather mobile, i.e., delocalized, in the rather open hydrous oxide structure. These two features, namely, the fact that the insoluble lattice may be charged and that mobile counterions may be present, are important aspects of hydrous oxide electrochemistry in general. It is clear, for instance, that with hydrous oxide electrodes in aqueous media there are now two sources of protons (or indeed any other species capable as functioning as counterions), i.e., the oxide itself and the solution phase. The acidic properties of a hydrous material may also be regarded as a hydrolysis process[38] in which protons are expelled from coordinated water molecules due to the high positive charge on the central metal ion, viz.,

$$Pt(OH)_4 \cdot 2H_2O = [Pt(OH)_6]^{2-} + 2H^+ \qquad (3)$$

According to this view the acidity of the hydrous material should decrease, as is known for solution species,[39] with decreasing charge on the central metal ion. Eventually a nondissociated hydrous layer, or possibly even a basic, i.e., cationic, film (in which protons are attracted in to neutralize coordinated hydroxide ions, or the latter are simply lost by desorption), will be attained at low central metal cation charge values.

The reversible potential of oxide systems, where both reduced and oxidized forms are insoluble, are almost invariably regarded (in terms of the RHE scale) as being pH independent, and this assumption is undoubtably valid in many cases, a typical example being the widely used Hg/HgO reference electrode.[40] However, deviations were noted in potential sweep experiments with oxide films on gold[23,41] and, more significantly (in view of the highly reversible character of the charge storage peaks), in cyclic voltammetry work with hydrous iridium oxide systems.[4,9,24,30] In the case of gold[23] two reduction peaks were noted with oxide films formed at high potentials ($E >$ ca. 2.1 V) under potentiostatic conditions ($t >$ ca. 0.5 min). The first peak was attributed to the reduction of a layer of compact oxide (possibly with a minor degree of hydration), viz.,

$$Au_2O_3 + 6H^+ + 6e^- = 2Au + 3H_2O \qquad (4)$$

while the second, usually much larger peak (corresponding to reduction of the outer hydrous material), showed a variation of ca. $1/2(2.303RT/F)$ V/pH unit (RHE scale) i.e., $3/2(2.303RT/F)$ V/pH unit (SHE scale).† According to this result the ratio of

---

† The Standard Hydrogen Electrode (SHE) scale is the well-known, internationally recognized, primary reference based on the following electrode, $Pt/H_2(a = 1)/H^+(a = 1)$, arbitrarily assigned a potential value of zero. As such an electrode is virtually impossible to construct, it is convenient in practice to use the Reversible Hydrogen Electrode (RHE) scale in which the zero point is taken as that of a hydrogen electrode in the cell electrolyte (hydrogen ion activity = a_{H^+}), i.e., the electrode system $Pt/H_2(a = 1)/H^+(a_{H^+})$. The two scales coincide for solutions of pH = 0; while the potential of the SHE is pH independent, that of the RHE drops (relative to the SHE) by a factor of $2.303RT/F$ volts per unit increase in solution pH.

hydrogen or hydroxide ions to electrons involved in the electrode process is 3/2. Since the most stable oxide of gold involves Au(III), the above result was attributed to the presence of an anionic, octahedrally coordinated gold species, $[Au_2(OH)_9]^{3-}$ or $[Au_2O_3(OH)_3 \cdot 3H_2O]^{3-}$, present at the surface in some type of polymeric or aggregated form. Application of the Nernst equation to the reduction reaction

$$[Au_2(OH)_9]^{3-} + 6e^- + 9H^+ = 2Au + 9H_2O \text{ (acid)} \quad (5)$$

or

$$[Au_2(OH)_9]^{3-} + 6e^- = 2Au + 9OH^- \text{ (base)} \quad (6)$$

must obviously yield the observed potential/pH dependence.

From a thermodynamic viewpoint, oxide electrodes are frequently regarded as metal/insoluble salt electrodes[40,42] in which the activity of the metal ion is modified by interaction with the ligands, which in this case are the OH^- ions. Thus, reaction (6) may be regarded as a combination of the following:

$$[Au_2(OH)_9]^{3-} = 2Au^{3+} + 9OH^- \quad (7)$$

$$2Au^{3+} + 6e^- = 2Au \quad (8)$$

The dissociation constant for the complex anion, whose activity may be represented as a_c, is given by the expression

$$K_d = \frac{a_{Au^{3+}}^2 \cdot a_{OH}^9}{a_c} \quad (9)$$

Applying the Nernst equation to (8) gives

$$E = E^0 - \frac{RT}{6F} \ln \frac{a_{Au}^2}{a_{Au^{3+}}^2} \quad (10)$$

$a_{Au} = 1$ and on substituting for $a_{Au^{3+}}^2$ from (9), it can readily be shown that

$$E = E^0 - \frac{2.303RT}{6F} \log \frac{P^9}{K_d a_c} - \frac{3}{2} \frac{2.303RT}{F} \cdot pH \quad (11)$$

P being the ionic product of water. Apart from demonstrating that potential/pH values greater than zero (RHE scale) are of thermodynamic significance, this approach provides a simple interpretation of the unusual decrease in redox potential with increasing pH for hydrous oxide systems in general. The electrode can be viewed as an Au/Au^{3+} electrode in which the Au^{3+} ions are contained within a hydrous matrix where their activity is influenced, as described by Eqs. (7) and (9), by the hydroxide ion activity of the solution. In the case of the uncharged oxide (Au_2O_3), which may also be treated as a hydroxide, $Au_2(OH)_6$ or $Au(OH)_3$, the pH effect is less marked as the equivalent of Eq. (7) is as follows:

$$Au(OH)_3 = Au^{3+} + 3OH^- \qquad (12)$$

and

$$K_d = \frac{a_{Au^{3+}} \cdot a_{OH^-}^3}{a_c} \qquad (13)$$

Obviously the potential/pH dependence here will be zero (on the RHE scale) as the numbers of electrons and OH^- ions in the overall redox reaction are equal.

The hydrous oxide/gold transition outlined here is only quasireversible: thus although the peak potential for the hydrous oxide to metal transition is not highly dependent on sweep rate,[41] the reverse process occurs only with great difficulty. This type of behavior is discussed more fully later in the case of platinum, but certainly no limitation or objection of this type arises in the case of relatively thin hydrous oxide layers on iridium[4] or rhodium[29] where the peak potentials for the major anodic and cathodic process virtually coincide. The same usual potential/pH behavior is observed in cyclic voltammetry experiments with these systems as with gold, i.e., the ratio of H^+ (or OH^-) to e^- is again 3/2. Assuming that the Ir(III) and Au(III) species have the same composition, Burke and Whelan[4] recently proposed the following reactions in the case of iridium:

$$2[Ir(OH)_6]^{2-} + 2e^- = [Ir_2(OH)_9]^{3-} + 3OH^- \text{ (base)} \qquad (14)$$

and

$$2[Ir(OH)_6]^{2-} + 2e^- + 3H^+ = [Ir_2(OH)_9]^{3-} + 3H_2O \text{ (acid)} \qquad (15)$$

Originally these iridium species were represented[4] as $[IrO_2(OH)_2 \cdot 2H_2O]^{2-}$ and $[Ir_2O_3(OH)_3 \cdot 3H_2O]^{3-}$; however, in view of the known structure of the hexahydroxoplatinum(IV) compound,[37] the hydroxy representation may be preferable (the distinction between the structures for the same metal oxidation state is probably not important as the conversion for instance of $[IrO_2(OH)_2 \cdot 2H_2O]^{2-}$ to $[Ir(OH)_6]^{2-}$ would involve very rapid proton transfer reactions). The oxidized states in Eqs. (14) and (15) are clearly more anionic (or acidic), the number of excess OH^- ions associated with each Ir ion being 2 in the oxidized and 1.5 in the reduced state.

To date the acid-base behavior of several hydrous oxide systems, Au,[41] Ir,[4] Rh,[42] Ni,[43] and Pt[27] has been discussed in considerable detail. It was also suggested recently that, as will be discussed later, anionic species may be involved in the onset of anodic oxidation of metals such as Pt,[44] Pd,[28] Ni,[45] and Au[28] where a significant decrease in the initiation or onset potential (RHE scale) for surface oxidation is observed on increasing the solution pH.

IV. STRUCTURAL ASPECTS OF HYDROUS OXIDES

Structural data for hydrous oxides tend to be very incomplete: indeed so little work has been done on hydrous oxides of the noble metals that the main purpose of this section is simply to illustrate how, from a structural viewpoint, microdispersion can occur in these systems using, as examples, such extensively investigated materials as aluminum hydroxide and the two battery compounds, manganese oxide and nickel oxide. The structure of oxides, hydroxides, and oxyhydroxides (though unfortunately with little emphasis on the hydrous materials) have been discussed by Wells.[46] The complexity of the area is clear from the work done on the aluminum-water system; this is a relatively simple case since only one (the trivalent) oxidation state is involved. According to Alwitt[1] there are three forms of $Al(OH)_3$, two of the $AlO \cdot OH$, and two of Al_2O_3. These are simply the well-characterized phases; the formation of amorphous or poorly crystalline deposits in this case is common, and slow interconversion between phases, involving recrystallization and hydration/dehydration effects, further illustrates the structural complexities of this type of system. In the case of the well-characterized hydroxides of both aluminum[1] and nickel,[2,43] planar structures are most common. The basic $Al(OH)_3$ structure consists of six coplanar, octahedrally OH^- coordinated, Al^{3+} ions arranged in a repeating ring pattern, with each Al^{3+} ion in the ring being linked to its neighbors by pairs of OH^- groups, $Al{<}^{OH}_{HO}{>}Al$, the latter being evenly distributed in planes above and below the Al^{3+} plane. Structural differences arise here due to (1) variations in the manner in which layers superimpose upon one another and (2) variations in the amount of interlamellar water in the more hydrated phases. The latter seems to be the important factor distinguishing hydrous from anhydrous hydroxides or oxyhydroxides. Relevant examples here are pseudoboehmite and boehmite, $Al(OH)_3$,[1] α-$Ni(OH)_2$ and β-$Ni(OH)_2$,[2] and γ-$NiO \cdot OH$ and β-$NiO \cdot OH$[2]; in each case the first member of the pair contains substantial quantities of water (plus electrolyte in the case of γ-$NiO \cdot OH$) in the regions between the layers.

Any type of open, porous structure, allowing intimate contact between the solid and aqueous phase, is likely to yield a hydrous phase. In the interaction between water and aluminum, for instance, transmission electron microscopy studies[47] indicate that the initial product consists of very fine fibrils, ca. 3.5 nm thick; similarly, Selwood[48] has postulated a threadlike or strand structure to account for the unusually high degree of dispersion evident from magnetic susceptibility studies of various hydrous metal oxide systems. A further obvious possibility here[49] is that open, three-dimensional, cage-type structures (as found, for instance, in zeolites) may be involved.

One of the most widely investigated systems illustrating the complexity of oxide structures is found in the case of manganese dioxide. The structure of this widely used battery material was summarized recently by Burns and Burns.[19] The Mn^{4+} ion is assumed to be octahedrally coordinated by six oxygen ligands, and by sharing edges and vertices the MnO_6 octahedra can combine to form almost a limitless number of phases. The basis of most of these structures are semi-infinite strands composed of edge-sharing octahedra. These can align side-by-side to give planar or layer structures similar to aluminum hydroxide. However, in many cases the number of strands aligned side-by-side is quite small, but corner sharing occurs as illustrated in Fig. 2 to yield tunnel structures. The presence of such tunnels—especially when they are relatively

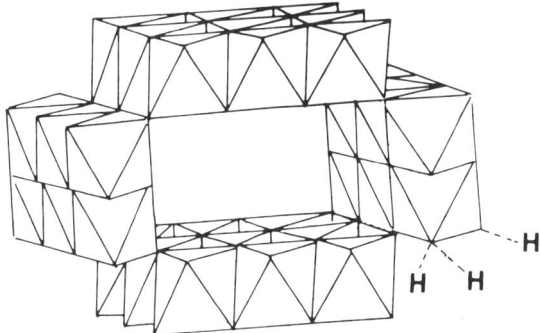

Figure 2. Representation of the Psilomelane structure (treble chains of MnO_6 octahedra joined to double chains of same, resulting in extended tunnel structure) of manganese dioxide. In aqueous media the terminal groups may have coordinated H_2O or OH (or their dissociated equivalents, OH^- or O^-) present.

large—allows extensive interaction between the solid and liquid phases.

It is realized currently in the battery oxide field that the structures encountered in many real oxide samples are probably vastly more complex than those outlined here. The extent of any planar region in a sample may be quite small and the orientation of one relative to another may be quite random. In the MnO_2 system, one of the most active varieties from a charge storage viewpoint is γ-MnO_2. This material is prepared electrolytically, and it has been suggested[19] that its high activity is related to its chaotic structure, e.g., microparticulate, porous form (with high specific surface area), phase intergrowths, and multidimensional tunnels; all these allow penetration of, and extensive interaction with, the aqueous phase. It is interesting to compare the MnO_6 octahedra in an extended regular structure, e.g., pyrolusite or hollandite, with that of a defect structure—as an example of the latter we may take the octahedra at the front end of the model represented in Fig. 2. In any crystal the ideal lattice is found only in the bulk; the surface (because of the absence of bonding on one side) always constitutes a defect. It is clear from Fig. 2 that all the oxygen atoms present in MnO_6 species in the bulk are bridging (Mn-O-Mn) species. However, at the surface of the lattice there are a number of terminal oxygen species—the precise number for any given octahedron varies with the bonding of the latter in the surface layer, but it appears to range from 1 to 4. Such terminal groups may exist as OH_2, OH, or O^-, e.g., $(-O-)_2Mn(OH)_4^{2-}$ or $(-O-)_2Mn(OH)_4^{2-} \cdot 2H^+$. In the active battery-type material the number of imperfections present is so large that these terminal species probably play quite an important role.

V. TRANSPORT PROCESSES IN HYDROUS OXIDE FILMS

Hydrous oxy-hydroxide films can be classified as space distributed redox electrodes consisting of redox centers uniformly distributed in a layer of thickness L, limited on one side by the metal electrode and on the other by the aqueous solution. A theoretical description of the process of charge transport through an electroactive layer of finite thickness has only been attempted in recent years. This is largely due to the fact that a space distributed redox electrode system is more complicated than the classical model of a simple surface redox reaction between two adsorbed species, the latter situation being the subject of numerous theoretical approaches.[50-55] Apart from the transmission of electrons, which is accompanied by the diffusion of protons or hydroxide ions via a Grotthus-type mechanism through the surface layer, other factors such as the transport of counterions (M^+) or solvent molecules through the hydrous oxide material must also be considered. Consideration must also be given to the fact that interactions may occur between the space distributed redox centers and that the film can exhibit a certain degree of inhomogeneity.

Laviron[56,57] and Saveant et al.[58] recently proposed a simple multilayer model to describe charge transport through a redox coating of finite thickness. In this approach the film is assumed to be divided into p sublayers each of thickness $\varepsilon = L/p$. The concentration of redox centers A and B in each sublayer is assumed constant, the sum being given by Γ_Σ, the concentration of active sites. It is further assumed for simplicity that the layer is homogeneous and that the electroactive centers are distributed evenly inside it. Further simplifying assumptions are that no interaction forces exist between the electroactive centers and that the electron hopping process between the latter sites is rate limiting. This latter assumption has been shown to be reasonable from recent experiments with hydrous iridium oxide films in aqueous acid solution.[4] The charge transfer reaction through the film is considered to occur via the following sequence (k denoting the rate constant for the electron exchange process):

$$A_1 + \bar{e} \rightleftharpoons B_1$$

$$B_1 + A_2 \overset{k}{\rightleftharpoons} A_1 + B_2$$

$$\vdots \qquad \vdots$$

$$B_{j-1} + A_j \overset{k}{\rightleftharpoons} A_{j-1} + B_j \qquad (16)$$

$$B_j + A_{j+1} \overset{k}{\rightleftharpoons} A_j + B_{j+1}$$

$$B_{p-1} + A_p \overset{k}{\rightleftharpoons} A_{p-1} + B_p$$

In this reaction sequence the subscript denotes the layer. If p and ω denote the surface concentrations of A and B centers, respectively, with $p_j + \omega_j = \Gamma_\Sigma$ for all j, then one has that

$$\frac{dp_j}{dt} = -kp_j\omega_{j-1} + kp_{j-1}\omega_j + kp_{j+1}\omega_j - kp_j\omega_{j+1} \qquad (17)$$

This expression simplifies to

$$\frac{dp_j}{dt} = k\Gamma_\Sigma[p_{j-1} - 2p_j + p_{j+1}] \qquad (18)$$

If a_j denotes the volume concentration of species A in layer j then Eq. (18) takes the form

$$\frac{da_j}{dt} = k\Gamma_\Sigma \varepsilon^2 [\varepsilon^{-2}(a_{j-1} - 2a_{j-1} - 2a_{j+1})] = k\Gamma_\Sigma \varepsilon^2 \frac{d^2a}{dx^2} \qquad (19)$$

The latter expression is simply a finite difference[59] form of the diffusion equation with the diffusion coefficient given by $D = k\Gamma_\Sigma \varepsilon^2$.

The voltammetric response of spatially distributed redox centers in a layer of finite thickness to an applied linear or cyclic potential sweep has been described by Laviron,[56,57] Andrieux and Saveant,[58] Peerce and Bard,[60] and Aoki et al.[61,62] The most comprehensive analytical approach to date has been that presented by Aoki et al.,[61,62] who developed a theoretical approach to describe the finite diffusion voltammetric response for reversible, totally irreversible, and quasireversible charge transfer reactions. The mathematical formulation of the problem is as follows. The charge storage reaction through the layer of thickness L is assumed to be of the form

$$A \rightleftharpoons B + ne^- \qquad (20)$$

where either the redox centers (A, B) are essentially rigidly fixed in the matrix and the electrons hop between the fixed sites, or the redox sites have sufficient local mobility to come in close contact and exchange the electron in a manner similar to electron exchange between redox couples in solution. The transport problem is described by the diffusion equation, i.e., the equations to be solved are

$$\frac{dC_A}{dt} = D\frac{d^2 C_A}{dx^2}, \qquad \frac{dC_B}{dt} = D\frac{d^2 C_B}{dx^2} \qquad (21)$$

subject to the initial condition

$$C_A(0, L) = C^0, \qquad C_B(0, L) = 0 \qquad (22)$$

and the boundary conditions

$$\frac{dC_A(L, t)}{dx} = \frac{dC_B(L, t)}{dx} = 0 \qquad (23)$$

$$\frac{i}{nFA} = D\frac{dC_A(0, t)}{dx} = -D\frac{dC_B(0, t)}{dx} \qquad (24)$$

Application of the Laplace transformation method[63] to this bounded diffusion problem results, in the case of reversible charge transfer through the layer of surface area A (at a sweep rate v), in a current-potential relation of the form

$$i = nFAC_0(DvnF/RT)^{1/2} f(\omega, \xi) \qquad (25)$$

where

$$f(\omega, \xi) = \omega^{1/2} \int_0^\infty \theta_2(0/Z) \frac{e^{\omega z - \xi}}{(1 + e^{\omega z - \xi})^2} dz \qquad (26)$$

In these latter expressions one has

$$\omega = (nF/RT)v(L^2/D)$$
$$\xi = \frac{nF}{RT}(E - E^0) \qquad (27)$$

and $\theta_2(0/u)$ is the Theta function[64] defined by the expression

$$\theta_2(0/u) = \frac{1}{(\pi u)^{1/2}}\left[1 + \sum_{k=1}^{\infty} (-1)^k e^{-k^2/u}\right] \qquad (28)$$

If the quantity ω (where $\omega^{1/2} = L/X_D$, X_D being the diffusion layer thickness) is large enough so that only the first term in Eq. (26) is considered, then the current–potential relation takes the form

$$i = nFAC^0(\pi nFDv/RT)^{1/2} \int_0^{\infty} \frac{e^{z-\xi}}{(1 + e^{z-\xi})^2} z^{-1/2} dz \qquad (29)$$

This expression is similar to the well-established integral equation describing semi-infinite diffusion of solution phase species originally proposed by Nicholson and Shain.[65] On the other hand, when ω is small the current–potential expression reduces to

$$i = \frac{n^2 F^2 A C^0 L}{RT} \frac{e^{\xi}}{(1 + e^{\xi})^2} \qquad (30)$$

This latter expression is similar to that obtained for a simple surface process if the term $C^0 L$ is replaced by the surface concentration Γ. It is possible to integrate Eq. (25) numerically to obtain various diagnostic criteria of the form

$$i_p = 0.446 nFA(C^0 D/L)\omega^{1/2} \tanh(0.56\omega^{1/2} + 0.05\omega) \qquad (31)$$

and

$$E_p = E^0 + 0.555(RT/nF)$$
$$\times \{1 + \tanh[2.41(\omega^{0.46} - 1.20) + 1.20(\omega^{0.46} - 1.20)^3]\} \qquad (32)$$

It is clear that these latter expressions are considerably more complicated than the corresponding expressions applicable either to surface or semi-finite diffusion processes. The extension of this type of analysis to irreversible and quasireversible processes has been recently described.[62]

Laviron[55] has recently noted that linear potential sweep or cyclic voltammetry does not appear to be the best method to determine the diffusion coefficient D of species migrating through a layer of finite thickness since measurements are based on the shape of the curves, which in turn depend on the rate of electron exchange with the electrode and on the uncompensated ohmic drop in the film. It has been established that chronopotentiometric transition times or current-time curves obtained when the potential is stepped well beyond the reduction or oxidation potential are not influenced by these factors.[55] An expression for the chronopotentiometric transition has been derived for thin layer cells.[66] Laviron[55] has shown that for a space distributed redox electrode of thickness L, the transition time (τ) is given implicitly by an expression of the form

$$\frac{2i_0\tau^{1/2}}{\pi^{1/2}nFAD^{1/2}\Gamma_\Sigma}\left[1 + 2\pi^{1/2}\sum_{k=1}^{\infty} i\,\mathrm{erfc}(kLD^{-1/2}\tau^{-1/2})\right] = 1 \quad (33)$$

where i_0 denotes the applied current and the function $i\,\mathrm{erfc}\,u$ is defined by an expression of the form

$$i^k\,\mathrm{erfc}\,u = \int_u^\infty i^{k-1}\,\mathrm{erfc}\,u\,du \quad (k = 0, 1, 2, \ldots) \quad (34)$$

and $\mathrm{erfc}\,u$ denotes the error function complement defined by an expression of the form

$$\mathrm{erfc}\,u = \frac{2}{\pi^{1/2}}\int_u^\infty e^{-x^2}\,dx = 1 - \mathrm{erfu} \quad (35)$$

and erfu denotes the well known error function. Note that for large values of i_0, τ becomes small and consequently the sum in Eq. (33) tends toward zero so that the classical Sand equation[68] for the transition time under semi-infinite diffusion conditions is obtained, viz.,

$$2i_0\tau^{1/2}/\pi^{1/2}nFAD^{1/2}\Gamma_\Sigma = 1 \quad (36)$$

The exact form of the current-time response obtained as a result of application of a potential step is given by[69,70]

$$i = nFAD^{1/2}\Gamma_\Sigma(\pi t)^{-1/2}\left[1 + 2\sum_{k=1}^{\infty}(-1)^k e^{-k^2L^2/Dt}\right] \quad (37)$$

The expression outlined in Eq. (37) reduces to the classical Cottrell equation

$$i = nFAD^{1/2}\Gamma_\Sigma(\pi t)^{-1/2} \qquad (38)$$

when the quantity $Dt/L^2 \to 0$. The diffusion coefficient can easily be determined from either of the latter equations if the quantities Γ_Σ or L are known. Γ_Σ can either by determined voltammetrically or by integrating the current–time transient. The thickness L can be obtained using the expression $L = \Gamma_\Sigma V$, where V denotes the molar volume of the coat. The rate constant for the charge transfer process k can be deduced from the value of the diffusion coefficient using the equation[70]

$$k = D\lambda^{-2}\Gamma_\Sigma^{-1} \qquad (39)$$

where λ denotes the mean distance between electroactive sites.

Charge transfer (electron and proton) diffusion coefficients have been determined for a number of hydrous oxide films. Reichman and Bard[71] reported a value of ca. 10^{-7} cm^2 s^{-1} for the proton diffusion coefficient for hydrated WO_3 layers. MacArthur[72] obtained values in the region 2×10^{-9} to 4.5×10^{-11} cm^2 s^{-1} for nickel hydroxide electrodes. In contrast, a marked difference in the charge transfer diffusion coefficient was obtained for reduction and oxidation of hydrous iridium oxide films in aqueous acid solution—values of 1.5×10^{-9} and 1×10^{-7} cm^2 s^{-1} being obtained for the reduction and oxidation processes, respectively.[4] In this latter case the diffusion coefficients were assumed to correspond to rate-limiting electron transfer through the layer. It is instructive to compare such values with proton diffusion coefficients reported for anhydrous oxides such as $FeOOH$,[73] $AlOOH$,[74] and MnO_2.[75,76] Such coefficients are typically of the order of 10^{-18} cm^2 s^{-1}, this latter value being indicative of the compact, nonporous nature of these oxide layers. In contrast the diffusion coefficient for protons or hydroxide ions in water is typically 10^{-4}–10^{-5} cm^2 s^{-1}. Proton diffusion coefficients for hydrous oxide systems thus fall between these two extremes—the latter conclusion being in good accord with the open, highly dispersed nature of hydrous oxide layers.

It must be noted that it is very difficult to identify unambiguously the rate-limiting transport process occurring within the film. Murray[77] has recently noted that the observed diffusion coefficient may correspond either to counterion motion or solvent motion, electron self-exchange between neighboring redox sites, or the segmental motion of polymer chains antecedent of these latter processes. It is indeed possible that the rate-determining factor can depend on the nature of the polymeric film, the type of solvent used, or the identity of the counterion. If one also considers the diverse ideas encountered in the literature to rationalize diffusion in polymers, and that in the electrochemical situation one must consider chemical transformation of the polymeric film, then a realistic measure of the magnitude of the problem can be obtained.

VI. THEORETICAL MODELS OF THE OXIDE–SOLUTION INTERPHASE REGION

The oxide–solution interphase is a complex region, significantly different in many respects from such classical electrochemical double layer systems as the mercury—or silver iodide—solution interphase.[78,79] Such features include, for example, the presence of an extended space charge layer in semiconducting oxides and the presence of a reversible double layer within the compact Helmholtz region which is due to the transfer of potential determining H^+ or OH^- ions across the interphase occurring via a variety of adsorption and dissociation reactions at a hydroxylated oxide surface. The acid–base properties of oxide surfaces in contact with aqueous solutions has been discussed in a previous section. Healy et al.[80,81] have recently noted that oxides generally exhibit very high charge density values (the latter being defined in terms of the quantity of potential determining H^+ or OH^- ions adsorbed). However, this latter observation is not reflected in the rather low electrokinetic potential and colloidal stability values recorded for oxide suspensions. Furthermore, differential capacitances of positively and negatively charged oxide surfaces are more symmetrical about the p.z.c. than for the classical Hg–solution and AgI–solution interphases. This indicates that cations and anions are adsorbed to similar extents on oppositely charged oxide surfaces.

Several theories have been postulated recently for the structure of the oxide–solution interface.[81] These approaches may be conveniently classified into two general categories—classical and nonclassical models.

1. Classical Models[82–88]

The Gouy-Chapman-Stern-Grahame (GCSG) model is the classical approach to interfacial behavior and has been used extensively in modelistic descriptions of the Hg- and AgI-solution interfaces. Three planes of charge are proposed in this formulation: (1) the plane of surface charge, i.e., that of the adsorbed potential determining ions; (2) the plane of the centers of specifically adsorbed ions, the inner Helmholtz plane (I.H.P.), and (3) the outer Helmholtz plane (O.H.P.). The basic GCSG model for the oxide-solution interface originally developed by Blok and de Bruyn[84] was extended by Levine and Smith,[85] who allowed for non-Nernstian behavior of the surface potential and considered discreteness of charge effects on the adsorption process. These workers obtained theoretical charge-capacitance and potential-pH curves in reasonable agreement with published experimental data. Wright and Hunter[86,87] have used a GCSG-based model to interpret the zeta-potential behavior of oxides. They compared theoretical potential-concentration variations at fixed electrolyte pH and obtained a general correspondence, although calculated surface charge densities were much less than those found experimentally.

GCSG-based models can qualitatively account for observed interfacial properties provided certain values are chosen for various fundamental parameters (e.g., specific adsorption potential, inner layer integral capacity) which are associated with each model. However, Healy et al.[81] have pointed out that it is not at all certain that the values assigned to some of these parameters (especially a rather large inner layer integral capacitance) are reasonable. Also explanations proposed to account for the magnitude of these parameters have not as yet received experimental verification. However, the major defect associated with the GCSG-based models is that the surface charge values observed in certain cases exceed that developed due to the ionization of all surface hydroxyl groups.

2. Nonclassical Models[89–100]

Various nonclassical models have been proposed to account for the very high surface charge densities observed at oxide-solution interphases. These models include the porous double layer model,[89-92] the gel layer model,[93-95] the transition layer theory,[96,97] and the polyelectrolyte or site-binding models.[98-100]

The porous double layer (PDL) approach[89-92] assumes that the oxide surface is porous to both potential-determining ions and counterions, i.e., the two separated planes of adsorbed ions and counterions of the GCSG model are replaced by a model in which the counterions reside in the same region as the potential-determining ions. This picture accounts at least qualitatively for the general trends observed in interfacial properties. First, the more permeable the surface layer is the higher the charge density that can be accommodated in the region. Second, because sorption is not restricted to the oxide surface itself, charge densities exceeding those corresponding to the ionization of the total number of hydroxyl groups on the surface are reasonable. Finally, because counterions may also penetrate the surface to a certain extent, charge densities and zeta potential values at the solution side of the interface remain low. The PDL model can, therefore, generally account for the experimentally observed low electrokinetic potentials and colloidal stability provided that the degree of counterion penetration is extensive. Perram et al.[93-95] have proposed an alternative gel layer model for the oxide–solution interphase. These workers based their analysis on the assumption that there is a homogeneous gel-like layer of definite thickness (2-5 nm) at the interface containing metal hydroxy groups $M(OH)_n$ into which various ions can adsorb. The gel-layer model accounts for the high surface charge in terms of ion penetration into subsurface regions of the oxide. Trasatti[101] has recently noted that this latter model is consistent with experimental facts but that the presence of a gel layer has not been proved by radiotracer experiments.[102] Dignam[96,97] has combined the basic ideas of porous and hydrolyzed gel-like layer theories to propose a transition layer model for the oxide–solution interface. The essential feature of this model is the existence of a transition layer of atomic dimensions within which is established an equilibrium between oxide and aqueous ionic species.

Yates, Levine, and Healy[99] and Davis et al.[100] have formulated a so-called polyelectrolyte or site-binding model for the oxide–solution interphase. In this approach the oxide surface is represented by a two-dimensional array of positive, negative, and neutral discrete charged sites exhibiting acid-base properties. Potential determining ions and counterions are then considered to react or bind with the acid-base sites. It is assumed that the adsorbed supporting electrolyte cations and anions are distributed in two ways: first, as interfacial ion pairs formed with oppositely charged surface groups, and second, in a diffuse layer to balance the remaining unneutralized surface charges. In this model the high surface charge and low zeta-potential are rationalized in terms of ion-pair formation between surface ionized groups and counterions. In order to obtain reasonable correlation with experimental data Yates et al.[99] assumed a rather high inner zone integral capacity ($\sim 140\ \mu F\ cm^{-2}$) and a rather low outer zone capacity ($\sim 20\ \mu F\ cm^{-2}$) for the compact layer. These values were rationalized in terms of a porous hydrated oxide surface layer at the interface which is penetratable by supporting electrolyte ions, where the centers of charges of the potential-determining and supporting electrolyte ions are located on two planes which are parallel to the interface and lie within the porous layer. Hence in this case the IHP is assumed to be actually imbedded in the porous layer at a given distance (ca. 1 Å) from the plane defining the charge center due to the potential-determining ions. Consequently, depending on the thickness of the porous layer, a wide separation between the IHP and the diffuse layer in solution becomes possible.

While it appears that no totally satisfactory model for the oxide-solution interface has emerged to date, progress is obviously being made in this area. Dispersion, hydration, hydroxylation, and acid-base properties (and, in particular, the variation of the latter with change in the oxidation state of the central metal ion) are all factors which must be combined before a model of general validity is obtained.

VII. PLATINUM

1. Monolayer Oxidation

Anodic oxidation of noble metals in general, and platinum in particular, has received a great deal of attention, especially at the monolayer level.[14-15] The selection of platinum as a substrate of choice for most of the fundamental work in this area is due to its marked resistance to corrosion, high electrocatalytic activity, and apparently clear separation between the potential regions for hydrogen and oxygen adsorption. The electrochemical behavior of platinum in aqueous media is summarized by the cyclic voltammogram shown in Fig. 3.

In such diagrams four main potential regions are usually distinguished on the anodic sweep: 0–0.4 V, ionization of adsorbed hydrogen; 0.4–0.8 V, double layer charging (this process occurs over the entire potential range but is the sole process over the intermediate region quoted here); 0.8–1.5 V, oxidation of, i.e., deposition of hydroxy and oxy species at, the metal surface; above 1.5 V, oxygen gas evolution plus continued growth of the surface oxide layer (formation of a thick, yellow, hydrous oxide film occurs under constant polarization conditions[20-22] at 2.1 V < E < 2.5 V). On the subsequent cathodic sweep the corresponding regions are as follows: the peak at 0.8 V, reduction of the monolayer oxide film; 0.5–0.4 V, double layer discharging (again this process occurs over the entire cathodic sweep); 0.4–0 V, adsorbed hydrogen formation

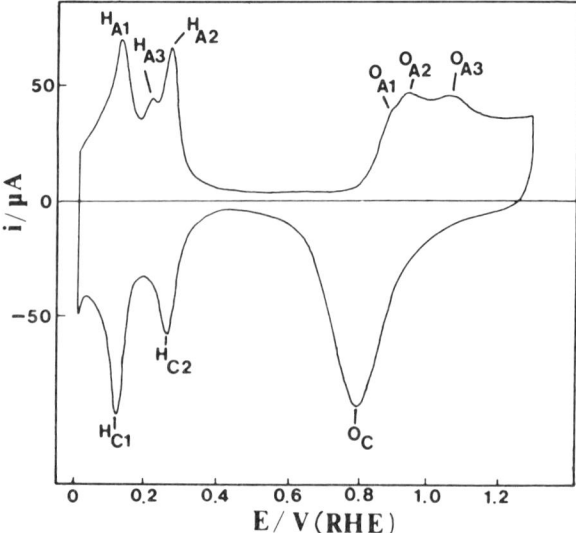

Figure 3. Potentiodynamic i/E profile for a platinum surface in high-purity 0.5 mol dm^{-3} H$_2$SO$_4$ at 25°C, V = 0.1 V s^{-1} (from Ref. 103, with permission).

(plus reduction of any hydrous platinum oxide formed under highly anodic conditions); below 0 V, hydrogen gas evolution. The potential ranges quoted here are approximate and refer mainly to acidic media.

The complexity of some of the surface processes is illustrated by the fact that as many as five peaks, corresponding to five different states of adsorbed hydrogen, have been reported[17] for the lower region of the anodic sweep. The main peaks in the hydrogen region display reversible characteristics at moderate sweep rates, i.e., the peak potentials are virtually independent of sweep rate and the values for corresponding anodic/cathodic pairs virtually coincide. A useful feature of the hydrogen region, from the experimental viewpoint, is the fact that the charge associated with the adsorption or desorption of a monolayer of hydrogen, ca. 210 μC cm^{-2} (the value is somewhat dependent on the orientation of the exposed crystal plane[15]), can be used to measure the real area or roughness of a platinum surface. By comparing the charge for oxide formation or reduction (Q_0) with the charge for monolayer hydrogen conversion ($^M Q_H$) the degree of surface oxidation can be assessed; it is generally assumed that each platinum atom at the surface can chemisorb one hydrogen or one oxygen atom, and that the condition $Q_0/^M Q_H = 2.0$ corresponds to monolayer oxide coverage.

Before outlining the theories proposed to account for monolayer oxide growth on platinum above ca. 0.8 V, it is worth noting that this layer is not the most stable oxidation product of the metal in question. The hydrous material, which is probably platinic acid $H_2Pt(OH)_6$ or one of its salts, is much more stable—especially in base,[27] where little reduction of this substance is observed even at potentials corresponding to hydrogen gas evolution. The formation of this material during the anodic sweep is evidently inhibited by the large activation energy barrier associated with the extraction of a platinum atom from a regular surface lattice position to form some type of loosely bonded adatom state (Pt*). Platinic acid formation

$$Pt^* + 4OH^- + 2H_2O = H_2Pt(OH)_6 + 4e \qquad (40)$$

requires that the ligands eventually have access to six coordination sites at each platinum atom—a situation that is clearly not very feasible when the latter is imbedded in the surface of a metal lattice.

The formation of monolayer oxide films on platinum under anodic conditions has been extensively investigated by Conway and co-workers.[17] It was demonstrated[103] that with a highly purified aqueous 0.5 mol dm^{-3} H$_2$SO$_4$ electrolyte three poorly resolved peaks (O$_{A1}$–O$_{A3}$) could be identified in the anodic sweep below monolayer OH (i.e., one electron per Pt atom) coverage. These authors also designated the broad flat region from $Q_{OH} = 1$ ($E \approx 1.1$ V) to the oxygen gas evolution stage as "O$_{A4}$" (the same type of nomenclature, O$_{A1}$–O$_{A4}$, was subsequently used by Conway in discussing the anodic behavior of other metals, e.g., gold[104] and iridium[53]). Hysteresis was always observed between the regions O$_{A2}$, O$_{A3}$ and O$_{A4}$ and the single peak on the subsequent cathodic sweep—the only region showing signs of reversibility was that associated with the start of the O$_{A1}$ peak (up to ca. 0.88 V at a sweep rate of 100 mV s^{-1}). Based on the recorded charge values, the three peaks were assigned to the following processes:

$$4\text{Pt} + \text{H}_2\text{O} = \text{Pt}_4\text{OH} + \text{H}^+ + e^-, \quad E_p(\text{O}_{A1}) \sim 0.89 \text{ V} \quad (41)$$

$$\text{Pt}_4\text{OH} + \text{H}_2\text{O} = 2\text{Pt}_2\text{OH} + \text{H}^+ + e^-, \quad E_p(\text{O}_{A2}) \sim 0.95 \text{ V} \quad (42)$$

$$\text{Pt}_2\text{OH} + \text{H}_2\text{O} = 2\text{PtOH} + \text{H}^+ + e^-, \quad E_p(\text{O}_{A3}) \sim 1.05 \text{ V} \quad (43)$$

The formulas in these equations represent only surface lattice occupancy ratios, not chemically distinct species.

The hysteresis in the oxide formation/removal process at anodic limits above ca. 0.9 V was attributed to a slow postelectrochemical step in the surface oxidation reaction. It was assumed that interaction between neighboring OH, or PtOH, species induced place-exchange reactions in the surface layer (adsorbed OH groups switching positions with metal atoms in the surface layer of the metal lattice) resulting in significant formation of "OHPt" species. The fact that the three peaks on the anodic sweep are quasireversible in character (the peak potentials being virtually independent of sweep rate—a feature that seems in conflict with the marked hysteresis between the oxide formation/removal reactions) was attributed to the continued existence, at any part of the sweep, of species such as OH$_{ads}$ which exhibit microscopic reversibility. For an anodic limit up to 1.0 V the species formed at the interface were assumed to be OH$_{ads}$ and its place-exchanged equivalent OHPt; at higher anodic limits, up to 1.1 V, an increased thickness of OHPt is assumed to be present; only at higher anodic limits was it postulated that PtO species were present in the surface layer.

The above approach by Conway and co-workers has been questioned by a number of authors. Ross,[105] for instance, has shown that the fine structure at the initial stages of surface oxidation depends strongly upon the solution composition and is absent in the presence of a hydrogen fluoride electrolyte—according to this author the fine structure in the voltammogram is due to anion adsorption in solutions of low pH. Bagotzky and Tarasevich[18] observed only two very broad peaks on the anodic sweep in base, the first of which they attributed to OH and the second to O adsorption. They also postulated that in acid media the main product of oxidation at any point of the anodic sweep is O_{ads}. Attempts have been made to simulate the experimentally observed current/voltage response obtained under cyclic voltammetry conditions for the platinum oxide formation/removal reactions by Appleby,[106] and Bagotzky and Tarasevich[18]; the fact that both approaches yield reasonable agreement between experimental and recorded voltammograms, using significantly different reaction schemes and assumptions as to the origin of the irreversibility, illustrates the absence of a diagnostic role in this approach as used to date.

One feature that seems to have been largely ignored in earlier work is the fact that the species formed in the initial stages of the platinum oxidation reaction may be charged, rather than neutral, in character. Evidence for this is found in the work of Vetter and Berndt,[107] whose charging curve data are reproduced here in Fig. 4.

It is clear from this diagram that the potential (E_i) for the onset of oxidation (dotted line in Fig. 4) decreases with increasing pH by ca. $3/2$ ($2.303RT/F$) V/pH unit; since the authors were apparently unaware at that time of the significance of oxide acidity, they automatically assumed a conventional $2.303RT/F$ variation—which is obviously in poor agreement with the experimental data in this initial oxidation region. The unusual decrease in E_i with increasing pH is now widely accepted; Conway and co-workers[104] attributed it to the effect of anion adsorption while Bagotzky and Tarasevich[18] attributed it to a change in oxidation mechanism with variation in pH. According to Burke and Roche[44] the effect is due to formation of an initial acidic or anionic product according to

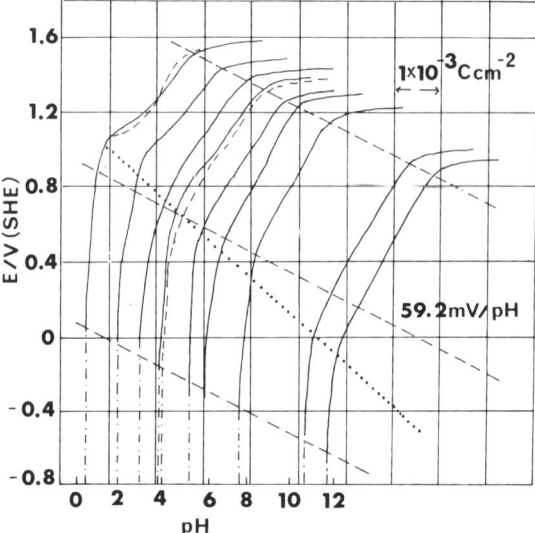

Figure 4. Anodic charging curves ($i = 5.0$ μA cm^{-2}) for platinum in aqueous buffer solutions at different pH values, $T = 25°C$. The dotted line of slope $3/2(2.303RT/F)$ V/pH unit has been added (original diagram is from Ref. 107, with permission).

the following reaction schemes:

$$2Pt + 2H_2O = [Pt_2O \cdot OH]^- + 3H^+ + 2e^- \quad \text{(acid)} \quad (44)$$

$$2Pt + 3OH^- = [Pt_2O \cdot OH]^- + H_2O + 2e^- \quad \text{(base)} \quad (45)$$

The product of this reaction, $O_{ads}^- \cdots H - O_{ads}$, is assumed to be stabilized to a significant extent by hydrogen bonding between species coordinated at adjacent Pt sites. Apart from inducing hysteresis, place-exchange reactions are also assumed to destroy the anionic character of the surface oxide species—the return to conventional $2.303RT/F$ behavior at the onset of oxygen gas evolution is clear from Fig. 4. The high OH^- ion activity in base stabilizes the $[Pt_2O \cdot OH]^-$ species; it effectively lowers the Pt^{1+} activity by driving the dissociation equilibrium of the surface hydroxy species, viz.,

$$[Pt_2O \cdot OH]^{1-} + H_2O = 2PtOH + OH^- \quad (46)$$

to the left-hand side. Thus, from a simple Nernstian viewpoint (which should be applicable to the initial stage of oxidation as this apparently occurs under reversible conditions[17]), the equilibrium potential for the process

$$Pt^{1+} + e^- = Pt \quad (47)$$

must be considerably lower in base than in acid [it is clear from Eqs. (44) and (45) that $\delta E/\delta pH = 3/2(2.303 RT/F)$]. Because the oxidation starts at a higher potential in acid, and the voltage is changing at the same rate in cyclic voltammetry experiments carried out under similar conditions, the currents in the reversible region must be greater at low pH (the point of completion of the largely anhydrous O_{ads} monolayer at 1.4–1.5 V is virtually pH independent)—clearly the arrest in the charging curve for acid just after the onset of oxidation has the same origin. Thus changes in the acidic or anionic character of the species involved in the early stages of surface layer formation exert a strong influence on the current/voltage behavior in this region.

The mechanism of monolayer oxide formation on platinum above 1.0 V is still a topic of considerable controversy. Visscher and Devanathan[108] suggested that the rate of reaction was controlled by the entry of metal ions into the film and derived rate equations on the assumption that this was predominantly a field-assisted process. Vetter and Schultze,[109,110] Schultze,[111] and Ord and Ho[112] proposed a model involving migration of platinum and oxygen ions across the oxide layer under the influence of a high electric field—this approach has been extensively employed in recent times by Damjanovic and co-workers.[113–115] Such an approach, however, is not universally accepted. As Belanger and Vijh[14] have pointed out, it is in conflict with the finding of a limiting coverage (oxygen/platinum ratio) of just above 2.0 at ca. 2.2 V by Biegler and co-workers.[116,117] This point may not be quite valid because (as outlined later) hydrous oxide growth commences in this region, i.e., there may be a change in the nature of the film at this potential, perhaps making it more susceptible to dissolution (Visscher and Blijlevens[118] have shown that the limiting coverage is dependent on the acid concentration). A possibly more serious objection is the fact that quite large currents may be passed through a platinum electrode in the thin oxide layer region provided a suitable simple redox species is present in solution[119]—even the occurrence of the oxygen gas evolution reaction, which is quite vigorous above 1.60 V, suggests that the surface layer is not highly resistive—as would be required if a large electric field (which has been suggested[110] to be as high as 10^7 V cm^{-1}) were to be developed across the surface layer. A more detailed criticism of the high-field theories of oxide growth as applied to platinum can be found in the work of Gilroy.[120,121] The latter author has, as an alternative, proposed a low-field theory based on the assumption that surface oxidation occurs, from the onset, by a nucleation mechanism, the process being controlled by the rate at which growth sites are initiated. The basis of this approach has already been reviewed elsewhere.[122]

2. Hydrous Oxide Growth on Platinum

(i) dc Growth

Early work on thick oxide growth on platinum under dc conditions has been summarized by Nagel and Dietz[123] and James.[20] In acidic media a distinctly yellow surface oxide layer can be produced by polarizing the metal at potentials in the region of 2.1–2.5 V. According to James[20] the initial thick film is formed on platinum, even under quite vigorous dc polarization conditions, only with difficulty; however, once such growth occurs it is easier to repeat the process after reduction of the initial film. This activation, which appears to be related to an expansion of the outer layers of the metal lattice (including the generation of defects centers in this region), decays with time—evidently the disturbed outer layers of the metal rearrange gradually to a more stable structure. The need for active centers to initiate thick film growth under dc conditions was discussed by James,[20] who attributed the beneficial effects of grinding with Emery paper (a technique used deliberately in subsequent work by Balej and Spalek[21] to reduce the period of induction required to obtain thick oxide growth on fresh platinum) to the generation of surface strain. It was also observed[21,124] that no thick oxide growth could be observed on platinum surfaces after vigorous annealing. While most of the dc hydrous oxide growth on platinum to date has been carried out in acid media (and Kozawa[125] found no cathodic peak for the reduction of hydrous oxide after strong anodizing in base), Shibata[126] has recently used a base electrolyte as a medium for preparing hydrous films of low foreign anion content. The distinction between anhydrous (also known as superficial, α, and type I) and hydrous (phase oxide, β, or type II) oxides is usually clear from voltammetric data. Cathodic charging or potential sweep curves for hydrous oxide coated platinum electrodes in acid show evidence of two oxide reduction processes, one at 0.6–0.7 V and the other at 0.2–0.4 V. The first is usually attributed to the reduction of a monolayer-type oxide film at the metal surface and is not of primary interest here. The second is due to the reduction of the hydrous oxide and the charge for the latter varies with the duration of the preanodization and may, depending upon the preanodization conditions, greatly exceed that for the monolayer reduction. According to James,[20] the hydrous oxide reduction peak recorded under cathodic sweep conditions may split after growth under prolonged, intense polarization into a doublet—suggesting the presence of two types of hydrous material. The influence of the nature and concentration of the electrolyte on the oxide reduction process was examined recently by Shibata and Sumino.[126] In some dilute acids, especially in the absence of agitation, it was not possible to distinguish voltammetrically between the two reduction steps. However, in strong base the hydrous oxide reduction step was displaced so far in the cathodic direction that it could no longer be recorded electrochemically due to overlap with hydrogen gas evolution (the view expressed by these authors in a series of fairly recent publications,[22,126] namely, that the hydrous film resides between the monolayer and the metal surface, seems quite invalid). Both James[20] and Shibata[127] reported considerable difficulty in forming thick films on platinum in acid at temperatures less than ca. 20°C. Balej and Spalek[21] observed formation of hydrous oxide in aqueous H_2SO_4 only within the potential range 2.1–2.5 V (RHE), the rate of formation decreasing with increasing acid concentration. These authors also claimed that hydrous material was produced at constant potential according to a parabolic rate law.

Shibata's experiment, involving the use of periodically interrupted cathodic current for the reduction of thick oxide films,[22] is quite significant. The smaller potential variations observed during the reduction of the hydrous, as compared with the anhydrous, component of the surface layer clearly demonstrate that the reaction in the former case is far more reversible in character. On the basis of the temperature dependence of the cathodic peaks, James[20] also appears to have concluded that the hydrous oxide reduction process is not highly irreversible in character—despite the large difference between the formation and reduction potential for this material.

Formation and reduction of a hydrous oxide film on platinum resulted in an increase in the charge for monolayer oxide reduction in subsequent experiments.[20] Obviously the metal surface was roughened as a result of formation and subsequent reduction of the phase oxide. However, the roughness developed with this technique tended to decay under open-circuit conditions (a process that was accelerated by occasional potential sweeps for monitoring purposes). Such decay was more rapid than with platinum black deposits prepared from conventional chloroplatinic acid baths— evidence perhaps that in the former case the active metallic layer is of a more highly strained, finely divided character.[128]

Most authors to date seem to agree with the proposal of Altmann and Busch[129] that the hydrous oxide may be described as $PtO_2 \cdot nH_2O$. Attempts to analyze the film by X-ray techniques[130] have not been successful—the inability to record a diffraction pattern for the hydrous material is probably due to its amorphous character. Electron diffraction studies[128] suggested that the dc grown thick film is poorly crystalline PtO_2. On the basis of XPS (or ESCA) studies Hammond and Winograd[131] have postulated that vigorous anodization of platinum in dilute acid media leads to the formation of $PtO_2 \cdot nH_2O$, and in more concentrated acid solutions $PtO_x(anion)_y$, species. The existence of a Pt(IV) species in the yellow film produced by vigorous dc anodization was confirmed by Allen et al.[132]

(ii) Growth under Potential Cycling Conditions

An account of early work in this area, i.e., production of a yellowish brown layer of hydrated PtO_2 at a platinum electrode in sulfuric acid solution on imposition of an ac component on a dc current, has been given by Altmann and Busch.[129] More recent work by Burke and Roche[133,134] has demonstrated that quite thick films can be produced in both acid and base on cycling the electrode potential between suitable limits. In the initial work[133] oxide growth in acid was carried out using a triangular sweep technique, typically 2.82-0.50 V at 4.6 V s^{-1} (the optimum upper and lower limit at this sweep rate), and a thickness corresponding to an oxide charge capacity of ca. 200 mC cm^{-2} was attained after cycling for ca. 100 min. Unlike the dc case, abrasive pretreatment was not required and the reproducibility of oxide growth was virtually unaffected by electrode history. Similar results were recorded in base,[134] the optimum upper and lower limits here (2.70 and 0.44 V, respectively) being slightly lower than in acid—possibly a reflection of a slight acidic character in the inner oxide layer. One of the most marked effects of solution pH, however, was on the ease of hydrous oxide reduction. The thick brown layer on platinum in acid was readily reduced (Fig. 5) under the cathodic sweep conditions at a potential of ca. 0.3 V. Similar reduction was not observed in base (the origin of the *minor* peak at ca. 0.3 V for a hydrous-oxide-coated anode in base [135] will be discussed later); in agreement with the results for dc grown films by Shibata and Sumino[126] the unreduced brown film appeared to be stable at the metal surface even under hydrogen gas evolution conditions. As outlined earlier (Section

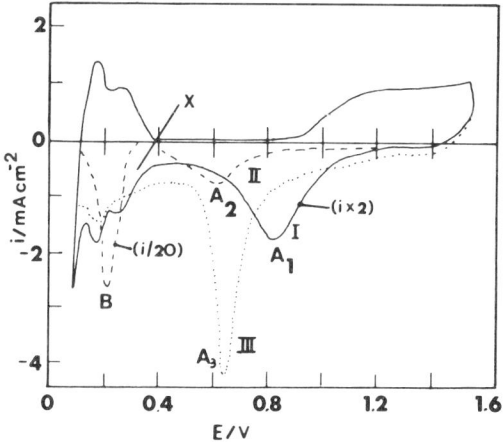

Figure 5. Curve I (———) is a typical cyclic voltammogram for polycrystalline platinum in 1.0 mol dm^{-3} H$_2$SO$_4$ at 25°C, $V = 0.3$ V s^{-1}. Curve II (- - - - -) is a reduction sweep for the same electrode coated with a thick hydrous oxide layer (at point × on this sweep the recorder sensitivity was considerably reduced). Curve III (· · · · ·) is a similar reduction sweep (2.0-0 V, 40 mV s^{-1}) recorded without regrowing the hydrous oxide layer (from Ref. 27, by permission).

III), this result may be rationalized by assuming that the layer produced on cycling is anionic, i.e., the reaction in question is assumed to be

$$[Pt(OH)_6]^{2-} + 4e^- + 6H^+ = Pt + 6H_2O \qquad (48)$$

There is an important distinction in these hydrous systems between redox processes involving oxide/metal, e.g., Pt(IV)/Pt, and oxide/oxide, e.g., Ir(IV)/Ir(III), transitions in that the former are not easily reversed. As outlined earlier, both Shibata[22] and James[20] have suggested that the reduction process is quasireversible. Detailed examination of the hydrous oxide reduction peak[27] revealed a number of odd features, e.g., while it is quite sharp it is asymmetric—the return to the base line toward the end of the reduction process is quite steep. Furthermore, while the peak height is proportional to sweep-rate (again an indication of the quasireversibility of the process involved), the peak maximum potential shifted anodically on decreasing the sweep rate—especially at low values of the latter. Such behavior has been interpreted recently[135] in terms of crystallization overpotential effects. By analogy with the treatment outlined earlier for gold, the platinum/hydrous oxide system may be treated as a metal-insoluble salt electrode involving the metal/metal ion reaction

$$Pt^{4+} + 4e^- = Pt^* \qquad (49)$$

Owing to slow incorporation of the metal atoms (Pt*) produced in this reaction into regular lattice sites at fast rates of reduction, the activity of Pt* will be large and the equilibrium potential for Eq. (49) more negative than for the case where regular bulk platinum is formed directly. The latter situation is more likely to arise at slow sweep rates where the concentration or activity of Pt* at the surface is low—hence the anodic drift in peak potential with decreasing sweep rate. Crystallization overpotential in the hydrous oxide reduction process, i.e., the inability of discharged metal atoms to attain rapidly a regular lattice arrangement, is obviously also the basis of the use of this technique to generate finely divided platinum black surfaces of high electrocatalytic activity.

The asymmetry of the reduction peak may also be attributed to crystallization overpotential. At the peak the rate of production of Pt*, and hence the crystallization overpotential (η_{cry}) is at a maximum. Just beyond the inflection η_{cry} decays owing to the decrease in current; more of the applied potential is available to drive the reduction process—which then goes rapidly to completion. The equivalent behavior during an anodic sweep is well known in the study of metal passivation phenomena.[136]

The generation of hydrous films on platinum at low potentials on the anodic sweep seems quite feasible from a thermodynamic viewpoint—especially in base. The inhibition here is evidently related to the need for six hydroxide ions to have access to coordination sites at the same platinum atom—an improbable condition for a metal atom in a regular surface site. Evidence will be presented later indicating that such hydrous oxide formation can, to a very limited extent, precede monolayer formation in the case of gold—the atoms of the latter involved in formation of the hydrous material are presumably at low lattice coordination sites on the surface. There is some evidence from recent single-crystal studies (see Section XIV) for this type of behavior in the case of platinum—it is obviously difficult to detect with polycrystalline substrates as with only a small fraction of a monolayer involved optical techniques would need to be extremely sensitive and electrochemical procedures are hampered by the fact that the redox behavior of the hydrous material coincides with that for adsorbed hydrogen.

In the cycling procedure the barrier to hydrous oxide growth is overcome by virtue of the fact that at the lower end of the sweep an appreciable number of displaced (Pt*) atoms are produced by reduction of quite a significant portion of the inner compact film. Too low a value is detrimental as not only does this permit loss of Pt* activity but there is a strong possibility of reducing some of the hydrated or partially hydrated material that would eventually be incorporated into the thick outer layer. The minimum upper limit required for hydrous growth on cycling is ca. 1.70 V [27]; this is clearly above the oxygen monolayer level and significant oxygen penetration (to give a highly disturbed metal layer on reduction) and formation of some PtO_2 species (more likely perhaps than PtO to yield hydrated species on reduction) are probably important features in the development of the dispersed hydrous oxide layer. The complex interrelationship between thermodynamic and kinetic factors in the hydrous growth reaction is clear from the shift[27] in optimum upper limit in acid from ca. 2.8 V at 4.64 V s^{-1} to ca. 2.2 V at 100 V s^{-1}; at very high upper limit values the stability of the inner compact layer is evidently so high that at reasonably fast sweep rates reduction is inhibited at the lower end of the sweep. Formation of hydrous oxide layers of thickness (measured in terms of the charge required for reduction) in excess of 2500 mC cm^{-2} (geometric area) is readily achieved on cycling[27]; in terms of the quoted roughness value of ca. 5.0, this corresponds to over 1000 monolayers of oxygen. The rate of hydrous oxide growth decreases with increasing cycling time or thickness, probably due to gradual accumulation of poorly hydrated material at the compact oxide/hydrous oxide interface. Apart from direct analysis of this type of film[35] indicating a gradation in the degree of hydration, there is also electrochemical evidence for an intervening layer in the case of thick films grown on platinum in base.[135] On a cathodic sweep the latter showed a minor reduction peak (corresponding to ca. 2% of the net oxide reduction charge as measured in acid) at ca. 0.25 V in 1.0 mol dm^{-3} NaOH. After such reduction the main hydrous oxide film (which is not easily reduced at high pH) was only weakly adherent—obviously the reducible material (which is clearly not the normal inner monolayer) must reside between this film and the metal surface.

Little hydrous oxide growth was observed under cycling conditions[27] within the pH range 4.0-9.0; it was suggested that hydrous oxide growth requires some type of rearrangement of the initial, amorphous surface oxide layer, a reaction requiring H$^+$ adsorption at low pH and interaction with OH$^-$ at high pH values. With both acid- and base-grown films reduction results in the generation of a finely divided, platinum black-type surface. The procedure may be used as a method for controlling the degree of activation or roughening of platinum surfaces, and possibly for reactivating platinum impregnated fuel cell electrodes. Very little platinum is lost from the surface due to dissolution in this type of cycling/activation procedure.[27]

Work of a similar nature, involving hydrous oxide growth on platinum under square-wave perturbation conditions in acid, has been reported recently by Chialvo *et al.*[137] Changes in real surface area were monitored by measuring the hydrogen monolayer charge before and after the hydrous oxide growth and reduction processes. The optimum limits observed in this case (especially the lower value of 0.0 V, where even the hydrated film in acid is not thermodynamically stable) were significantly different from that observed by Burke and Roche.[27] This difference is most likely due to the much higher cycling rate, ca. 1.8 kHz, used by Chialvo *et al.*

VIII. PALLADIUM

The oxygen adsorption behavior of palladium[14-16] has received far less attention than that of platinum and the main reason for this is the complication associated with the unusual capacity of palladium to adsorb large quantities of hydrogen. The metal forms a nonstoichiometric hydride of equilibrium composition[138] $PdH_{0.69}$ at $p_{H_2} = 1.0$ atm, $T = 25°C$. Since the adsorption of hydrogen under cathodic conditions occurs under potential sweep conditions in the same region as hydrogen adsorption, it is quite difficult to determine accurate values for $^M Q_H$ for palladium; Woods[15] has suggested that charge values associated with monolayer ($^M Q_0 = 424$ μC cm^{-2}) or submonolayer oxide coverage is a convenient method for determining real surface area in this case.

The current-voltage response for palladium in the monolayer oxide formation/removal region under cyclic voltammetry conditions[15] is not very different from that of platinum. One significant difference between the two metals, however, is the greater susceptibility of palladium to dissolution in acid media; Rand and Woods,[139] for instance, found that Pd dissolved ca. 36 times faster than Pt under potential cycling conditions. Furthermore, while dissolution with the latter occurs mainly during the course of oxide film reduction in the cathodic sweep,[140] dissolution of the former was observed during the course of both the anodic and cathodic sweeps.[141] Recent work by Bolzan *et al.*[142] has shown that this activity for dissolution under potential cycling conditions decreases dramatically over the first few cycles. One possible interpretation of the latter is that oxygen penetration into the metal lattice—interesting evidence for which has been obtained from combined LEED and UPS studies[143]—at the anodic end of the sweep lowers its reactivity with respect to dissolution. The general shape of the voltammogram in acid in fact seems to vary considerably, in the oxide formation region, on going from the oxygen-free (first cycle) to the oxygen-containing (second and subsequent cycles) state.[142]

It seems generally accepted[15] that the onset potential for palladium oxidation in both charging curve and potential sweep experiments is lower in base than in acid. This drop is quite steep (ca. 109 mV/pH unit, SHE scale[28]) at low pH (0-2.0)—similar behavior to that outlined earlier for platinum.[44] However, over most of the pH scale (2.0-14.0) a linear drop of ca. 72 mV/pH unit (SHE scale) was recorded. The latter result is in agreement with the data of both Bagotzky and Tarasevich[18] and El Wakkad and El Din.[144] This difference between E_i values at high and low pH was noted in cyclic voltammetry experiments (again as in the case of platinum) carried out at reasonably fast sweep rates; on decreasing the latter in the range below ca. 50 mV s^{-1} the acid and base values eventually coincided at an intermediate level of ca. 0.74 V—this evidently corresponding to the behavior of the neutral, place-exchanged film.

The variation of E_i with pH for palladium is similar in magnitude to that reported for nickel[45] and thermally prepared RuO_2[8] under cyclic voltammetry conditions, and hydrous iridium oxide under open-circuit conditions.[145] The species involved in these cases is evidently less acidic (or anionic) than that produced initially in the oxidation of platinum, i.e., there is a lower overall fraction of the surface PdOH species present in ionized (PdO^-) form in the palladium case. It is possible to write equations to account for the reported value, e.g.,

$$Pd + H_2O = [PdO_{0.2}(OH)_{0.8}]^{0.2-} + 1.2H^+ + e^- \quad (50)$$

where on average only ca. one in five of the surface OH species is ionized. Why palladium should differ from platinum in this respect is not easily explained. Possibly it is influenced in some way by the significant rate of dissolution in the former case during the anodic sweep. Such a process generates surface defects, and OH species discharged at the latter will be of a partially place-exchanged character—a situation that generally favors formation of a more neutral film. There is an interesting analogy between the charged species written in Eq. (50) for the case of palladium and the $AuOH^{0.65-}$ species for the case of gold in the work of Angerstein-Kozlowska et al.[146]

Very little work has been carried out on hydrous oxide growth on palladium under dc conditions. According to Rand and Woods[147] the oxide coverage, at the end of the PdO plateau (i.e., at ca. 1.7 V), increases again with increasing potential—the effect being quite dramatic at ca. 2.0 V where a limiting coverage value is not observed. The layer formed under such conditions is a visible, dark gray, phase oxide whose reduction behavior under potential sweep conditions is considerably different from that of monolayer films, the electrode surface at the end of the process being considerably roughened. Kim et al.[148] have used ESCA techniques to examine the nature of the oxide films formed under dc conditions, the main product on anodization at potentials in the monolayer region appeared to be PdO and possibly small amounts of PdO_2. With thicker films formed at higher potentials a mixture of PdO and PdO_2, with considerable quantities of excess oxygen (probably present as bound water and hydroxy species), was detected.

Thick hydrous (β-type) oxide, yellowish brown in color, may be formed on palladium under potential cycling conditions in either acid or base.[28] As outlined earlier for platinum, such films can be reduced under potential sweep conditions only in acid media, Fig. 6; the reaction in question was assumed to be

$$[Pd(OH)_6]^{2-} + 4e^- + 6H^+ = Pd + 6H_2O \tag{51}$$

Quantitative work on hydrous oxide growth in base was restricted by the fact that the film produced in the latter decomposed— apparently to PdO [the Pd(II) state is significantly more stable than Pd(IV)[149]]—on transfer to acid. The optimum upper and lower limits for hydrous oxide growth on palladium in acid were found to be ca. 2.82 and 0.48 V (5-100 V s^{-1}), respectively; a similar film could be produced on palladium in base using approximately the same limits. Thick film growth on palladium is surprisingly rapid; for instance, a layer corresponding to a reduction charge of ca. 300 mC cm^{-2} was generated in 1.0 mol dm^{-3} H_2SO_4 using the above limits at 0.5 V s^{-1} after only ca. 10 cycles (to attain the same value with platinum using almost identical limits, but at a slightly faster cycling rate—4.62 V s^{-1}, involved about 600 cycles). Unlike the layer on platinum, the hydrous film on palladium in acid is subject to chemical dissolution; this can create errors in coulometric estimations of thickness at slow sweep rates (below ca. 60 mV s^{-1}).

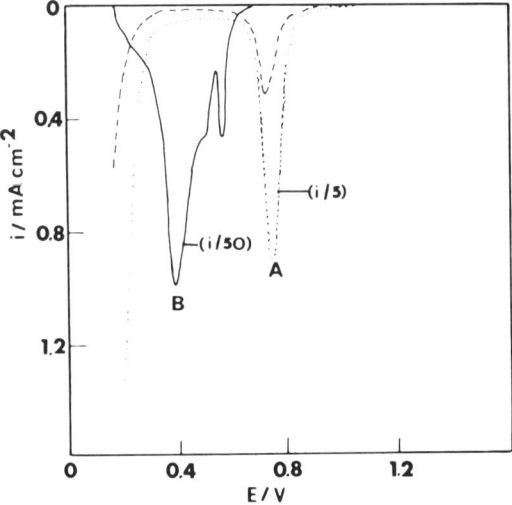

Figure 6. Reduction sweeps (1.50-0.15 V, 60 mV s^{-1}) for a Pd electrode in 1.0 mol dm^{-3} H$_2$SO$_4$ at 25°C: (- - - - -) uncycled electrode; (———) cycled electrode (0.48-2.82 V at 0.5 V s^{-1} for 1 min); (· · · · ·) second sweep for the same electrode after reduction but without regrowing the hydrous oxide layer (from Ref. 28, with permission).

IX. GOLD

1. Monolayer Behavior

In recent reviews of monolayer oxide formation on gold it was claimed[14-16] that oxidation of the metal commences at ca. 1.35 V in acid media with a shift to less anodic potentials on increasing the pH. Recent investigation[28] has shown that at fast sweep rates this decrease is virtually linear with pH with a slope of ca. −18 mV per unit increase in pH (RHE scale). This value for $\delta E_i/\delta \text{pH}$ is lower than that for platinum—a result confirmed by the significantly smaller drop in onset potential for gold as compared with platinum in the charging curve data for a wide range of buffer solutions reported by Vetter and Berndt.[107] Evidently a somewhat less anionic species is produced in the initial stages of gold oxidation. It has been suggested by quite a number of authors[150-156] that some preliminary oxidation of gold occurs at significantly less anodic potentials. While these currents below the main monolayer region have been attributed[15] to impurity effects, the situation was reassessed recently[157] in view of the fact that hydrous oxides may be involved in this region.

A typical example of a voltammogram for gold in $Ba(OH)_2$ solution (the CO_3^{2-} content of the latter being reduced to a minimum[146]) is shown in Fig. 7. Two reversible peaks of quite small charge (allowing for background current, and assuming the anodic charge between 1.2 and 1.5 V to represent a monolayer, the coverage for each of the two minor peaks in the region of 0.5-0.6 V scarcely exceeds 2% of the available gold surface) can be seen in the double layer region, one at ca. 0.53 V and the other at ca. 0.60 V. Alternating current impedance studies have confirmed[157] the presence of faradaic processes in this region, and since these peak potentials are quite close to the values recorded for the reduction of hydrous films grown on gold under vigorous dc conditions[41] it may be assumed that formation of traces of hydrous material precedes regular oxide growth at gold in base. It is quite clear that hydrous film formation on gold, as on platinum, is thermodynamically more favorable, especially in base, than its compact equivalent. Little hydrous material is normally produced on the anodic sweep because most of the gold atoms are embedded in the surface lattice. Such

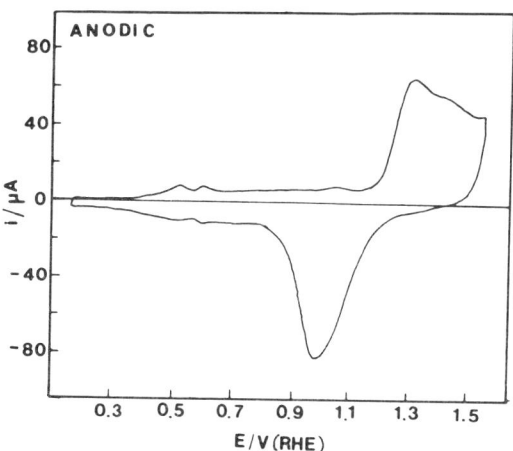

Figure 7. Cyclic voltammogram for a pure gold electrode in 0.2 mol dm^{-3} $Ba(OH)_2$ solution at 25°C, $V = 50$ mV s^{-1} (from Ref. 146, with permission).

a reaction can only occur at unusual defect sites where the metal atoms are present in a very low lattice coordination state—presumably as a type of adatom (Au*) species. These minor peaks,[157] like those for the reduction of the equivalent thick hydrous layers,[41] shift to more anodic potentials on decreasing the solution pH, $\delta E/\delta \text{pH} = 3/2(2.303 RT/F)$ V/pH unit (pH-independent scale).

Conway and co-workers[146] have interpreted monolayer oxide growth on gold in terms of the same model as for platinum, i.e., four peaks corresponding to four different overlay structures. Peak broadening was attributed to repulsive interactions and much attention was devoted to the effect of anions which undoubtedly influenced the shape of the voltammogram, causing for instance a shift of material between different adsorption states. There is a significant difference between this view of the behavior of gold in the monolayer region and that given in the earlier work of Dickertmann et al.[158] (which has been criticized by Conway and co-workers[146] on the basis of insufficient care in the production and retention of a clean surface). The latter group attributed the monolayer peaks on gold in acid to the effect of different crystallographic orientation at the surface of polycrystalline gold samples. Conway and co-workers[146] have pointed out that with gold in base there is a region (0.5–1.2 V) prior to monolayer formation where a reversible i/V profile is observed. The currents in this region are well above double-layer charging values, the charge involved constituting ca. 35% of that required for an OH monolayer. Of the two possibilities listed—formation of 35% of a normal OH monolayer or formation of an "OH" monolayer with an electrosorption valence of 0.35, effectively giving rise to an anionic species ($OH^{-0.65}$) at the surface—the latter was considered as being the most likely. Since the potential range in question here is between the expected reduction values for anhydrous and hydrous gold oxide species, it may well be that a variety of partially hydrated, discrete anionic species, of a more localized nature, are produced on gold in base in this region.

2. Hydrous Oxide Growth

Thick hydrous oxide films on gold can readily be produced, as in the case of platinum, by either dc or potential cycling techniques.

Polarizing the metal in acid above ca. 1.80 V results in the formation of a reddish brown oxide layer[159] which on reduction yields a relatively active form of the finely divided metal, i.e., a type of gold black. The reactions involved were investigated using potential sweep techniques by Lohrengel and Schulze,[160] who found that rapid growth in acid required a potential in excess of 2.0 V; a multilayer film was clearly formed under such conditions as two, and occasionally three, peaks were recorded on the cathodic sweep. These authors postulated that while the inner layer was both a poor electronic and ionic conductor, the reverse conditions prevailed for the thicker outer film, which they also suggested was rather compact in character.

Burke and McRann[23] showed that these hydrous layers could be produced under potentiostatic conditions in base as well as in acid. They suggested that since the onset potential for hydrous oxide growth virtually coincided with that for oxygen gas evolution, and the latter is commonly assumed[161] to involve formation of unstable Au_2O_4 entities, changes in coordination state of gold cations in the outer region of the compact layer facilitated the conversion of anhydrous to hydrous oxide in this region. Hydrous oxide formation in base was much slower than in acid and a limiting value was more readily observed with the former (according to Lohrengel and Schultze[160] growth at constant potential, at values above 2.0 V, is linear with time in acid—the eventual limiting factor being loss of contact between flakes of hydroxide and the electrode surface). In this work it was observed that the separation in peak potential for the anhydrous and hydrous oxide reduction processes increased steadily with increasing solution pH—the slope for the former (ca. −65.6 mV/pH unit, pH-independent scale) being much lower than that for the latter (ca. −85.6 mV/pH unit). Such an effect was attributed to the much more acidic (or anionic) character of the hydrous material. Further details, including a simple thermodynamic treatment of the hydrous gold systems in terms of a metal/insoluble salt electrode, have since been published.[41] As pointed out earlier in the case of platinum, some caution is required here as the experimentally determined peak potential values may not be totally reversible in character (peak variation with sweep rate seems to be less marked with gold[41] as compared with platinum[27]). The hydrous film on gold was found to be slightly soluble in base,[41] presumably due to aurate formation; however, the effect was far less marked than for the corresponding behavior of hydrous palladium oxide films in acid.[135]

A brief account of the production of thick hydrous oxide films on gold in both acid and base using the potential cycling procedure was published recently.[28] The optimum lower limit was found to be ca. 0.75 V in base and ca. 1.0 V in acid—these values are significantly lower than the monolayer oxide reduction potential. Further work in base[162] suggests that while the optimum lower value is at 0.75 V, some hydrous growth is possible with this limit set at, but not far below, ca. 0.6 V. Clearly the freshly formed hydrous material is reduced below this value (Fig. 3a in Ref. 41 clearly shows that there is a hydrous oxide reduction peak in this region). This result also supports the suggestion made earlier that the minor reversible peak at 0.6 V for clean gold in base (Fig. 7) is associated with hydrous oxide material. No optimum upper limit was observed in either acid or base: using the optimum lower limit for the particular electrolyte, the rate of hydrous gold oxide formation per cycle increased steadily with increasing upper limit, with values for the latter ranging from ca. 2.1 to 2.8 V. Growth on cycling was again particularly marked in acid—presumably because the hydroxy species are generally more reactive at low pH; the initially amorphous deposit can adopt a more crystalline form, thereby exposing the underlying substrate to further attack.

Potential cycling was found to be more effective than the dc technique for producing hydrous films on gold in base.[162] However, the reduction profiles for deposits prepared by the former technique were found to be quite complex, with as many as six broad, poorly resolved peaks apparent in the range extending from ca. 1.0 to 0.2 V. The scan in question entailed reduction of both anhydrous (inner) and hydrous (outer) films; it clearly illustrated the overall complexity of the films generated on gold. An account of the changes in the surface of polycrystalline gold, including electron micrographs of the surface before and after reduction of a hydrous oxide layer produced by square wave treatment, has been reported recently by Chialvo and co-workers.[163]

X. IRIDIUM

1. Monolayer Growth

Much of the early work[14-16] on the initial stages of anodic oxide formation on iridium is of dubious significance as this metal readily undergoes irreversible oxidation,[12] thick oxide films being readily generated on cycling the potential over the normal potential range, e.g., 0–1.5 V, used in this area. Highly reversible peaks are observed in such cases and the observed increase in charge capacity was originally attributed[13] to the reversible uptake of oxygen whose depth of penetration into the metal was assumed to increase with each cycle. The obvious objection to this lattice expansion viewpoint was that no simultaneous increase in hydrogen adsorption capacity was observed under these conditions; in fact it has been demonstrated that there was a drop in the latter (in order[12,164] of ca. 25%) due to the cycling treatment while the double layer charging current observed between ca. 0.40 and 0.50 V [164] remained essentially constant. It is now generally accepted that cycling the potential of iridium in a wide variety of electrolytes between certain limits induces irreversible formation of a battery-type, porous, hydrous oxide layer[11] whose thickness increases with increasing number of cycles.

Two studies to date have been reported[12,53] where precautions have been taken to minimize hydrous oxide formation in the investigation of monolayer oxide growth on iridium. Rand and Woods[12] used a solution of 5 mol dm^{-3} H_2SO_4, in which the hydrous material is soluble, and observed hydrogen peaks on the anodic sweep at 0.09 and 0.25 V, with oxygen adsorption commencing above 0.4 V. Two very broad peaks were observed in the oxygen adsorption region, one at ca. 0.7 and the other at ca. 1.0 V. Similar behavior with regard to the latter was reported by Mozota and Conway[53]— this group restricted the maximum value of the upper limit in 0.5 mol dm^{-3} H_2SO_4 to a value of 1.3 V to avoid formation of the hydrous material. They assumed that a monolayer of OH species was formed in the region of 1.1–1.2 V, conversion to an IrO monolayer being completed at ca. 1.4 V. The two peak values on the anodic monolayer sweep are surprisingly close to those observed[11] with a hydrous-oxide-coated electrode—suggesting that some hydrous material is formed even on the first sweep. Against this viewpoint, however, is the fact that the cathodic counterpart of the hydrous oxide transition is missing in the monolayer case in the subsequent reduction sweep. The amount of hydrous material present on the first sweep is of course likely to be quite small, and possibly on attaining the higher oxidation state these hydrous species combine, in a postelectrochemical step, with nearby iridium atoms to form anhydrous material. Mozota and Conway, working with more dilute acid than Rand and Woods, observed four peaks in the hydrogen region: the peak potentials of the first three shifted to more anodic values (RHE scale) on transferring the electrode from acid to base.

Mozota and Conway[53] evidently regarded the lower oxide peak on the anodic scan for iridium in acid as a doublet and assumed that oxidation via OH adsorption involved the same type of overlay development, with again four anodic peaks, as discussed earlier for platinum. The onset of OH deposition was found to be lower in base than in acid—this the authors attributed to weaker anion adsorption in base, although obviously it could also be explained (as in the case of platinum[44]) in terms of formation of an initially anionic species. It is surprising, in view of the greater affinity of iridium for oxygen (as reflected in the lower value of E_i for oxidation) that monolayer OH coverage is only achieved at about the same value, 1.2 V, as on platinum; however, this—plus the broadness of the peaks—was attributed to greater repulsion between surface dipoles (IrOH being regarded as more polar than PtOH). The initial stages of oxidation were claimed to be more reversible than in the case of platinum, with the reverse condition holding at high coverages. This switch was rationalized on the basis of (1) the effect of electric field strength at the interface on anion adsorption (both the field strength and anion adsorption energy being lower at the start of the iridium oxidation process), and (2) the greater affinity of iridium for oxygen.

2. Hydrous Oxide Films

(i) Formation

The conditions required to produce a thick hydrous oxide layer on iridium under potential cycling conditions were investigated initially by Buckley et al.,[24] and, more recently, by Conway and Mozota.[30] In acid media (the behavior in base is outlined later in a separate section) the optimum upper and lower limits were found to be ca. 1.60 and 0.01 V, respectively. Unlike platinum or gold, the hydrous film cannot be generated on iridium by vigorous dc polarization for the simple reason that both the metal itself,[165] and the hydrous oxide layer,[11] dissolve under such conditions. Oxide growth on cycling was observed with the upper limit just above ca. 1.4 V—at this stage significant penetration of oxygen into the second, or possibly even the third, layer of metal atoms of the bulk lattice is probable. Increasing the upper limits favors hydrous oxide growth until, above 1.6 V, dissolution causes loss of oxide species at the interface.

Enhanced oxide growth on cycling requires a minimum lower limit of ca. 0.15 V (critical values for both upper and lower limits are influenced by a range of kinetically and thermodynamically important factors, e.g., sweep rate, solution composition, temperature, degree of oxygen penetration, and oxide stability). Evidently most, though probably not all, of the compact monolayer material must be reduced. A disrupted layer of metal atoms is assumed to be produced at the electrode surface (perhaps partially separated from the bulk metal lattice by a region of unreduced monolayer material) and on the next anodic sweep some of these are converted to the hydrated form, i.e., formation of octahedrally coordinated IrO_6-type hydrous species is possible since some of the reduced atoms (Ir*) are present in the highly reactive adatom state. Assuming a typical rate of charge capacity increase[11] per cycle of 13.3 $\mu C\ cm^{-2}$ (with a one-electron reaction involved in the charge storage process) and a monolayer hydrogen charge capacity for iridium[166] of ca. 210 $\mu C\ cm^{-2}$, then only ca. 6% of the outer layer of iridium atoms is converted to the hydrous state in any one cycle. This figure is only approximate as no correction has been applied for the roughness factor and the value is dependent on the operating conditions—it does indicate, however, that formation of a relatively thick hydrous oxide layer requires an appreciable number of cycles. Reducing the lower limit from ca. 0.15 to 0.01 V enhances the rate of hydrous oxide formation, probably due to increased production of iridium adatoms. Further reduction below ca. 0.01 V is detrimental not because it reduces the hydrous oxide material (the latter in acid is evidently resistant to reduction under hydrogen gas evolution conditions[164] at −0.10 V) but presumably complete reduction of the compact layer favors the transformation of iridium atoms from the adatom state to regular lattice sites.

On repetitive cycling continued conversion of the metal to the hydrous oxide, via repeated generation and partial reduction of inner compact oxide at the metal surface, continues—the charge capacity of the outer film increases almost linearly with the number of cycles in the initial stages[11]; however, the rate eventually decreases to quite low values as (1) the inner region of the outer layer becomes less porous (limiting access of water to the metal/oxide interface—in addition a significant proportion of the outer layer does not participate in the redox process at high film thicknesses[4]) and (2) loss of oxide in the form of a stream of blue colloidal particles occurs, on prolonged cycling at relatively slow sweep rates, from the outer regions of the thick film. Among other factors which affect the hydrous oxide growth reaction (the original literature[11,30,145,167] can be consulted for details) are sulfuric acid concentration, pH, the presence of chloride ions and temperature.

(ii) Characterization

In the original work on this system Rand and Woods[12] postulated that cycling an iridium electrode in acid resulted in irreversible oxide formation at the metal surface—a result which was shortly confirmed by X-ray emission spectroscopy.[168] However, they assumed the material to be a porous form of the anhydrous oxide, IrO_2, attributing the density value (11.68 g cm^{-3}) of the latter to the surface layer; they also postulated[12] that the charge storage properties of the film were due to gain and loss of oxygen by the anhydrous film. Buckley and Burke[11,24] claimed from the onset that the layer formed was a hydrated oxide or oxyhydroxide, such as $IrO_2 \cdot 2H_2O$, and that the high reversibility of the charge storage process was due to the role of the proton as a counterion—the analogy with some well-known battery oxide electrodes, e.g., $Ni(OH)_2/NiOOH$ and $MnOOH/MnO_2$, was pointed out. The equation which they proposed[24] to explain the main charge storage reaction, involving a redox transition between the Ir(III) and Ir(IV) states in the surface layer, viz.,

$$IrO_2 + H^+ + e^- = IrOOH \tag{52}$$

is now widely accepted—although it had to be modified recently[4] to account for potential/pH effects.

Preliminary electron microscopy work by Mitchell et al.[168] showed that the film formed on cycling in acid was quite rough and not very adherent. The grain size was quite small (<50 nm), and while electron diffraction patterns suggested a hexagonal structure the data were not of high quality, probably owing to the amorphous character of the film material. Using combined ellipsometric and reflectometric techniques Gottesfeld and Srinivasan[169] showed that potential cycling resulted in the formation of a slightly absorbing, relatively thick film (250 nm after 15 h of cycling), which they suggested was a hydrated hydroxide. Using a microbalance to estimate the change in weight associated with the dissolution of a dried hydrous oxide layer, whose charge capacity has previously been determined by cyclic voltammetry, McIntyre et al.[5] demonstrated that the charge transfer process in the film involved one electron per iridium site (not two as suggested by other authors[168]). By combining coulometric with optical data they found that the apparent density of the hydrous material was ca. 2.0 g cm^{-3}, i.e., ca. one-sixth of that of anhydrous IrO_2. They showed, using electron microscopy, that the film was porous, with microvoids in the order of 2.5 nm diameter and suggested that the particles present were composed of a network of interlinked octahedral oxyhydroxide units. A similar model was outlined by Glarum and Marshall,[170] who assumed that all sites in the polymer network were accessible to the electrolyte phase. A chain structure was envisaged, with a certain degree of random cross-linking; the bonding was considered to be dynamic, with continual forming and breaking of linkages in the polymer. The model outlined recently by Burke and Whelan[4] is quite similar to the latter except that, in order to account for potential/pH effects, it was necessary to assume that the sites in the film were anionic, with counterions present in a delocalized manner, probably in the solution present throughout the hydrous material. In these terms the latter may be viewed as a type of acidic gel at low pH or, more generally, as a cationic, inorganic ion-exchange resin in gel form. As outlined later in connection with conductivity data the oxidized material was viewed[4,9] as being in a slightly mixed (IV and VI—predominantly the former) oxidation state.

Hackwood et al.[171] have carried out combined thermal and gas evolution analysis of sputtered iridium oxide films—judging from the quoted density values these were probably not as dispersed or hydrated as films produced by cycling. Some water was lost on heating in vacuum at 120°C; an amorphous to crystalline transition occurred at 300°C, the system releasing considerable energy without decomposing; finally, at ca. 700°C, there was another change associated with dehydration as the material assumed the anhydrous, rutile form of IrO_2.

In discussing data on hydrous iridium oxide, Conway[164,172,173] used the term "charge enhancement factor" (CEF). This is the ratio Q_O/Q_H where Q_O is the charge associated with the main transition of the hydrous film and Q_H is the corresponding charge in the hydrogen region—measured in the absence of oxide. For relatively thin films and slow sweep rates it can be regarded as a measure of oxide loading.

In a recent ESCA study[172] of films produced by potential cycling it was found that the initial layer (CEF = 1) was not IrO_2 but the hydrated hydroxide—some water was assumed to be chemisorbed on the surface beneath the $Ir(OH)_4$ species. With thicker films (CEF = 80) some peaks attributable to Ir(VI) were also evident; appreciable incorporation of SO_4^{2-} anions from solution was observed. It was not possible to examine the reduced state as the dried film was rather readily oxidized in air. Electron microscopy combined with X-ray emission analysis[173] showed that surprising differences in morphology arise in oxide deposits generated electrochemically on iridium and ruthenium. In the case of iridium the oxide was classified as an amorphous, hyperextended material. X-ray emission analysis showed a much higher O/Me ratio in the oxide formed on iridium as compared with ruthenium.

(iii) Electrochemical Behavior

Two fundamental properties influencing the electrochemical behavior of acid-growth hydrous iridium oxide films are (1) the acidity of the hydrated material and the variation of this property with change in the iridium oxidation state,[4] and (2) the difference in electrical conductivity between the reduced and oxidized form of the surface film.[170,174] According to Burke et al.[9,38] both properties are strongly influenced by the dispersed hydrated nature of the iridium oxide layer. The unusual pH variation of the reversible potential of the redox transitions observed with the latter (see Fig. 8), with a linear shift of ca. $3/2(2.3RT/F)$ V/pH unit, is now well established[9,30,145,175,176]; a similar shift was recorded recently for base-grown hydrous iridium oxide films.[177] A detailed thermodynamic treatment of this phenomenon has been published recently by Burke and Whelan,[4] and in view of the structure proposed by Scott[37] for the equivalent Pt(IV) compound, i.e., $H_2Pt(OH)_6$, the reactions assumed to be involved, viz.,

$$2\{(H_f^+)_{2n}\cdot[Ir(OH)_6]_n^{2n-}\} + 2ne^- + 2nH_s^+ = (H_f^+)_{3n}\cdot[Ir_2(OH)_9]_n^{3n-} + 3nH_2O \quad \text{(acid)} \quad (53)$$

and

$$2\{(M_f^+)_{2n}\cdot[Ir(OH)_6]_n^{2n-}\} + 2ne^- = (M_f^+)_{3n}\cdot[Ir_2(OH)_9]_n^{3n-} + 3nOH^- + nM_s^+ \quad \text{(base)} \quad (54)$$

are presented here as hydroxy species. Both oxidized and reduced forms are assumed to be polymeric network structures of a rather attenuated character—the basic unit in the reduced form being dimeric [the latter assumption was made earlier in the case of gold

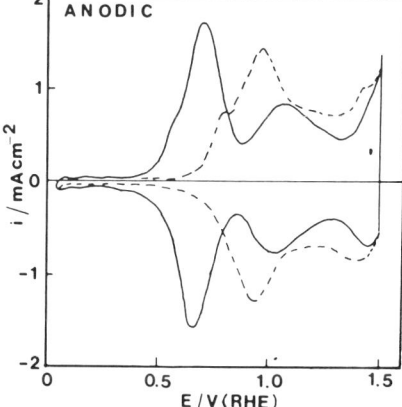

Figure 8. Cyclic voltammograms recorded for a hydrous-oxide-coated iridium electrode (40 mV s^{-1}) in 1.0 mol dm^{-3} H$_2$SO$_4$ (- - - - -) and 0.1 mol dm^{-3} NaOH (———), $T = 25°C$. The film was produced by repetitive potential cycling in the acid medium (from Ref. 186, with permission).

and is supported by the fact that even in solution Fe(III) salts are dimeric in character[178]. Possible structures for the basic units in the oxidized (A) and reduced (B) form are given below:

$$\begin{array}{cc} \text{HO}\diagdown\overset{\text{OH}}{\underset{\text{Ir}}{|}}\diagup\text{OH} & \text{HO}\diagdown\overset{\text{OH}}{\underset{\text{Ir}}{|}}\overset{\text{H}}{\diagup}\overset{\text{OH}}{\underset{\text{O}}{\diagdown}}\diagup\text{OH} \\ \text{HO}\diagup\underset{\text{OH}}{|}\diagdown\text{OH} & \text{HO}\diagup\underset{\text{OH}}{|}\diagdown\underset{\text{H}}{\text{O}}\diagup\underset{\text{OH}}{|} \\ (\text{A}) & (\text{B}) \end{array} \qquad (55)$$

It is obvious that the structure of the Ir(IV) compound cannot be the normal rutile-type structure of IrO_2—there is a much higher oxygen content in the hydrous material, even in dry state (the unusually high level of oxygen in electrochemically generated iridium, as compared with ruthenium, oxide films was pointed out recently by Birss et al.[173]). Support for the view that Ir(IV) oxide is acidic is also provided by the work of Ardizzone et al.,[179] who found an unusually low value (ca. 0.5) for the point of zero charge for thermally prepared IrO_2—the surface of the latter, like the corresponding RuO_2 compound,[8] is assumed to be hydrated.

There has been considerable controversy in the literature as to whether cations[35,180,181] or anions[34,182-184] act as charge balancing ions, maintaining electroneutrality in the film in the course of electron insertion or withdrawal reactions. The scheme represented in Eqs. (53) and (54) should resolve this problem: reduction is accompanied by insertion of a proton into an OH group coordinated to the central metal ion so that the latter effectively loses an OH^- ligand, i.e., there is a certain validity in both viewpoints. From (53) and (54) there appear also to be chemical and structural changes associated with the redox processes—these would probably be easily accommodated in the loosely linked polymer. Ion transfer processes in these films, even at high pH, probably involve largely proton switching, i.e., Grotthus-type transfer, in the hydrous material.

An account of a comparison of the potential/pH behavior of hydrous and anhydrous iridium oxide films was published recently.[145] The open-circuit response for the hydrous material, in the half charged state, was ca. $1.25(2.3RT/F)$ V/pH unit; this lower value, as compared to the response to the same film on cycling, was attributed to a decrease in acidity, due to desorption of OH^- ions or uptake of protons—especially by Ir(IV) centers in the metastable film, during the rather long equilibration times required before and during the open-circuit measurements.

The ac impedance measurements by Glarum and Marshall,[170] and faradaic experiments by Gottesfeld,[174] have clearly shown that the hydrous film in the reduced state has a low electronic conductivity while in the oxidized state it is highly conductive. While Gottesfeld[174] has provided a qualitative interpretation of this behavior in terms of conventional band theory, Burke and Whelan[9] have outlined a more chemical approach involving electron hopping between neighboring centers. With all these centers in the reduced state, which is assumed to be highly stoichiometric, there is a barrier hindering electron transfer between adjacent strands in the polymer network (because of the labile nature of the latter[170] it is assumed that there are always unconnected segments present in such films). The higher conductivity of the oxidized material was attributed to the presence of Ir(VI) sites (which behave as acceptor states) in the predominantly Ir(IV) film—direct evidence for these Ir(VI) species was obtained recently[172] in an ESCA investigation of the oxidized state.

Although the main charge storage process of the hydrous material is regarded as being quite reversible—as it certainly is at low sweep-rate and film thickness values—substantial departures have been observed at the other extreme, i.e., with thick films and fast sweep rates.[4,30] The effect is most marked in the anodic sweep where the main peak moves so far in the anodic direction that it eventually overlaps with the current increase associated with the oxygen gas evolution process. During the anodic sweep the poorly conducting Ir(III) state is altered to the highly conducting Ir(IV) state; peak currents are very large and proportional to film thickness. At the fast rates of conversion involved in the oxidation of the surface layer the departure of the peak potential from its reversible value was assumed to be due to a slow electrochemical step. The unusually high Tafel slope value observed for this process (ca. 200 mV decade^{-1}) was tentatively ascribed[4] to the involvement of a slow chemical or structural alteration in the hydroxy complex during the course of its oxidation.

Much lower currents were observed during the reduction sweep and the values did not increase as rapidly with increasing sweep rate. Clearly the formation of the poorly conducting material in the inner region of the film impeded electron transfer[4,30]—with thick films the cathodic charge capacity value dropped substantially with increasing sweep rate. This inhibition of the reduction process is also clear from the electrochromic behavior of the system; the change in color from the blue (oxidized) to transparent (reduced) state is distinct only at low film thickness. With thicker films the layer remains blue throughout the cycle as the inhibition of the electron transfer during the reduction sweep leaves a large residue of Ir(IV) centers within the film even under quite reducing conditions.

In sulfuric, though not in phosphoric, acid media[185] a shoulder is usually observed at ca. 0.6 V on the cathodic side of the main anodic peak (its cathodic counterpart can also be clearly observed under certain conditions[4,167]). According to Burke and Whelan[4] it is probably an III/IV transition in species containing anions other than OH^- within the coordination sphere. Conway and Mozota[30] attributed the large currents above the main anodic peak mainly to charging of the rather entensive oxide-solution interface. However, Burke and Whelan,[4] who assumed that the dispersed, hydrated oxide material scarcely constituted a distinct phase, assumed that the currents in this region were due to changes in stoichiometry in a mixed valence state, Ir(IV)/Ir(VI), deposit. A peak—again showing a shift of ca. $3/2(2.303RT/F)$ V/pH unit— can be observed[9] above the main anodic peak on transferring an acid grown film to solutions of higher pH. Such a peak was also observed recently[177] following transfer of a base-grown film to acid. The increase in the region just prior to oxygen gas evolution is presumably due to further formation of Ir(VI) species.

(iv) Growth in Base

Although hydrous iridium oxide deposits grown in acid show little change in redox activity—apart from the potential/pH shift— on transfer to base, it was demonstrated by Buckley *et al.*[24] that the film could not be grown on cycling a clean iridium electrode in such a medium (1.0 mol dm^{-3} NaOH, 0.01 to 1.50 V, 30 V s^{-1}). There have been subsequent reports[186,187] that some growth can be achieved in dilute base (0.1 mol dm^{-3} NaOH). In a recent detailed investigation of this topic[177] it was confirmed that hydrous oxide formation was possible in dilute, though not in strong, base. Some unusual effects were observed; e.g., after subjecting the electrode in base to preliminary cathodic pretreatment, which probably involved removal of any residual oxide film, no growth was observed. This result, and others, were explained by the assumption that local pH changes within the pores of even a very superficial oxide layer played a crucial role in the conversion of anhydrous to hydrous oxide. The evidence for the pH change was quite striking; on cycling in dilute base the main charge storage peak is at ca. 0.65 V, but owing to the production of protons within the oxide during the anodic sweep a peak for acid conditions, at ca. 0.98 V, was also observed (no peak response is observed for intermediate pH values as there is little buffering in this region). The charge ratio of the "acid" to the "base" peak on the anodic sweep increased quite considerably with increasing sweep rate. A similar shift was not observed on the cathodic sweep—the solution within the pores in this case is made just slightly more basic.

It was concluded from the above behavior that hydrous oxide growth in base requires exposure of the film at some stage of the cycle to an acidic environment. The low hydroxide ion activity of the latter is assumed to destabilize partially the hydroxy complex, activating it with respect to rearrangement as suggested earlier in connection with the slow rate of hydrous oxide formation on gold in base.[23] Such rearrangement removes the passivating influence of the highly amorphous initial hydrous deposit, allowing access of water and further transition to occur in the inner oxide region.

The important role of rearrangement also appears to be responsible for the difference in texture and reactivity between acid- and base-grown films. The latter were much more reactive (they dissolve readily in 1.0 mol dm^{-3} H_2SO_4 or NaOH), probably because of their more amphorous character. The acid-grown films are probably more crystalline (the larger grain size would explain their greater stability[177]) and open in texture—local pH gradient effects were not evident in dilute base with such films.

There are a number of reasons as to why hydrous oxide growth is not observed on cycling in strong base. Local pH changes, and therefore hydrous rearrangement, would not occur in these more highly buffered solutions. Conway and Mozota[30] attributed the lack of growth to the absence of strongly adsorbing anions which assist the place-exchange reaction in the formation of the initial film at the metal/oxide interface; however, this should apply, possibly to an even greater extent, to the more dilute base where hydrous oxide growth is in fact observed (if not in quite as rapid a rate as in acid). Hydrous oxide dissolution is another possibility[24]; the objection by Woods[186] to this viewpoint on the basis that no loss of charge capacity is observed on cycling in 0.1 mol dm^{-3} NaOH is clearly invalid as in such dilute base hydrous oxide growth, not dissolution, is apparently the main reaction.

XI. RHODIUM

1. Hydrous Oxide Growth

According to Burke and O'Sullivan[29] hydrous oxide growth on rhodium can be achieved in base, though not in acid—the metal dissolves on cycling in the latter medium—on repetitive cycling over the range 0–1.55 V at a sweep rate of ca. 30 V s^{-1}. The effect of the usual variables—e.g., sweep limits, sweep rate, base concentration, and solution temperature—was investigated; a noteworthy feature of this work was the absence of stirring dependence, clearly demonstrating, for the first time, that hydrous oxide growth on cycling does not involve a precipitation/dissolution mechanism. As in the case of iridium, film growth on rhodium had little effect on the charge in the hydrogen or double layer region; the main effect was the development of a large, reversible charge storage peak (Fig. 9) at ca. 1.37 V.

The electrochromic transition described for this system by Burke and O'Sullivan[29]—pale yellow for the reduced state and dark green for the oxidized state, the latter color persisting throughout the cycle at high film thicknesses—was not observed by Woods.[186] Gottesfeld[188] reported that the color of the oxidized state in this system varied with thickness, ranging from brown, to green and even purple—it was suggested that the colors observed in this case were due to a combination of film adsorption and interference effects.

In a further examination of the hydrous rhodium oxide system Burke and O'Sullivan[42] showed that thick films grown by fast cycling procedures were only partially active with regard to charge storage

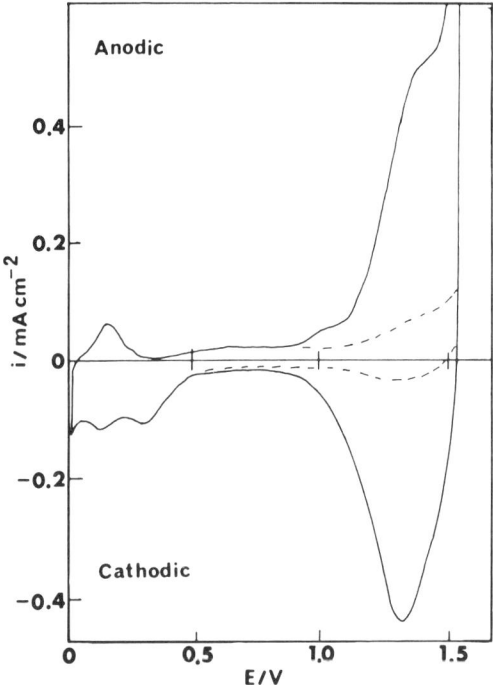

Figure 9. Cyclic voltammograms (31 mV s^{-1}) recorded for an uncycled (-----) and a cycled (0.01–1.55 V, 15 V s^{-1} for 40 s) rhodium electrode in 1.0 mol dm^{-3} NaOH at 25°C (from Ref. 29, with permission).

behavior. The charge capacity of such films could be significantly increased by slow cycling, possibly as a result of decreasing the extent of cross linking within the film—thereby increasing the degree of hydration. Severe polarization of these films (either under steady-state or cycling conditions) introduced a degree of heterogeneity as shown by the presence of an additional charge storage peak at ca. 1.0 V; such treatment also modified the electrochromic behavior of the system. Both the charge storage and electrochromic properties of these films were irreversibly lost on heating to 170°C.

The influence of solution pH on the main cathodic charge storage peak for the hydrous material, and a minor cathodic peak at ca. 0.3 V in base (the latter was attributed to reduction of an inner, compact oxide layer on the metal surface), was also examined.[38,42] Both shifted anodically, the former by ca. 82.5 and the latter by 65.8 mV/pH unit (SHE scale), with decreasing pH. Such behavior is quite similar to that of gold[23]; the charge storage reaction for the hydrous layer is assumed to be quite similar to that outlined earlier for iridium—Eqs. (53) and (54).

2. Behavior of Rh/Pt Alloys

This system seems to be the only alloy to date whose hydrous oxide growth behavior under potential cycling conditions has been investigated.[189,190] Burke and O'Sullivan[189] demonstrated that with an alloy containing 10% by weight of rhodium in platinum both components corroded on cycling between certain limits (0-1.5 V) in 1.0 mol dm^{-3} NaOH. However, while the platinum corrosion product was found to be soluble under these conditions the rhodium one was not—in fact the hydrous film developed on the surface in this case was apparently derived almost totally from the minor component in the alloy.

The behavior of a 1:1 Pt-Rh alloy on cycling in various buffer solutions was examined recently by Aston et al.[190] The corrosion behavior is more complex here, as shown by the fact that hydrous film growth is quite slow even on pure rhodium at pH values less than ca. 13.0.[29] The hydrous rhodium oxide layer (whose redox behavior Aston et al. attributed to the model outlined by Burke and O'Sullivan[42]) was observed on cycling at pH = 13.0. For the lower pH values some surface enrichment by rhodium at the alloy itself was noted at the early stages of cycling; however, this changed on prolonged cycling to yield a platinum-rich surface. The diagram given for pH = 10.9 does not show the hydrous rhodium oxide charge storage region; it is not possible, therefore, to assess whether such material is produced in this case. It is possible that on prolonged cycling the platinum enrichment at the surface of the underlying metal could be due, not to preferential rhodium dissolution, but to replating of dissolved platinum (especially if the latter were trapped within the pores of an insoluble hydrous rhodium oxide layer). The behavior of such alloy systems is obviously fairly complex but an awareness of the problems involved is essential for the interpretation of voltammetric data (particularly important in electrocatalysis studies is the possibility of surface enrichment by the more insoluble component of the hydrous oxide deposit generated in preliminary cycling work employed to characterize or clean the alloy electrode surface).

XII. RUTHENIUM

Studies of oxide formation on ruthenium under anodic condition have been reported by Hadzi-Jordanov et al.,[191-193] Burke et al.,[194,195] and Michell et al.[196] Burke and Mulcahy[194] suggested that instead of a layer of adsorbed OH or O species[191] the film produced on ruthenium may be regarded as a phase oxide of composition ranging from Ru_2O_3 to RuO_2, i.e., the processes taking place, especially above ca. 1.0 V, may not be restricted solely to monolayer-type phenomena. The marked pH dependence of the charge involved in the anodic sweep with initially oxide-free ruthenium surfaces[195] is indicative of the occurrence, even under single sweep conditions, of multilayer oxide growth, the latter being particularly marked in either strongly acidic or strongly alkaline media.

The generation of a phase oxide on ruthenium under anodic conditions is also suggested by current/time profiles[194] recorded at 0 V for the reduction of anodic films generated at different formation potentials (E_f). As the latter were increased over the range 0.15-1.4 V the reduction profiles altered from a simple exponential decay ($E_f = 0.15$-0.3 V) to one showing a distinct plateau ($E_f = 0.5$-1.0 V) and, finally (at $E_f = 1.2$-1.4 V), to a bell-shaped curve characteristic of nucleation-controlled phenomena. The shape of the reduction profile, and the charge required for complete removal of the film formed at 1.2 V, were found to be significantly dependent upon the time the electrode was held at the latter. Clearly the behavior of the system is complex, as coverage, thickness, and possibly even the nature of the film itself vary with both potential and time.

To examine the behavior of the ruthenium metal/solution interface using potential sweep techniques[191-195] it is necessary to reduce any oxide present from previous experiments either by holding the potential constant at 0 V for ca. 2 min or by scanning at very low sweep rates (ca. 13 mV s^{-1}). There is significant overlap at low potentials between currents due to the hydrogen and oxygen (O or OH) deposition/removal process[192-194] (this creates an additional complication in that true surface area values—and hence oxide thicknesses—cannot readily be assessed for this metal). The initial stages of oxidation in acid, up to a potential of ca. 0.6 V, apparently occur in a virtually reversible manner.[192] A peak is usually observed at ca. 1.0 V; however, its origin is uncertain as it is not clear whether at this stage the film is of monolayer or multilayer character. Apart from possible oxidation of the underlying metal, currents in this region are probably associated with increasing oxide stoichiometry as adsorbed OH and O species are converted to oxyruthenium entities, e.g., Ru_2O_3, RuO_2, and RuO_3, as well as their hydrated analogs. Repeated cycling of ruthenium in acid between 0 and 1.3 V (or higher values[192,196]), or holding the electrode potential at values in the latter region,[194] results in formation of an oxide film that is not readily reduced. Such an oxidized surface exhibits a voltammogram of surprisingly reversible character (Fig. 10), quite like that recorded for thermally prepared RuO_2 deposits on titanium—even to the extent of showing two very broad peaks, one at ca. 0.7 V and the other at ca. 1.2 V on both the anodic and cathodic sweeps. According to Hadzi-Jordanov et al.[192] the spreading of the charge over the entire voltammetric range is due to Ru(III)/Ru(IV) transitions in a material where the individuality of such redox species is destroyed by interaction effects.

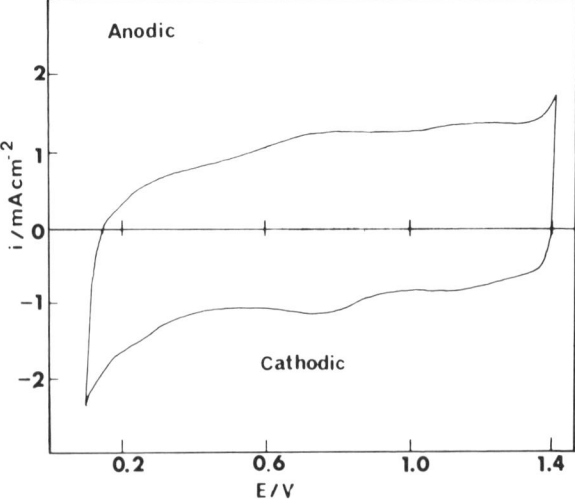

Figure 10. Cyclic voltammogram recorded under continuous potential cycling conditions for a ruthenized platinum electrode in 0.5 mol dm^{-3} H_2SO_4 at 25°C, $V = 50$ mV s^{-1} (from Ref. 191, with permission).

In a more recent investigation of hydrous oxide growth on ruthenium in acid Birss et al.[173] found that a ruthenium oxide film, grown by cycling, had a thickness as measured by scanning electron microscopy 50-100 times greater than that estimated from charge capacity values. This discrepancy was attributed to (1) microporosity in the hydrous oxide film, (2) limited electrochemical access to material inside the film, and (3) electrochemical redox activity being restricted only to the surface of pores. The film formed on cycling was found to have a smooth shiny appearance due to the presence of a hard outer oxide crust. Removal of the latter with the aid of a steel needle or scalpel revealed a much looser oxide layer beneath. Scanning electron microscopy work showed that the latter, in films formed by the fast cycling techniques, possessed a remarkable columnar structure—the latter feature was absent in films produced by oxidation using constant current, or programmed change in current, techniques. The films in the latter cases were found to be more amorphous, with a higher O/Ru ratio than the material formed on cycling—although this ratio was always substantially lower than that corresponding to the nominally stoichiometric "RuO_2"—taking the value for thermally prepared RuO_2 samples as standard (there is of course some evidence[197] that the outer layers of this high surface area material is oxygen-rich). Transmission electron microscopy studies indicated that there were channels or pores in the columnar material of diameter ca. 10-15 nm.

It was suggested that the columnar structure present in the oxide formed on cycling is the result of anodic etching accompanied by the formation of a thin layer of oxide on the structure resulting from the latter. The columnar material is assumed to contain unoxidized metal, coated with oxide—this assumption was made to explain the low (O/Ru) ratio of this type of deposit. The retention of the scratch marks on the extremely hard outer crust on these deposits suggested that the oxide deposit forms by migration of OH^- or O^{2-} ions into the film rather than by outward migration of metal ions.

The behavior of ruthenium in base was examined by Burke and Whelan.[198] The metal dissolves even more readily in base than in acid but, by comparing the behavior of the metal in unstirred solution with data for ruthenium solution species at a platinum substrate in base as reported by Lam et al.,[199] the following peak assignments were made: Ru(III)/Ru(IV), $E = 0.45$ V; Ru(IV)/Ru(VI), $E = 1.0$ V; Ru(VI)/Ru(VII), $E = 1.4$ V (these voltage values are approximate). In marked contrast to thermally prepared RuO_2, the oxide generated electrochemically on the metal in base was a poor catalyst for oxygen gas evolution, the preferred reaction being dissolution. In a further extension of this work Burke and Healy[8] showed that highly reversible peaks occur at the same potential in the case of thermally prepared RuO_2-type films on titanium in base (the peak at ca. 1.4 V being quite distinct, the lower two being quite broad). In addition, these peaks for the two lower transitions were shown to display the usual effect of acidic character, i.e., the peak potentials rose on decreasing the solution pH by a value of ca. $1.25(2.3RT/F)$ V/pH unit (SHE scale). This suggests that the broad peak at ca. 0.7 V in the case of the multicycled ruthenium electrode in acid (Fig. 10) is due to a Ru(III)/Ru(IV) transition while the higher peak at ca. 1.2 V is due to a Ru(IV)/Ru(VI) transition. It may be noted that similar peaks, at approximately the same potential values, have been reported by other authors[200,201] for thermally prepared RuO_2 electrodes in acid media.

XIII. SOME NONNOBLE METALS

1. Iron and Cobalt

Because of their extensive technological applications, the anodic behavior of most of the nonnoble metals has been extensively investigated. Only a brief outline, with emphasis on the role of hydrous oxide material, is given here. The conditions necessary for the production of a thick hydrous oxide film on iron in base (repetitive cycling between -0.50 and 1.25 V at ca. 3.3 V s^{-1} at 25°C) have been outlined by Burke and Murphy.[202] As outlined in Fig. 11, four peaks [with maxima at ca. -0.04, 0.13, 0.30, and 0.55 V (RHE)] were observed with the bare metal on the first anodic sweep, and only two (-0.05 and -0.20 V) on the subsequent cathodic sweep. On repetitive cycling the anodic peak at $+0.3$ V and the cathodic peak at -0.05 V became greatly enhanced and the presence of an electrochromic effect was noted. Further indication that a dispersed hydrous oxide film was produced under these conditions

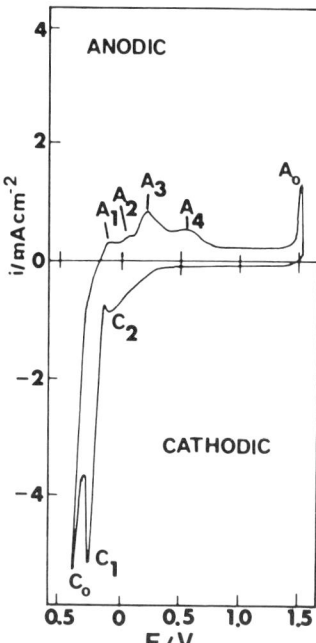

Figure 11. Cyclic voltammogram (-0.35–1.65 V, 40 mV s^{-1}) for an uncycled iron electrode in 1.0 mol dm^{-3} NaOH at $25°C$.

was provided by the large potential/pH shift (ca. -89.6 mV/pH unit, SHE scale) subsequently reported for this system.[38]

In a more recent investigation of this system the present authors[203] have assigned, in a provisional manner, the peaks for the bare metal as follows. The first anodic peak is possibly due to a combination of adsorbed hydrogen oxidation and formation of a layer of adsorbed hydroxy (FeOH) species. The second is probably due to conversion of both Fe and FeOH to a film of Fe(II) oxide or hydroxide material, the outer region of the latter evidently being hydrated. The third anodic peak, i.e., the one that becomes enhanced on cycling, is most likely due to an Fe(II)/Fe(III) conversion in hydrous material present at the oxide/solution interface. By analogy with the process occurring in the case of iridium in base, Eq. (14), the reaction involved here may be formulated in simple terms as

$$[Fe_2(OH)_9]^{3-} + 3H_2O + 2e^- = [Fe_2(OH)_6 3H_2O]^{2-} + 3OH^- \quad (56)$$

However, since the composition of the hydrous material is currently unknown, the observed potential/pH dependence for the peak in question could (by analogy with nickel[204]) also be represented as

$$[Fe(OH)_{3.5} \cdot nH_2O]^{0.5-} \cdot (Na^+)_{0.5} + e^-$$
$$= Fe(OH)_2 \cdot nH_2O + 0.5Na^+ + 1.5OH^- \qquad (57)$$

The final anodic peak in the case of the bare metal was attributed to an Fe(II)/Fe(III) transition in a layer of compact oxide material at the inner region of the surface deposit. Of the two cathodic peaks, one is clearly the reverse of the hydrous oxide transition in the anodic sweep while the nature of the second is still unclear.

The reaction of major technological importance in the case of iron is the corrosion reaction and the slow step here with regard to dissolution may well be the conversion of the anhydrous oxide or hydroxide to its hydrous equivalent. The Fe(II) species appears to be involved as in the pH range where dissolution is significant (<9.0) the metal dissolution and passivation processes occur just above the second anodic peak and significantly below the region for the Fe(II)/Fe(III) transition in the hydrous material at lower pH. At high pH the high hydroxide activity in solution inhibits this conversion, the complexes formed being too stable to allow rearrangement (OH$^-$ ion penetration here is assumed to be a field-assisted reaction). At lower pH both OH$^-$ and H$^+$ ions may be absorbed by the compact material, the proton catalyzing the process by converting bridging oxygen to terminal oxygen species (thus effectively opening up the lattice)

$$-Fe-O-Fe- + H^+ \cdots OH^- = 2-Fe-OH \qquad (58)$$

Once hydrated, the oxide material is assumed to lose much of its resistance to dissolution. According to this view passivation, i.e., the sharp drop in the rate of dissolution at higher potential, is attributed to the effect of the increasing positive electric field at the interface on the ability of protons to enter the compact layer (it is, however, not possible at present to eliminate the alternative possibility, namely, that the inhibition of hydration of the compact film is influenced by some formation of Fe_3O_4 species in the latter). The role of protons, and indeed hydrous material in general, in the iron dissolution/passivation reaction is still controversial[205] (an account of recent work in this area has been published by Murphy et al.[206]), and it appears that greater care must be taken to distinguish between hydrous and anhydrous material when examining the composition, structure, and indeed the precise location of the actual passivating layer. A brief account of the enhanced electrocatalytic activity of iron for oxygen gas evolution, following hydrous oxide growth, has recently been published.[207]

Hydrous oxide growth on cobalt in base can readily be achieved[208] by potential cycling (typically 0.0-1.2 V, 33 mV s^{-1}). Conversion of Co(O) to Co(II) occurs over the region 0-0.5 V (RHE) on the anodic sweep—at low film thicknesses both oxide formation and metal dissolution are evidently involved here. Co(II) to Co(III) conversion appears to occur over the region 1.0-1.25 V (RHE); the processes involved, however, are complex as more than one pair of peaks may be observed in this region. With relatively thick hydrous oxide layers a Co(III)/Co(IV) transition, with an associated electrochromic effect, was observed at ca. 1.48 V (RHE), i.e., just prior to oxygen gas evolution. Some slight enhancement of the oxygen gas evolution rate, relative to that observed with an uncycled cobalt substrate, was observed following hydrous oxide formation at the metal surface.

2. Nickel and Manganese

In view of their major application in aqueous battery systems more work has been carried out on the structural aspects of the oxides of these two metals than any of the systems discussed earlier. Details of the structure and reactivity of the nickel oxide battery materials can be found in recent reviews by Briggs[209] and Oliva *et al.*[2] Both hydrous and anhydrous phases exist for both the Ni(II) hydroxide and Ni(III) oxyhydroxide systems. Most interesting are the comments of Le Bihan and Figlarz,[210] and McEwen,[211] with regard to turbostatic structures: the latter are found in materials where the ordering of the oxide is quite limited, i.e., the systems consist of highly ordered nuclei linked in a disordered manner—the latter feature should certainly enhance mass transfer processes and may well be involved in many other hydrous oxide systems.

According to Burke and Twomey[45] nickel metal in base oxidizes just above 0 V (RHE) to form an initially anionic Ni(II) species. The Ni(II)/Ni(III) reaction, which is the main process of interest in battery systems, occurs at ca. 1.4 V (RHE). Thick oxide films may be produced[45,212] on the metal by repetitive potential cycling [typically 0.5-1.55 V (RHE), 31 mV s^{-1}]. With such deposits cyclic voltammograms usually display[213] one anodic ($\delta E/\delta$pH = -86.5 mV/pH unit, SHE scale) and two cathodic ($\delta E/\delta$pH is -86.5 mV/pH unit for one and -59.2 mV/pH unit for the other) peaks in the region of the Ni(III)/Ni(II) transition; however, two anodic peaks have been noted[45] at ca. 1.4 V under certain circumstances. Since structural data for the Ni(OH)$_2$ and NiO·OH systems show that these can exist in both anhydrous and hydrated forms, it was assumed that the film produced on cycling was heterogeneous with an inner layer of anhydrous material (which may well be a coating of hydroxide on oxide) and an outer hydrous oxide deposit. The ratio of hydrous to anhydrous material present may well vary during the course of a cycle—this would explain the absence of the peak for the anhydrous material under most conditions on the anodic sweep (such oxide transitions have also been proposed by Macagno et al.[214]). The reactions assumed to be occuring in the components of different activity in these films are as follows:

$$\text{NiO·OH} + \text{H}^+ + e^- = \text{Ni(OH)}_2 \text{ (anhydrous)} \quad (59)$$

$$[\text{Ni(OH)}_{3.5} \cdot n\text{H}_2\text{O}]^{0.5-} (\text{Na}^+)_{0.5} + e^-$$
$$= \text{Ni(OH)}_2 \cdot n\text{H}_2\text{O} + 0.5\text{Na}^+ + 1.5\text{OH}^- \text{ (hydrated)} \quad (60)$$

Again it must be borne in mind here that these processes are occurring in an extended array of oxynickel species. A Nernstian treatment of the unusual potential/pH shift for this system has been published[43]; a brief account of the interaction between the acid-base and redox properties of this system has also been given by Bourgault and Conway.[215] Points requiring further clarification in this system include the somewhat irregular potential/pH variation noted in strong base[213] and the role of the higher oxidation state, Ni(IV), in the oxidized material.[43] The E/pH variation of the Ni(III)/Ni(II) oxide electrode (produced by potential cycling) on open-circuit, in the half-charged state, was found to be somewhat unusual[216]: with relatively thick films a constant value of ca. -78 mV/pH unit (quite similar to that noted earlier with iridium[145]) was observed. In the case of the battery-type material, produced by cathodic electroprecipitation, the anhydrous component in the film appeared to be of increasing importance (at least in terms of open-circuit behavior) with increasing deposit thickness. In cyclic voltammetry experiments with relatively thin films of such material[213] only one anodic and one cathodic peak was observed. The E/pH variations of these transitions were again rather unusual, the value for the former being characteristic of a hydrous deposit while that for the latter being typical of an anhydrous deposit. Such a result again suggests a hydrous overlayer on an anhydrous deposit, the outer material dehydrating at the anodic, and rehydrating at the cathodic, end of the sweep.

Two surprising features of the electrochemistry of the manganese system are the small amount of attention devoted to the behavior of the metal itself and the limited use of cyclic voltammetry in the study of the charge storage reaction of the oxides. The metal itself tends to oxidize spontaneously in an aqueous environment, but, as usual, thick oxide films are more readily prepared[217] by potential cycling in base [$-0.40-1.65$ V (RHE), $40\,\text{mV}\,\text{s}^{-1}$]. The voltammogram recorded for such a substrate (Fig. 12) is somewhat similar to that observed for hydrous iridium oxide in so far as one pair of peaks, evidently corresponding to an Mn(IV)/Mn(III) transition, is observed—again with considerable charge in the anodic sweep at potentials above the main peak (a feature also noticeable in the case of Ni[213]). The degree of hysteresis between the anodic and cathodic peaks is also more marked in the case of Mn as compared with Ir.

The loss of electrochromic behavior of these oxide films on manganese with increasing surface layer thickness, and the drop in charge capacity with increasing sweep rate with films of appreciable thickness, suggest that these deposits are not homogeneous—again there may well be relatively anhydrous microparticles of compact oxide present coated with overlayers of more reactive hydrous material. Cyclic voltammetry studies[218] carried out with base-grown oxide films on manganese over the pH range 9.0-14.0 showed unusual peak potential/pH variation. As outlined in Fig. 13 the observed variations at slow sweep rates (where peak potential values were independent of the latter) were ca. $137\,\text{mV/pH}$ unit (SHE scale) for the anodic sweep and ca. $90\,\text{mV/pH}$ unit for the cathodic sweep. The following reaction scheme was proposed to explain these results:

(a) Main anodic transition

$$[\text{Mn}_2(\text{OH})_9]^{2-} + 4.6\text{OH}^-$$
$$= [\text{Mn}_2\text{O}_{4.6}(\text{OH})_{4.4}]^{5.6-} + 4.6\text{H}_2\text{O} + 2e^- \quad (61)$$

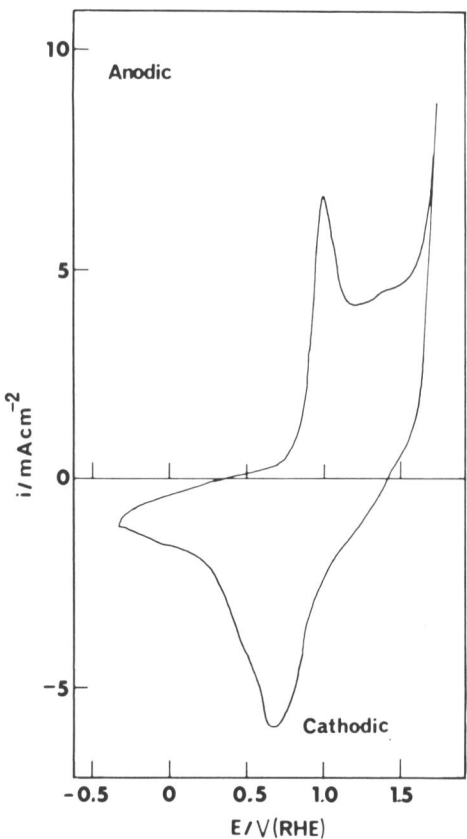

Figure 12. Cyclic voltammogram (-0.30–1.70 V, $40\,\text{mV}\,\text{s}^{-1}$) for a hydrous-oxide-coated manganese electrode in $1.0\,\text{mol}\,\text{dm}^{-3}$ NaOH at 25°C (from Ref. 218, with permission).

(b) Postelectrochemical decomposition step

$$[\text{Mn}_2\text{O}_{4.6}(\text{OH})_{4.4}]^{5.6-} + 4.6\text{H}_2\text{O}$$
$$= [\text{Mn}(\text{OH})_6]^{2-} + 1.6\text{OH}^- \qquad (62)$$

(c) Main cathodic transition

$$2[\text{Mn}(\text{OH})_6]^{2-} + 2e^- = [\text{Mn}_2(\text{OH})_9]^{3-} + 3\text{OH}^- \qquad (63)$$

The observed scheme is clearly speculative—its main function is to account for the observed electrochemical data. The greater poten-

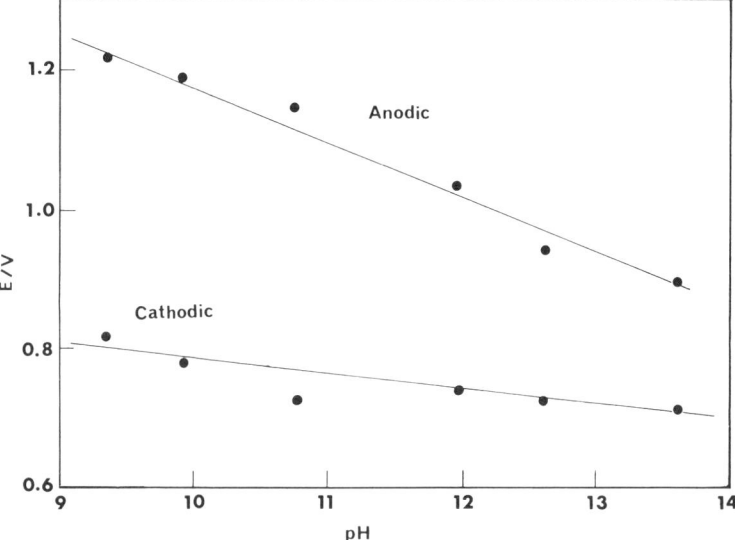

Figure 13. Variation of peak potential with solution pH at a slow sweep rate (ca. 10 mV s^{-1}). Slope values, mV/pH unit (SHE scale): Anodic, 137; cathodic, 87 (from Ref. 218, with permission).

tial/pH response in the case of manganese, as compared with iridium, on the anodic sweep may be attributed to the greater acidity of the manganese hydroxy complex (for the same oxidation state the small first-row transition metal ion is expected to generate a more intense, local, positive electric field). The noninteger values shown in the product of the primary oxidation step, Eq. (61), may be rationalized on the basis that the product obtained here is a polymeric species composed of a mixture of complexes, e.g., $[Mn_2O_4(OH)_5]^{5-}$ and $[Mn_2O_5(OH)_4]^{6-}$, differing slightly in their degree of hydrolysis. The postelectrochemical decomposition step was postulated to account for the significant difference in potential/pH response between the anodic and cathodic sweeps. The possible importance of this unusual behavior of hydrous manganese oxide (an effect which recent experiments in this laboratory have shown to be relevant also to the behavior of anodically deposited, battery-type MnO_2) in terms of the interpretation of the role of manganese as an oxygen gas releasing agent in plant photosynthesis has also been outlined.[218]

3. Tungsten

There is considerable interest in the electrochemistry of tungsten trioxide as the redox behavior of thin films of this material is accompanied by a dramatic, highly reversible, color change (transparent to blue on reduction) which can be used as the basis of an electrochromic display device (a comprehensive review of the use of hydrous oxide materials in this area has been published recently by Dautremont-Smith[219]). Studies of electrochromism in both evaporated and chemically produced tungstic oxide (WO_3) films[220] have shown that extensive reduction, leading to electrocoloration, requires the presence of water within the amorphous oxide phase. Schlotter and Pickelmann[221] have described the active oxide material as a dried gel or colloid (which they referred to as a Xerogel) in which hydroxylated oxide clusters are randomly linked with W-O-W chains; they emphasized the fact that the voids in this material are not to be confused with the regular voids and tunnels found in crystalline tungstic oxides. They also demonstrated that the active oxide exhibited both acidic and cation ion-exchange behavior.

Electrochemical processes for the production of tungstic oxide films on the metal in acid solution include both dc anodization[222] and repetitive potential cycling.[223] Subjecting freshly grown films to repetitive coloration/bleaching cycles usually increases the water content and enhances both the optical and electrical responses of the film. The degree of microdispersion of the film is evidently increased by such treatment which, however, adversely affects the oxide stability. The hydrated film dissolves readily[223] in aqueous media at pH > 5.0; this is regarded as an oxide depolymerization process leading the formation of WO_4^{2-} monomers. Various other cations apart from H^+, e.g., Li^+, Na^+, and Ag^+, can function as counterions[224]—these entering the solid on reduction of W(VI) to W(V) centers in the hydrous oxide

WO_3(hyd.), transparent $+ xM^+ + xe^-$

$$= WO_3^{x-}[M^+]_x(\text{hyd.}), \text{blue} \qquad (64)$$

For use in display devices it appears that these films may have to be operated in moist, nonaqueous solvents to achieve adequate service life.

XIV. CONCLUSION

The study of hydrous oxide systems is clearly of interest in many areas of electrochemistry. The unusually high level of activity of hydrous oxide-coated metal anodes for such industrially important reactions as oxygen[49,169,173,225] and chlorine[164,173,226] gas evolution has frequently been demonstrated. Such enhanced electrocatalytic activity, compared with metal substrates anodized in the conventional manner, appears to be largely a matter of increased contact at the porous oxide surface.[49] In some instances, however (mainly in work on oxygen gas evolution[135,227]—the corresponding process for chlorine seems to be more strongly influenced by diffusion[228]), a decrease in Tafel slope (which suggests a change in mechanism) has been reported on altering from a virtually anhydrous, to a thick hydrated, surface oxide layer. Such a change may be due to the decrease in lattice oxygen (M-O-M) coordination bonds. The oxycations at the surface of a highly hydrated oxide have more H_2O and HO ligands: the latter may either participate directly in the oxygen evolution process, or rearrange more readily to allow formation of a reactive higher oxidation state, i.e., facilitate the setting up of an electrocatalytic cyclic redox process. This enhanced activity is probably not of immediate technological utility as these hydrous films currently lack adequate long-term stability under severe operating conditions. Studies of hydrous oxide systems have, however, already provided useful insight into the mode of operation of the industrially more important anhydrous oxide (i.e., dimensionally stable) anodes. Apart from forming the basis of the Reactive Surface Group Theory[8] of oxide electrocatalysis, work on these systems has shown that the proposed use of the redox potential of oxides as a guide to their electrocatalytic behavior[229] needs to be applied with caution[204] as the transition potential, leading to the generation of the more reactive, higher oxidation state, alters with solution pH—even in terms of the RHE scale.

With regard to models of the oxide-solution interface, the general agreement between the microdispersed or hyperextended viewpoint (as developed largely on the basis of current-voltage studies) and the various nonclassical models (based largely on adsorption studies) is quite striking. Obviously there is a need in the latter case to take into account variations in acid-base strength with change in oxidation state of the central ion at the oxymetal centers. Such variation in the acid-base behavior of oxides is relevant also in other areas, e.g., heterogeneous catalysis, ion-exchange, and soil science: it is worth noting that the idea that oxide systems frequently bear a variable net negative charge is generally accepted in the latter field.[230]

The role of hydration in oxide battery systems seems to be quite important. In several of these systems, e.g., those involving MnO_2[231] and $NiOOH$,[216] the redox potential of the cathode shows a pronounced pH dependence, suggesting that at least the outer regions of these microcrystalline oxide deposits are of a dispersed, hydrated character. It has been suggested[216] that, because of their more open structures, such materials play a most important role in charge storage reactions. An optimum battery oxide material may represent a compromise between microdispersion (which would favor rapid mass transfer) and a degree of compactness adequate to ensure stability and electronic conductivity (an account of dispersion due to defect formation, without emphasis on the change in oxide acidity accompanying the redox transition in the case of battery-type MnO_2, has been published recently by Ruetschi[232]). These comments on the battery materials also apply to those oxides, e.g., IrO_2 and WO_3, that have been rather extensively investigated recently as the basis of electrochromic display devices. Oxide hydration is evidently an undesirable complication in the case of oxides designed for pH-monitoring purposes—mainly because of the variable nature of the potential/pH response under open-circuit conditions.

The role of hydrous oxides in metal oxidation reactions has already been briefly outlined here. Where such materials are involved the thermodynamic data summarized in most current Pourbaix diagrams are inadequate.[233] In such diagrams, at present, lines relating to redox processes where both the oxidized and reduced form of the couple are insoluble are generally assumed to involve a conventional, $2.303(RT/F)$ V/pH unit, potential variation with pH. It is only necessary to consider one example, namely, the difficulty of reducing hydrous oxide films on platinum in base[27] (where there is evidently little inhibition of kinetic origin), to appreciate the need to extend such diagrams to take hydrous oxide behavior into account.

The fact that some hydrous oxide growth can precede compact monolayer oxide formation in certain circumstances under potential sweep conditions seems to be a matter of fundamental significance. There have been a number of quite unusual voltammograms reported recently for various noble metals where specific faces of single crystals (freshly prepared, exposure to oxygen or an oxidizing environment being avoided) were examined in highly purified aqueous solutions. The recorded very sharp, reversible, anodic and cathodic peaks (usually observed within the range 0.1–0.5 V (RHE) in the case of platinum[234]) were attributed initially by Clevalier et al.[235,236] to hydrogen deposition/removal processes. However, Scherson and Kolb[237] recently reported similar sharp peaks in the case of gold, the peaks in this case occurring at a potential [ca. 1.0 V (RHE) in acid solution] where hydrogen adsorption is quite improbable. While the latter authors attributed the unusual peaks in question here to anion adsorption effects, it is clear from the potentials involved (both in the case of Pt and Au) that reversible hydrous oxide formation, probably at high energy adatom sites generated during the high-temperature pretreatment,[238] could well be responsible. Competition between OH^- and other anions for coordination at metal sites (especially at low pH) may account for the anion sensitivity of these reactions.

It is clear from this brief survey that the behavior of hydrous oxides is of quite widespread interest. There is at the present time a need for more detailed data as to the structure and composition of these hydrous materials. As pointed out recently by Thomas,[239] the high level of disorder and the variability of composition may limit the applicability of X-ray crystallography. Among the suggested alternatives listed by this author are solid state, magic-angle-spinning NMR, X-ray photoelectron spectroscopy, high-resolution electron microscopy, and neutron diffraction.

ADDENDUM

There are many similarities (e.g., in charge storage and electrochromic behavior, and in transport mechanisms) between inorganic hydrous oxide films and the largely organic redox polymer-modified electrodes which are also of much interest at the present time. Of special interest is the appearance in certain instances[240] with the latter systems of a similar type of unusual pH shift (as mentioned in Section III for oxides) for the reversible potential of the charge storage reaction with change in solution pH. The general interpretation in both cases is similar; the number of electrons and protons (or hydroxide ions) involved in the electrode process differ; therefore, in either the reduced or the oxidized form (or possibly both) the polymer material exists in a charged state.

More important from a practical viewpoint (e.g., in the use of organic compounds in fuel cell systems) is the realization[241] that the oxidation of many organic compounds at noble metal anodes in aqueous media involves an interfacial cyclic redox mechanism based on metal/hydrous oxide couples. The organic species may be activated by interaction with adatoms, but the actual oxidation in many instances, especially in the case of gold, occurs only in the presence of the hydrous oxide material—the latter, i.e., the effective oxidizing agent, being generated in situ at quite low surface coverage.

While most of the hydrous oxides considered in this review appear to be anionic in character (the central metal ion being in a high, II to IV, oxidation state), cationic films can apparently be produced[242] on cycling Cu, Ag, or Au in aqueous media. In such films the metal ion is in the +1 oxidation state and, due to the low charge, the OH^- counterion is easily displaced not only by other anions but also by organic bases such as pyridine. The relevance of this type of behavior to the interpretation of certain aspects of surface enhanced Raman scattering (SERS) phenomena was pointed out recently[242] (SERS effects are particularly marked for the three metals in question here).

An extensive account of active MnO_2 was published recently.[243] This material is not extensively hydrated; however, hydration accompanying rearrangement of the lattice on reduction to (or beyond) the $MnO_{1.5}$ state may well contribute[244] to the nonrechargeability of alkaline Zn/MnO_2 battery systems. The use of high-vacuum techniques, e.g., XPS,[245] to determine the composition of hydrous materials in general is fraught with the problem of dehydration (and lattice rearrangement accompanying same). It may be pointed out also that it is important, when examining hydrous films, to grow the material under appropriate conditions; severe polarization may lead, e.g., in the case of Pt, to a thick film of compact, rather than hyperextended, oxide.

REFERENCES

[1] R. S. Alwitt, in *Oxides and Oxide Films*, Vol. 4, Ed. by J. W. Diggle and A. K. Vijh, Marcel Dekker, New York, 1976, pp. 169–254.
[2] P. Oliva, J. Leonardi, J. F. Laurent, C. Delmas, J. J. Braconnier, M. Figlarz, and F. Fievet, *J. Power Sources* **8** (1982) 229.
[3] G. Halpert, *J. Power Sources* **12** (1984) 177.
[4] L. D. Burke and D. P. Whelan, *J. Electroanal. Chem.* **162** (1984) 121.
[5] J. D. E. McIntyre, W. F. Peck, and S. Nakahara, *J. Electrochem. Soc.* **127** (1980) 1264.
[6] L. D. Burke and R. A. Scannell, *Plat. Metals Rev.* **28** (1984) 56.
[7] K. S. Kang and J. L. Shay, *J. Electrochem. Soc.* **130** (1983) 766.
[8] L. D. Burke and J. F. Healy, *J. Electroanal. Chem.* **124** (1981) 327.
[9] L. D. Burke and D. P. Whelan, *J. Electroanal. Chem.* **124** (1981) 333.
[10] H. Beer, *Chem. Ind.* **14** (1978) 491.
[11] D. N. Buckley and L. D. Burke, *J. Chem. Soc. Faraday Trans. I* **71** (1975) 1447.
[12] D. A. J. Rand and R. Woods, *J. Eectroanal. Chem.* **55** (1974) 375.
[13] W. Bold and M. Breiter, *Electrochim. Acta.* **5** (1961) 169.
[14] G. Belanger and A. K. Vijh, in *Oxide and Oxide Films*, Vol. 5, Ed. by A. K. Vijh, Marcel Dekker, New York, 1977, pp. 1–104.
[15] R. Woods, in *Electroanalytical Chemistry*, Vol. 9, Ed. by A. J. Bard, Marcel Dekker, New York, 1976, pp. 1–162.
[16] L. D. Burke, *Electrodes of Conductive Metallic Oxides*, Part A, Ed. by S. Trasatti, Elsevier, Amsterdam, 1980, pp. 141–181.
[17] B. E. Conway and H. Angerstein-Kozlowska, *Acc. Chem. Res.* **14** (1981) 49.
[18] V. S. Bagotzky and M. R. Tarasevich, *J. Electroanal. Chem.* **101** (1979) 1.
[19] R. G. Burns and V. M. Burns, in *Manganese Dioxide Symposium*, Vol. 2, Ed. by B. Schumm, H. M. Joseph, and A. Kozawa, The I.C. MnO_2 Sample Office, Box 6116, Cleveland, Ohio 44101, U.S.A. 1980, pp. 97–112.
[20] S. D. James, *J. Electrochem. Soc.* **116** (1969) 1681.
[21] J. Balej and O. Spalek, *Coll. Czech. Chem. Commun.* **37** (1972) 499.
[22] S. Shibata, *J. Electroanal. Chem.* **89** (1978) 37.
[23] L. D. Burke and M. McRann, *J. Electroanal. Chem.* **125** (1981) 387.
[24] D. N. Buckley, L. D. Burke, and J. K. Mulcahy, *J. Chem. Soc. Faraday Trans. I* **72** (1976) 1896.
[25] D. F. Pickett and J. T. Malloy, *J. Electrochem. Soc.* **125** (1978) 1026.
[26] G. W. Nichols, *Trans. Electrochem. Soc.* **62** (1932) 393.
[27] L. D. Burke and M. B. C. Roche, *J. Electroanal. Chem.* **164** (1984) 315.
[28] L. D. Burke, M. M. McCarthy, and M. B. C. Roche, *J. Electroanal. Chem.* **167** (1984) 291.
[29] L. D. Burke and E. J. M. O'Sullivan, *J. Electroanal. Chem.* **93** (1978) 11.
[30] B. E. Conway and J. Mozota, *Electrochim. Acta.* **28** (1983) 9.
[31] M. A. Anderson and A. J. Rubin, *Absorption of Inorganics at Solid–Liquid Interfaces*, Ann Arbor Science Publishers, Michigan, 1981.
[32] S. M. Ahmed, *J. Phys. Chem.* **73** (1969) 3546.
[33] K. Tanabe, *Solid Acids and Bases*, Academic, New York, 1970.
[34] G. Beni, C. E. Rice, and J. L. Shay, *J. Electrochem. Soc.* **127** (1980) 1342.
[35] J. D. E. McIntyre, S. Basu, W. F. Peck, W. L. Brown, and M. W. Augustyniak, *Phys. Rev. B.* **25** (1982) 7242.
[36] H. Angerstein-Kozlowska, J. Klinger, and B. E. Conway, *J. Electroanal. Chem.* **75** (1977) 45.
[37] H. G. Scott, *Acta Crystallogr. B.* **35** (1979) 3014.

[38] L. D. Burke, M. E. Lyons, E. J. M. O'Sullivan, and D. P. Whelan, *J. Electroanal. Chem.* **122** (1981) 403.
[39] J. Burgess, *Metal Ions in Solution*, Wiley, New York, 1978, p. 266.
[40] D. J. G. Ives, *Reference Electrodes, Theory and Practice*, Ed. by D. J. G. Ives and G. J. Janz, Academic, New York, 1961, pp. 322-392.
[41] L. D. Burke, M. E. Lyons, and D. P. Whelan, *J. Electroanal. Chem.* **139** (1982) 131.
[42] L. D. Burke and E. J. M. O'Sullivan, *J. Electroanal. Chem.* **129** (1981) 133.
[43] L. D. Burke and T. A. M. Twomey, in *Proceeding of the Symposium on the Nickel Electrode*, Ed. by R. G. Gunther and S. Gross, The Electrochemical Society, Pennington, New Jersey, 1982, pp. 75-96.
[44] L. D. Burke and M. B. C. Roche, *J. Electroanal. Chem.* **159** (1983) 89.
[45] L. D. Burke and T. A. M. Twomey, *J. Electroanal. Chem.* **162** (1984) 101.
[46] A. F. Wells, *Structural Inorganic Chemistry*, Fifth Edition, Clarendon Press, Oxford, 1984, pp. 496-652.
[47] W. J. Bernard and J. J. Randall, quoted in Ref. 1, p. 207.
[48] P. W. Selwood, *Magnetochemistry*, Interscience, New York, 1964, pp. 337-341.
[49] L. D. Burke and E. J. M. O'Sullivan, *J. Electroanal. Chem.* **117** (1981) 155.
[50] S. Srinivasan and E. Gileadi, *Electrochim. Acta.* **11** (1966) 321.
[51] H. Angerstein-Kozlowska, J. Klinger, and B. E. Conway, *J. Electroanal. Chim.* **75** (1977) 45.
[52] H. Angerstein-Kozlowska and B. E. Conway, *J. Electroanal. Chim.* **95** (1979) 1, 9.
[53] J. Mozota and B. E. Conway, *Electrochim. Acta.* **28** (1983) 1.
[54] E. Laviron, *J. Electroanal. Chem.* **100** (1979) 263; **101** (1979) 19; **105** (1979) 25.
[55] E. Laviron, in *Electroanalytical Chemistry*, Vol. 12, Ed. by A. J. Bard, Marcel Dekker, New York, 1982, pp. 53-57.
[56] E. Laviron, *J. Electroanal. Chem.* **112** (1980) 1, 11.
[57] E. Laviron, *J. Electroanal. Chem.* **122** (1981) 37.
[58] C. P. Andrieux and J. M. Saveant, *J. Electroanal. Chem.* **111** (1980) 377.
[59] A. J. Bard and L. A. Faulkner, *Electrochemical Methods*, Wiley, New York, 1980, pp. 675-697.
[60] P. J. Peerce and A. J. Bard, *J. Electroanal. Chem.* **114** (1980) 89.
[61] K. Aoki, K. Tokuda, and H. Matsuda, *J. Electroanal. Chem.* **146** (1983) 417.
[62] K. Aoki, K. Tokuda, and H. Matsuda, *J. Electroanal. Chem.* **160** (1984) 33.
[63] R. V. Churchill, *Operational Mathematics*, 3rd edn., McGraw-Hill, New York, 1972; G. E. Roberts and H. Kaufman, *Tables of Laplace Transforms*, Saunders, Philadelphia, 1966.
[64] E. T. Whittaker and G. N. Watson, *A Course of Modern Analysis*, 4th edn., Cambridge University Press, 1927, p. 464.
[65] R. S. Nicholson and I. Shain, *Anal. Chem.* **36** (1964) 706; **37** (1965) 178.
[66] C. R. Christensen and F. C. Anson, *Anal. Chem.* **35** (1963) 205.
[67] M. Abramowitz and I. A. Stegun, *Handbook of Mathematical Functions*, Dover, New York, 1972, p. 299.
[68] A. J. Bard and L. R. Faulkner, *Electrochemical Methods*, Wiley, New York, 1980, pp. 249-279.
[69] D. M. Oglesby, S. H. Omang, and C. N. Reilly, *Anal. Chem.* **37** (1965) 1312.
[70] W. J. Albery, M. G. Boutelle, P. J. Colby, and A. R. Hillman, *J. Electroanal. Chem.* **133** (1982) 135.
[71] B. Reichman and A. J. Bard, *J. Electrochem. Soc.* **126** (1979) 583.
[72] D. M. MacArthur, *J. Electrochem. Soc.* **117** (1970) 729.
[73] K. J. Gallagher and D. N. Phillips, *Trans. Faraday Soc.* **64** (1968) 785.
[74] Y. K. Wei and R. B. Bernstein, *J. Phys. Chem.* **65** (1959) 738.
[75] A. B. Scott, *J. Electrochem. Soc.* **107** (1960) 941.

[76] J. P. Gabano, *J. Electrochem. Soc.* **117** (1970) 147.
[77] R. W. Murray, *Phil. Trans. R. Soc. London* **A302** (1981) 253.
[78] R. M. Reeves, in *Modern Aspects of Electrochemistry*, Ed. by J. O'M. Bockris and B. E. Conway, Plenum Press, New York, 1974, Chap. 4.
[79] J. Lyklema and J. Th. G. Overbeck, *J. Colloid Sci.* **16** (1961) 595; J. Lyklema, *Trans. Faraday Soc.* **59** (1963) 418.
[80] G. R. Wiese, R. O. James, D. E. Yates, and T. W. Healy, in *International Review of Science, Electrochemistry, Physical Chemistry, Series Two*, Vol. 6, Consultant Ed., A. D. Buckingham, Volume Ed., J. O'M. Bockris, Butterworths, London, 1976, p. 54.
[81] D. N. Furlong, D. E. Yates, and T. W. Healy, *Electrodes of Conductive Metallic Oxides*, Part B, Ed. by S. Trasatti, Elsevier, Amsterdam, 1981, p. 367.
[82] S. M. Ahmed and D. Maksimov, *J. Colloid Interface Sci.* **29** (1969) 97.
[83] S. M. Ahmed, *Can. J. Chem.* **44** (1966) 1663, 2769.
[84] L. Blok and P. L. De Bruyn, *J. Colloid Interface Sci.* **32** (1970) 518, 527, 533.
[85] S. Levine and A. L. Smith, *Disc. Faraday Soc.* **52** (1971) 290.
[86] R. J. Hunter and H. J. L. Wright, *J. Colloid Interface Sci.* **37** (1971) 564.
[87] H. J. L. Wright and R. J. Hunter, *Aust. J. Chem.* **26** (1973) 1183, 1191.
[88] M. Borkovec and J. Westall, *J. Electroanal. Chem.* **150** (1983) 325.
[89] Th. F. Tadros and J. Lyklema, *J. Electroanal. Chem.* **17** (1968) 267.
[90] Th. F. Tadros and J. Lyklema, *J. Electroanal. Chem.* **22** (1969) 1.
[91] J. Lyklema, *J. Electroanal. Chem.* **18** (1968) 341.
[92] J. Lyklema, *Croat. Chem. Acta.* **43** (1971) 249.
[93] J. W. Perram, *J. Chem. Soc. Faraday II* **2** (1973) 993.
[94] J. W. Perram, R. J. Hunter, and H. J. L. Wright, *Chem. Phys. Lett.* **23** (1973) 265.
[95] J. W. Perram, R. J. Hunter, and H. J. L. Wright, *Aust. J. Chem.* **27** (1974) 461.
[96] M. J. Dignam, *Can. J. Chem.* **56** (1978) 595.
[97] M. J. Dignam and R. K. Kalia, *Surf. Sci.* **100** (1980) 154.
[98] T. W. Healy and L. R. White, *Adv. Colloid Interface Sci.* **9** (1978) 303.
[99] D. E. Yates, S. Levine, and T. W. Healy, *J. Chem. Soc. Faraday I* **70** (1974) 1807.
[100] J. A. Davis, A. P. James, and J. O. Leckie, *J. Colloid Interface Sci.* **63** (1978) 480.
[101] A. Daghetti, G. Lodi, and S. Trasatti, *Mater. Chem. Phys.* **8** (1983) 1.
[102] W. Smit, C. L. M. Holten, H. N. Stein, J. J. M. De Goeig, and H. M. J. Theelen, *J. Colloid Interface Sci.* **63** (1978) 120.
[103] H. Angerstein-Kozlowska, B. E. Conway, and W. B. A. Sharp, *J. Electroanal. Chem.* **43** (1973) 9.
[104] H. Angerstein-Kozlowska, B. E. Conway, B. Barnatt, and J. Mozota, *J. Electroanal. Chem.* **28** (1983) 1.
[105] P. N. Ross, *J. Electroanal. Chem.* **76** (1977) 139.
[106] A. J. Appleby, *J. Electrochem. Soc.* **120** (1973) 1205.
[107] K. J. Vetter and D. Berndt, *Z. Electrochem.* **62** (1958) 378.
[108] W. Visscher and M. A. V. Devanathan, *J. Electrochem. Soc.* **8** (1964) 127.
[109] K. J. Vetter and J. W. Schultze, *J. Electroanal. Chem.* **34** (1972) 131.
[110] K. J. Vetter and J. W. Schultze, *J. Electroanal. Chem.* **34** (1972) 141.
[111] J. W. Schultze, *Z. Phys. Chem. N.F.* **73** (1970) 29.
[112] J. L. Ord and F. C. Ho, *J. Electrochem. Soc.* **118** (1971) 46.
[113] L. B. Harris and A. Damjanovic, *J. Electrochem. Soc.* **122** (1975) 593.
[114] A. Damjanovic, A. R. Ward, B. Ulrick, and M. O'Jea, *J. Electrochem. Soc.* **122** (1975) 471.
[115] V. I. Birss and A. Damjanovic, *J. Electrochem. Soc.* **130** (1983) 1688.
[116] T. Biegler and R. Woods, *J. Electroanal. Chem.* **20** (1973) 1969.
[117] T. Biegler, D. A. J. Rand, and R. Woods, *J. Electroanal. Chem.* **29** (1971) 269.
[118] W. Visscher and M. Blijlevens, *J. Electroanal. Chem.* **47** (1973) 363.

119. A. T. Kuhn and T. H. Randle, *J. Chem. Soc. Faraday I* **81** (1985) 403.
120. D. Gilroy, *J. Electroanal. Chem.* **71** (1976) 257.
121. D. Gilroy, *J. Electroanal. Chem.* **83** (1977) 329.
122. M. Fleischmann and H. R. Thirsk, in *Advances in Electrochemistry and Electrochemical Engineering*, Vol. 3, Ed. by P. Delaney and C. W. Tobias, Interscience, New York, 1963, p. 123.
123. K. Nagel and H. Dietz, *Electrochim. Acta.* **4** (1961) 1.
124. S. Shibata and M. P. Sumino, *Electrochim. Acta* **17** (1972) 2215.
125. A. Kozowa, *J. Electroanal. Chem.* **8** (1964) 20.
126. S. Shibata and M. P. Sumino, *Electrochim. Acta* **26** (1981) 517, 1587.
127. S. Shibata, *Bull. Chem. Soc. Jpn* **40** (1967) 696.
128. S. Shibata, *Electrochim. Acta* **17** (1972) 395.
129. S. Altmann and R. H. Busch, *Trans. Faraday Soc.* **45** (1949) 720.
130. A. Kozawa, *J. Electroanal Chem.* **8** (1964) 20.
131. J. S. Hammond and N. Winograd, *J. Electroanal. Chem.* **78** (1977) 55.
132. G. C. Allen, P. M. Tucker, A. Capon, and R. Parsons, *J. Electroanal. Chem.* **50** (1974) 335.
133. L. D. Burke and C. M. B. Roche, *J. Electroanal. Chem.* **137** (1982) 175.
134. L. D. Burke and C. M. B. Roche, *J. Electrochem. Soc.* **129** (1982) 2641.
135. L. D. Burke and M. B. C. Roche, *J. Electroanal. Chem.* in press.
136. P. Lorbeer and W. J. Lorenz, *Corr. Sci.* **21** (1981) 79.
137. A. C. Chialvo, W. E. Triaca, and A. J. Arvia, *J. Electroanal. Chem.* **146** (1983) 93.
138. F. A. Lewis, *The Palladium Hydride System*, Academic, New York, 1967.
139. D. A. J. Rand and R. Woods, *J. Electroanal. Chem.* **35** (1972) 209.
140. D. C. Johnson, D. T. Napp, and S. Bruckenstein, *Electrochim. Acta* **15** (1970) 1493.
141. S. H. Cadle, *J. Electrochem. Soc.* **121** (1974) 645.
142. A. E. Bolzan, M. E. Martins, and A. J. Arvia, *J. Electroanal. Chem.* **172** (1984) 221.
143. K. Gossner and E. Mizera, *J. Electroanal. Chem.* **125** (1981) 347.
144. S. E. S. El Wakkad and A. M. Shams El Din, *J. Chem. Soc.* (1954) 3094.
145. L. D. Burke, J. K. Mulcahy, and D. P. Whelan, *J. Electroanal. Chem.* **163** (1984) 117.
146. H. Angerstein-Kozlowska, B. E. Conway, B. Barnett, and J. Mozota, *J. Electroanal. Chem.* **100** (1979) 417.
147. D. A. J. Rand and R. Woods, *J. Electroanal. Chem.* **31** (1971) 29.
148. K. S. Kim, A. F. Gossmann, and N. Winograd, *Anal. Chem.* **46** (1974) 197.
149. M. Pourbaix, *Atlas of Electrochemical Equilibria in Aqueous Solutions*, Pergamon, Oxford, 1966.
150. S. E. S. El Wakkad and A. M. Shams El Din, *J. Chem. Soc.* (1954) 3098.
151. G. M. Schmid and R. N. O'Brien, *J. Electrochem. Soc.* **111** (1964) 832.
152. R. S. Sirohi and M. A. Genshaw, *J. Electrochem. Soc.* **116** (1969) 910.
153. K. Ogura, H. Haruyama, and K. Nugasaki, *J. Electrochem. Soc.* **118** (1971) 531.
154. D. M. McArthur, *J. Electrochem. Soc.* **119** (1972) 672.
155. R. Cordova O., M. E. Martins, and A. J. Arvia, *J. Electrochem. Soc.* **126** (1979) 1172.
156. D. W. Kirk, F. R. Foulkes, and W. F. Graydon, *J. Electrochem. Soc.* **127** (1980) 1069.
157. L. D. Burke, M. M. McCarthy, and R. A. Scannell, *J. Electroanal. Chem.*, in press.
158. D. Dickertmann, J. W. Schultze and K. J. Vetter, *J. Electroanal. Chem.* **97** (1979) 123.
159. G. Gruneberg, *Electrochim. Acta* **10** (1965) 339.
160. M. M. Lohrengel and J. W. Schultze, *Electrochim. Acta* **21** (1976) 957.
161. A. I. Krasil'shchikov, *Russ. J. Phys. Chem.* **37** (1963) 273.
162. L. D. Burke and G. P. Hopkins, *J. Appl. Electrochem.* **14** (1984) 679.
163. A. C. Chialvo, W. E. Triaca, and A. J. Arvia, *J. Electroanal. Chem.* **171** (1984) 303.

[164] J. Mozota and B. E. Conway, *J. Electrochem. Soc.* **128** (1981) 2142.
[165] D. N. Buckley and L. D. Burke, *J. Chem. Soc. Faraday I* **72** (1976) 2431.
[166] R. Woods, *J. Electroanal. Chem.* **49** (1974) 217.
[167] M. S. Cruz, T. F. Otero, and S. U. Zanartu, *J. Electroanal. Chem.* **158** (1983) 375.
[168] D. Michell, D. A. J. Rand, and R. Woods, *J. Electroanal. Chem.* **84** (1977) 117.
[169] S. Gottesfeld and S. Srinivasan, *J. Electroanal. Chem.* **86** (1978) 89.
[170] S. H. Glarum and J. H. Marshall, *J. Electrochem. Soc.* **127** (1980) 1467.
[171] S. Hackwood, A. H. Dahem, and G. Beni, *Phys. Rev. B* **26** (1982) 471.
[172] J. Augustynski, M. Kondelka, J. Sanchez, and B. E. Conway, *J. Electroanal.* **160** (1984) 233.
[173] V. Birss, R. Myers, H. Angerstein-Kozlowska, and B. E. Conway, *J. Electrochem. Soc.* **131** (1984) 1502.
[174] S. Gottesfeld, *J. Electrochem. Soc.* **127** (1980) 1922.
[175] M. F. Yuen, I. Lauks, and W. C. Dautremont-Smith, *Solid State Ionics* **11** (1983) 19.
[176] S. Gottesfeld and J. D. E. McIntyre, *J. Electrochem. Soc.* **126** (1979) 742.
[177] L. D. Burke and R. A. Scannell, *J. Electroanal. Chem.* **175** (1984) 119.
[178] P. W. Selwood, *Magnetochemistry*, Interscience, New York, 1964, pp. 166-167.
[179] S. Ardizzone, D. Lettieri, and S. Trasatti, *J. Electroanal. Chem.* **146** (1983) 431.
[180] J. D. E. McIntyre, S. Basu, W. F. Peck, W. L. Browne, and W. M. Augustyniak, *Solid State Ionics* **5** (1981) 359.
[181] J. D. E. McIntyre, *J. Electrochem. Soc.* **126** (1979) 2171.
[182] G. Beni and J. L. Shay, *Appl. Phys. Lett.* **33** (1978) 567.
[183] S. Hackwood, G. Beni, W. C. Dautremont-Smith, L. M. Schiavone, and J. L. Shay, *Appl. Phys. Lett.* **37** (1980) 965.
[184] C. E. Rice, *Appl. Phys. Lett.* **35** (1979) 563.
[185] J. O. Zerbino and A. J. Arvia, *J. Electrochem. Soc.* **126** (1979) 93.
[186] R. Woods, *Israel J. Chem.* **18** (1979) 118.
[187] M. S. Ureta-Zanartu and J. H. Zagal, *Ext. Abs. Spring Meeting Montreal (1982)*, The Electrochemical Society, Pennington, New Jersey, 82-1, 1982, p. 1082.
[188] S. Gottesfeld, *J. Electrochem. Soc.* **127** (1980) 272.
[189] L. D. Burke and E. J. M. O'Sullivan, *J. Electroanal. Chem.* **112** (1980) 247.
[190] M. K. Aston, D. A. J. Rand, and R. Woods, *J. Electroanal. Chem.* **163** (1984) 199.
[191] S. Hadzi-Jordanov, H. Angerstein-Kozlowska, and B. E. Conway, *J. Electroanal. Chem.* **60** (1975) 359.
[192] S. Hadzi-Jordanov, H. Angerstein-Kozlowska, M. Vukovic, and B. E. Conway, *J. Electrochem. Soc.* **125** (1978) 1471.
[193] S. Hadzi-Jordanov, H. Angerstein-Kozlowska, M. Vukovic, and B. E. Conway, *J. Phys. Chem.* **81** (1977) 2271.
[194] L. D. Burke and J. K. Mulcahy, *J. Electroanal. Chem.* **73** (1976) 207.
[195] L. D. Burke, J. K. Mulcahy, and S. Venkatesan, *J. Electroanal. Chem.* **81** (1977) 339.
[196] D. Michell, D. A. J. Rand, and R. Woods, *J. Electroanal. Chem.* **89** (1978) 1046.
[197] K. S. Kim and N. Winograd, *J. Cat.* **35** (1974) 66.
[198] L. D. Burke and D. P. Whelan, *J. Electroanal. Chem.* **103** (1979) 179.
[199] K. W. Lam, K. E. Johnson, and D. G. Lee, *J. Electrochem. Soc.* **125** (1978) 1069.
[200] D. Galizzioli, F. Tantardini, and S. Trasatti, *J. Appl. Electrochem.* **4** (1974) 57.
[201] T. Doblhofer, M. Metikos, Z. Ogumi, and H. Gerischer, *Ber. Bunsenges. Phys. Chem.* **82** (1978) 1046.
[202] L. D. Burke and O. J. Murphy, *J. Electroanal. Chem.* **109** (1980) 379.
[203] L. D. Burke and M. E. Lyons, *J. Electroanal. Chem.* **198** (1986) 347.
[204] L. D. Burke and T. A. M. Twomey, *J. Electroanal. Chem.* **167** (1984) 285.
[205] S. C. Tjong and E. Yeager, *J. Electrochem. Soc.* **128** (1981) 2551.

[206] O. J. Murphy, T. E. Pou, and J. O'M. Bockris, *J. Electrochem. Soc.* **131** (1984) 2785.
[207] M. E. G. Lyons and L. D. Burke, *J. Electroanal. Chem.* **170** (1984) 377.
[208] L. D. Burke, M. E. Lyons, and O. J. Murphy, *J. Electroanal. Chem.* **132** (1982) 247.
[209] G. W. D. Briggs, in *Specialist Periodical Reports: Electrochemistry*, Vol. 4, Chemical Society, London, 1974, pp. 33-54.
[210] S. Le Bihan and M. Figlarz, *Electrochim. Acta* **18** (1973) 123.
[211] R. S. McEwen, *J. Phys. Chem.* **75** (1971) 1782.
[212] L. D. Burke and D. P. Whelan, *J. Electroanal. Chem.* **109** (1980) 385.
[213] L. D. Burke and T. A. M. Twomey, *J. Electroanal. Chem.* **134** (1982) 353.
[214] V. A. Macagno, J. R. Vilche, and A. J. Arvia, *J. Electrochem. Soc.* **129** (1982) 301.
[215] P. L. Bourgault and B. E. Conway, *Can. J. Chem.* **38** (1960) 1557.
[216] L. D. Burke and T. A. M. Twomey, *J. Power Sources* **12** (1984) 203.
[217] L. D. Burke and O. J. Murphy, *J. Electroanal. Chem.* **109** (1980) 373.
[218] L. D. Burke and M. J. Ahern, *J. Electroanal. Chem.* **183** (1985) 183.
[219] W. C. Dautremont-Smith, *Displays* **3** (1982) 3 and 67.
[220] R. Hurditch, *Electron. Lett.* **11** (1975) 142.
[221] P. Schlotter and L. Pickelmann, *J. Electron. Mater.* **11** (1982) 207.
[222] B. Reichmann and A. J. Bard, *J. Electrochem. Soc.* **126** (1979) 583.
[223] L. D. Burke and D. P. Whelan, *J. Electroanal. Chem.* **135** (1982) 55.
[224] O. Bohnke, C. Bohnke, G. Robert, and B. Carquille, *Solid State Ionics* **6** (1982) 121 and 267.
[225] L. D. Burke and E. J. M. O'Sullivan, *J. Electroanal. Chem.* **97** (1979) 123.
[226] J. Mozota, M. Vukovic, and B. E. Conway, *J. Electroanal. Chem.* **114** (1980) 153.
[227] E. J. M. O'Sullivan and L. D. Burke, in *Proceedings of the Symposium on Electrocatalysis*, Ed. by W. E. O'Grady, P. N. Ross, and F. G. Will, The Electrochemical Society, Pennington, New Jersey, 1982, pp. 209-223.
[228] L. D. Burke and J. F. O'Neill, *J. Electroanal. Chem.* **101** (1979) 341.
[229] A. C. C. Tseung and S. Jasem, *Electrochim. Acta* **22** (1977) 31.
[230] J. E. Odom, *Phil. Trans. R. Soc. London Ser. A* **311** (1984) 391.
[231] L. D. Burke and M. J. Ahern, *J. Electrochem. Soc.* **132** (1985) 2662.
[232] P. Ruetschi, *J. Electrochem. Soc.* **131** (1984) 2737.
[233] L. D. Burke and R. A. Scannell, *Equilibrium Diagrams Localized Corrosion* (*Proceeding of an International Symposium Honoring Marcel Pourbaix on his Eightieth Birthday*), Ed. by R. P. Frankenthal and J. Kruger, The Electrochemical Society, Pennington, New Jersey, 1984, pp. 135-147.
[234] S. Motoo and N. Furuya, *J. Electroanal. Chem.* **167** (1984) 309.
[235] J. Clavilier, R. Faure, G. Guinet, and R. Durand, *J. Electroanal. Chem.* **107** (1980) 205.
[236] J. Clavilier, *J. Electroanal. Chem.* **107** (1980) 211.
[237] D. A. Scherson and D. M. Kolb, *J. Electroanal. Chem.* **176** (1984) 353.
[238] R. Parsons, *J. Electroanal. Chem.* **118** (1981) 3.
[239] J. M. Thomas, *Phil. Trans. R. Soc. London Ser. A* **311** (1984) 271.
[240] C. Degrand and L. L. Miller, *J. Electroanal. Chem.* **117** (1981) 267.
[241] L. D. Burke and V. J. Cunnane, *J. Electrochem. Soc.* **133** (1986) 1657.
[242] L. D. Burke, M. I. Casey, V. J. Cunnane, O. J. Murphy, and T. A. M. Twomey, *J. Electroanal. Chem.* **189** (1985) 353.
[243] The Second Battery Material Symposium, Graz, September 1985, Ed. by K. V. Kordesch and A. Kozawa, The IBA Office, 14731 Sprengel Avenue, Cleveland, Ohio 44135, 1986.
[244] L. D. Burke and M. J. G. Ahern, Ref. 243, pp. 75-97.
[245] M. Peuckert, *J. Electroanal. Chem.* **185** (1985) 379.

5

Chemistry and Chemical Engineering in the Chlor-Alkali Industry

F. Hine

Nagoya Institute of Technology, Nagoya 466, Japan

B. V. Tilak and K. Viswanathan

Occidental Chemical Corporation, Research Center, Grand Island, New York 14072

I. INTRODUCTION

Chlor-alkali technology is one of the largest electrochemical industries in the world, the main products being, as the name implies, chlorine and sodium hydroxide (caustic) generated simultaneously by the electrolysis of sodium chloride. It is an energy-intensive process and is the second largest consumer of electricity among electrolytic industries next to aluminum. Chlorine and caustic are indispensable intermediates in the chemical industry, chlorine being used in the manufacture of polyvinyl chloride and chlorinated organic compounds which are precursors in the production of polymers such as polyesters and urethanes. Caustic soda, on the other hand, finds extensive use in mineral processing, the pulp and paper industry, and textile and glass manufacturing operations.

Basic and applied aspects related to chlor-alkali technology have been addressed in detail in several recent publications (see Refs. 1-10) without much discussion on the engineering features involved in these technologies. Hence, this chapter is devoted to the chemical engineering aspects with emphasis on the strongly emerging ion-exchange membrane cell technology. Underlying principles in diaphragm and mercury cell chlor-alkali technologies have been addressed for the sake of completeness. Thus, Section II is devoted to the basics involved in chlor-alkali production, Section III to the manufacturing processes, and Section IV to the electrode materials and electrode processes. Engineering design aspects related to amalgam decomposition and transport across separators are highlighted in Section V, and the engineering aspects related to the ion-exchange membrane technology are discussed in Section VI. Kinetic and electrocatalytic aspects involved in chlorine and hydrogen evolution reactions were not extensively addressed in Section IV and the reader is recommended to Refs. 5 and 11–15 for details.

II. CHEMICAL AND ELECTROCHEMICAL PRINCIPLES INVOLVED IN CHLOR-ALKALI PRODUCTION

About 75% of total chlorine capacity in the world is produced electrolytically using diaphragm and membrane cells, while 20% is made using mercury cells. In all these technologies, chlorine and caustic are produced simultaneously by electrolysis of sodium chloride (commonly termed "brine") following the overall reaction

$$2NaCl + 2H_2O \rightarrow 2NaOH + Cl_2\uparrow + H_2\uparrow \qquad (1)$$

the component electrochemical reactions in diaphragm and membrane cells being

$$2Cl^- \rightarrow Cl_2 + 2e \quad \text{(at the anode)} \qquad (2)$$

$$2H_2O + 2e \rightarrow H_2 + 2OH^- \quad \text{(at the cathode)} \qquad (3)$$

The main difference in these technologies lies in the manner by which chlorine and caustic are prevented from mixing with each other to ensure generation of pure products. Thus, in diaphragm cells, brine from the anode compartment flows through the separator to the cathode compartment, the separator material being either asbestos or polymer-modified asbestos composite deposited on a foraminous cathode. Flow through the separator is regulated by maintaining a finite hydrostatic head to achieve a given salt-to-caustic ratio in the catholyte. In membrane cells, on the other hand, an ion-exchange membrane is used as a separator and the Na^+ ions from the anode compartment are exchanged by the cation exchange membrane to form NaOH with the OH^- ions produced at the cathode.

While anolyte–catholyte separation is achieved in the diaphragm and membrane cells using separators and ion-exchange membranes, respectively, the mercury cells contain no diaphragm and the mercury cathode itself acts as a separator. Chlorine is generated at the anode and Na^+ is discharged at the cathode to form sodium amalgam which is passed into a second cell where it is reacted with water to form NaOH, H_2, and Hg as expressed by

$$2NaHg_x + 2H_2O \rightarrow 2NaOH + H_2 + 2xHg \qquad (4)$$

The regenerated mercury is cycled back to the electrolyzer. Owing to environmental hazards associated with mercury, these cells are being phased out gradually.

The primary anodic reaction in all these technologies is the discharge of chloride ions to form gaseous Cl_2 following Eq. (2). However, there are two parasitic reactions off-setting the anode efficiency. These reactions include generation of oxygen from the anodic discharge of water [Eq. (5)] and the electrochemical oxidation of OCl^- to chlorate [Eq. (6)].

$$2H_2O \rightarrow O_2 + 4H^+ + 4e \qquad (5)$$

$$6OCl^- + 3H_2O \rightarrow 2ClO_3^- + 4Cl^- + 6H^+ + 1.5O_2 + 6e \qquad (6)$$

If graphite anodes are used, another reaction leading to inefficiency is the oxidation of C to CO_2 as

$$C + 2H_2O \rightarrow CO_2 + 2H_2 \qquad (7)$$

At the cathode, water molecules are discharged to form H_2 gas and NaOH as described by Eq. (3). However, some of the caustic generated in the cathode compartment back-migrates to the anode compartment and reacts with dissolved Cl_2 to form hypochlorite and chlorate:

$$Cl_2(\text{dissolved}) + OH^- \rightarrow HOCl + Cl^- \qquad (8)$$

$$HOCl + OH^- \rightarrow H_2O + OCl^- \qquad (9)$$

$$2HOCl + OCl^- \rightarrow ClO_3^- + 2H^+ + 2Cl^- \qquad (10)$$

Reduction of OCl^- and ClO_3^- ions, following the reaction schemes

$$OCl^- + H_2O + 2e \to Cl^- + 2OH^- \quad (11)$$

$$ClO_3^- + 3H_2O + 6e \to Cl^- + 6OH^- \quad (12)$$

is thermodynamically favored and adversely influences the chlorine efficiency since it is related to the amount of dissolved chlorine in the anolyte[16] as

Chlorine current efficiency

$$= \frac{\% Cl_2}{\% Cl_2 + 2\% O_2 + Cl_2(\text{dissolved})} \quad (13)$$

the compositions being expressed in volume percent. For diaphragm cells, Cl_2(dissolved) can be shown to be equal to $(1/C_{OH}^-) \times [2.2546\, C_{ClO_3^-}(\text{anolyte}) + 1.0745\, C_{OCl^-}(\text{anolyte})]$, where C_{OH^-} refers to the catholyte caustic concentration, whereas in membrane cells, most of the dissolved chlorine is recovered by HCl additions.

In the case of mercury cells, the anodic inefficiency arises from the "recombination reaction," given by

$$2NaHg_x + Cl_2 \to 2Na^+ + 2Cl^- + 2xHg \quad (14)$$

which is significant at small interelectrode distances. It may be noted that the anodic current efficiency equals the cathode current efficiency when the feed brine is neutral.

The terminology employed in the chlor-alkali industry for comparing the performance characteristics of various cells is the energy consumption expressed in kilowatt hours per metric ton (kW h/M.T) of Cl_2 or NaOH. This is related to the cell voltage (E) and the current efficiency (η) as

Energy consumption (for Cl_2)

$$= \frac{756E}{\eta} (\text{d.c. kW h/M.T of } Cl_2) \quad (15)$$

Energy consumption (for NaOH)

$$= \frac{670.1E}{\eta} (\text{d.c. kW h/M.T of NaOH}) \quad (16)$$

III. MANUFACTURING PROCESSES

The unit operations in a commercial chlor-alkali plant can be generally classified as follows: (1) brine purification, (2) electrolytic cells, (3) H_2 and Cl_2 collection, and (4) caustic concentration and salt removal. In this section, the general process flowsheets for diaphragm, membrane, and mercury cell technologies are discussed with emphasis on the need for brine purification and the manner in which it is carried out.

1. Importance of Brine Purification

(i) Effect of Impurities in Brine on the Diaphragm and Membrane Cells

The feed material for the manufacture of chlorine is saturated brine, usually pumped from salt deposits, which contains about 1500 ppm Ca^{2+}, 40 ppm Mg^{2+}, 4000 ppm SO_4^{2-}, 40 ppm Fe^{2+}, and Fe^{3+} and others at ≤1 ppm levels. Ca, Mg, and Fe are harmful to the performance of porous separators and ion-exchange membranes, since these ions would precipitate as hydroxides in the separator (where the pH is >7) leading to an increase in the ohmic drop across the diaphragm and, hence, an increase in the cell voltage. (Decreased brine flow arising from the plugging of the diaphragm would also result in lowering of current efficiency.[17,18]) In addition, Ca^{2+} and Mg^{2+} ions percolating through the separators into the catholyte may deposit on the walls of caustic evaporators resulting in a decrease of the heat transfer coefficient, and an increase in maintenance costs associated with the descaling of the evaporators.

Membrane cells require ultrapure brine since the Ca and Mg hydroxides deposited on the membrane not only affect the mechanical integrity of the membrane[19] but also cause a decrease in cell performance. Hence, the membrane manufacturers recommend that the Ca and Mg levels in the feed brine be reduced to less than 0.04 and 0.08 ppm, respectively, depending on the membrane type and operating conditions to realize long life and good performance.

(ii) Effect of Impurities in Brine on the Amalgam Cathode

In mercury cells, impurities play a significant role during the formation of sodium amalgam and, hence, quality control of brine purity is an important unit operation.

In an amalgam-type chlor-alkali cell, Na^+ ions are discharged at the cathode to form dilute sodium amalgam in the potential range of -1.9 to -1.6 V (versus SHE), which is about 1 V positive to the reversible potential of Na/Na^+ couple, and considerably negative to the equilibrium potential of $H^+ \rightarrow H_2$ reaction. However, the hydrogen overvoltage on sodium amalgam and/or mercury is high and hence, sodium amalgam is formed with ease without any hydrogen evolution. Nevertheless, hydrogen evolution can be triggered by factors such as increased Na concentration in the amalgam and contamination of the cathode surface with impurities.

Two types of effects can be caused by impurities in brine:

1. Promoting the hydrogen evolution reaction during sodium amalgam formation either by direct discharge of hydrogen ions or by decreasing the hydrogen overpotential on the sodium amalgam.
2. Decomposing the sodium amalgam as it is being formed.

Thus, V, Cr, and Mo ions are extremely harmful even at ≤ 1 ppm level by promoting the hydrogen evolution reaction, whereas Mg and Ca ions form colloidal precipitate on the cathode surface on which hydrogen discharge is initiated.[20-25] Impurities such as Ag, Pb, and Zn form a relatively uniform amalgam and may not promote hydrogen formation on the amalgam cathode. Others such as Fe, Ni, and C do not dissolve in mercury, but form "amalgam butter" on which hydrogen discharge is favored. Thus, according to MacMullin,[25] V, Mo, Cr, Ti, Ta, and (Mg + Fe) are very harmful, Ni, Co, Fe, and W are moderately harmful, Ca, Ba, Cu, Al, Mg, and graphite are slightly harmful, and Ag, Pb, Zn, and Mn have no effect during sodium amalgam formation.

It may be noted that combination of two or more impurities is more harmful even though the impurity by itself has no effect on the amalgam formation, a typical example being Mg + Fe. Although Al has no effect, the colloidal $Al(OH)_3$ occludes other impurities and forms amalgam butter, which is undesirable. While

Ti and Ta are classified as harmful impurities, these metals are corrosion resistant in brine and, hence, have no significant influence on the primary cathodic process.

(iii) Brine Purification

Generally, crude salt contains Mg and Ca as chlorides and/or sulfates. These impurities are precipitated by the addition of caustic soda and soda ash to saturated brine at a pH of about 10.5, followed by settling, which is facilitated by coagulants such as starch or synthetic polymer and filtration. During this process $Mg(OH)_2$ and $CaCO_3$ carry along other impurities present in the brine.[26-28]

Since the solubility product of $Mg(OH)_2$ is in the range of 10^{-9}-10^{-10}, Mg^{2+} can be precipitated at a pH of about 11. However, this value varies[26] slightly in saturated brine as shown in Fig. 1.

Figure 2 illustrates differences in the settling rates of $Mg(OH)_2$ formed by $Ca(OH)_2$ addition to $MgCl_2$ solution and NaOH addition to $MgCl_2$ solution. While the sedimentation rate with $Ca(OH)_2$ addition is fast, the chlor-alkali industry prefers caustic addition for obvious reasons. For example, addition of $Ca(OH)_2$ results in

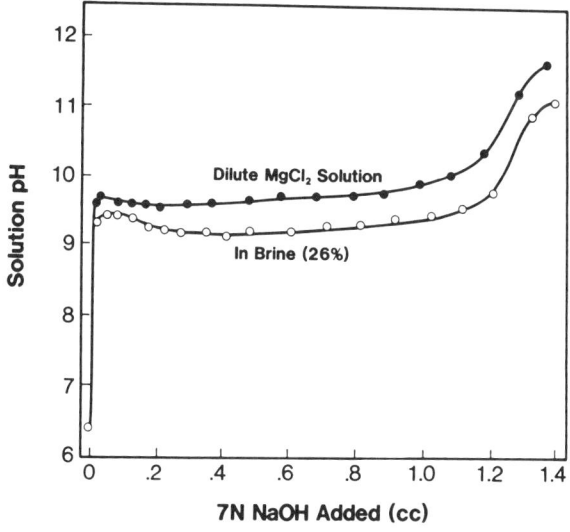

Figure 1. An example of the pH titration curve of saturated brine containing $MgCl_2$.

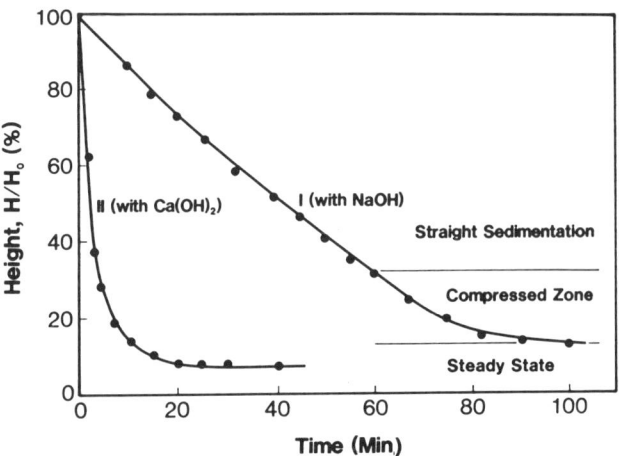

Figure 2. Sedimentation curves for $Mg(OH)_2$ precipitates.

the formation of $CaCl_2$, which has to be removed subsequently by Na_2CO_3 addition. Since the initial settling rate is slow with caustic addition, longer retention time is mandated. Only a small height needs to be devoted to the steady-state region to obtain a thick slurry which is removed from the bottom of the settling tank. The reason for the differences in the settling rates with $Ca(OH)_2$ and NaOH is attributed to the morphological characteristics of the $Mg(OH)_2$ precipitate.[27]

2. Diaphragm Cell Process

The general layout of a typical diaphragm-type chlor-alkali plant is shown in Fig. 3. Purified brine, fed into the anode compartment, passes through the separator to the cathode compartment where caustic is formed along with hydrogen. Degree of conversion of NaCl to NaOH is an important operating parameter in diaphragm cell process—50% decomposition being considered as an economic optimum. The catholyte containing ~15% NaCl and 12% NaOH, called "cell liquor," is fed to a multiple effect evaporator for concentrating the caustic to 50% by weight. Crystallized salt is separated from the concentrated caustic by gravity and/or by mechanical means. Caustic liquor is cooled further to 20°C, to

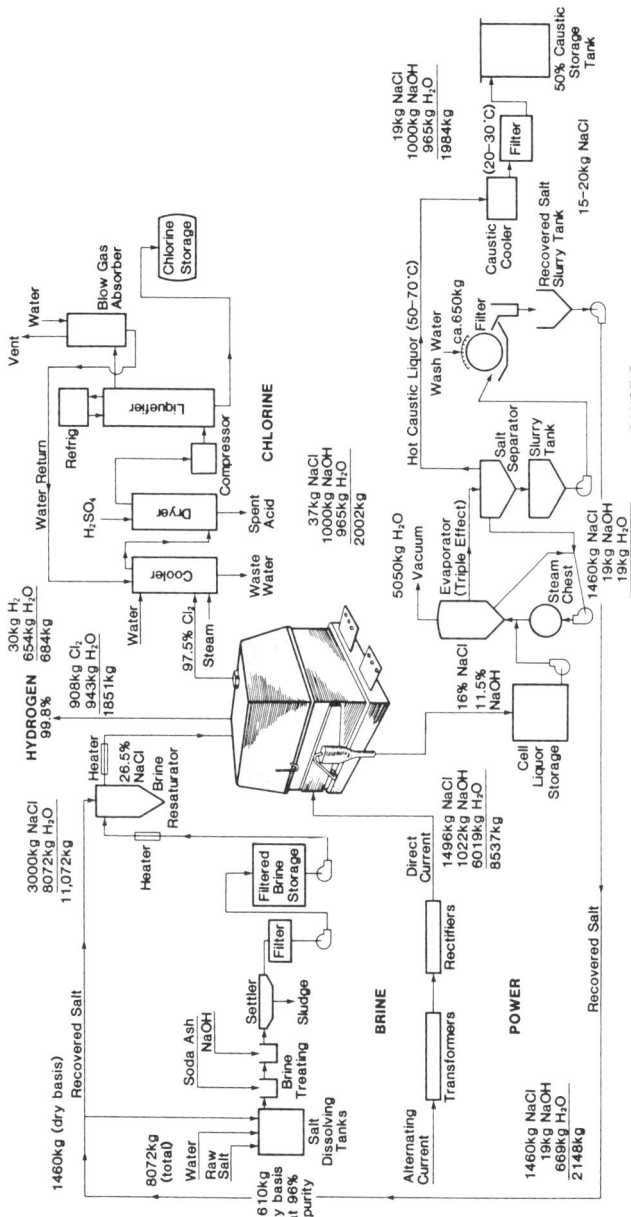

Figure 3. Flowsheet of typical diaphragm-type chlor-alkali plant.

remove additional salt, which is filtered. This caustic, which is 50% by weight, still contains ca. 1% NaCl and other impurities such as $NaClO_3$ which are generally acceptable for most end uses except in some specialized applications such as rayon manufacture. The diaphragm cell caustic can be purified further, if required, at a considerable cost.

As shown in Fig. 3, the cell gas is cooled by a heat exchanger to remove water and mist and is dried using sulfuric acid prior to liquefication. Chlorine gas can be cooled to about $-25°C$ or lower by indirect cooling with ammonia. However, ammonia has been replaced by safer fluorinated hydrocarbon gases such as chlorodifluoromethane, $CHClF_2$ (called R-22) and dichlorodifluoromethane, CCl_2F_2 (called R-12), which make it possible to cool chlorine by direct contact resulting in significant energy savings.

In modern chlorine plants, on the other hand, chlorine gas is compressed to 5-15 kg/cm² in a multistage turbocompressor at ambient temperature. Figure 4 is an example of the flowsheet of the high-performance liquefaction system, where the chlorine gas is compressed to 2.1 kg/cm² prior to cooling. Since the chlorine gas leaving the gas–liquid separator contains some hydrogen (0.5%-1.0%), the gas is mixed with dry air to avoid explosion, and is compressed and cooled further for liquefaction with an efficiency of over 99%. Waste gases are sent to remove Cl_2 by scrubbing with caustic. Other methods for recovering chlorine include using pres-

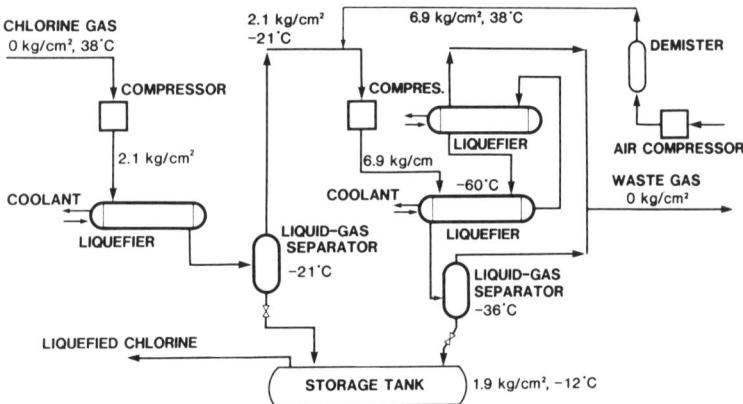

Figure 4. Flowsheet for chlorine liquefaction.

surized water² and carbon tetrachloride.²⁹ Hine *et al.* reported an electrochemical method[30,31] employing a mixed electrolyte containing HCl and $CuCl_2$. Chlorine is evolved at the graphite anode, and cupric chloro-complex ions are reduced to cuprous ions at the graphite cathode. The catholyte containing Cu^+ ions is sent to the gas absorption tower where cuprous ions are oxidized to cupric by the waste gases:

$$Cu^+ + \tfrac{1}{2}Cl_2 \rightarrow Cu^{2+} + Cl^- \tag{17}$$

3. Membrane Cell Process

A projected flowsheet of a membrane cell plant is shown in Fig. 5. The brine treatment is almost similar to that used for diaphragm cell operation except that the filtrate is further treated in a packed column containing an ion-exchange resin where other trace impurities such as Ca are also removed. The ion-exchange resin is regenerated periodically by passing muriatic acid solution followed by conversion with caustic soda.[32,33] Generally, the ion-exchange resin has amino-phosphonic functional groups, and the reactions

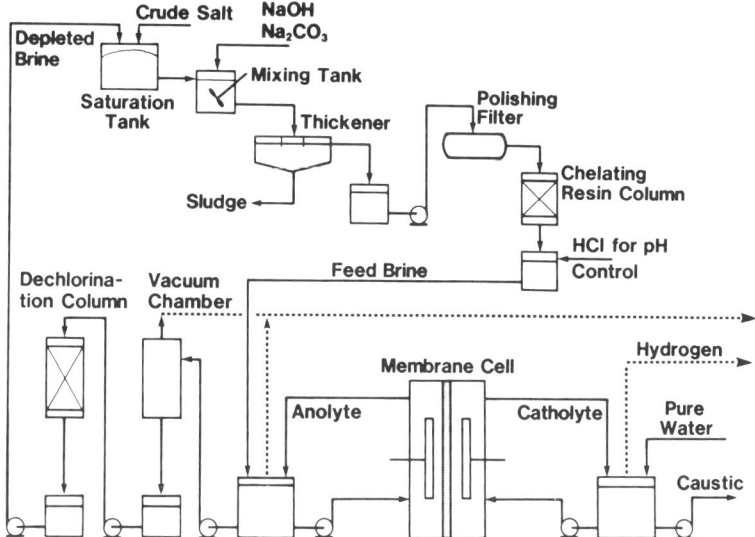

Figure 5. Flowsheet of the membrane-type chlor-alkali plant.

involved in the brine treatment are as follows:
Exchange reaction:

$$2RCH_2NHCH_2PO_3Na_2 + Ca^{2+}$$
$$= (RCH_2NHCH_2PO_3)_2CaNa_2 + 2Na^+ \qquad (18)$$

Regeneration:

$$(RCH_2NHCH_2PO_3)_2CaNa_2 + 4HCl$$
$$= 2RCH_2NHCH_2PO_3H_2 + CaCl_2 + 2NaCl \qquad (19)$$

Conversion:

$$RCH_2NHCH_2PO_3H_2 + 2NaOH$$
$$= RCH_2NHCH_2PO_3Na_2 + 2H_2O \qquad (20)$$

Since the ion-exchange resin is attacked by trace amounts of dissolved chlorine, it is essential that the depleted brine from the anode compartment is dechlorinated to decrease the chlorine content to ≤1 ppm prior to resaturation.

Purified water is fed into the cathode compartment where H_2 and OH^- are produced during electrolysis. Water is also transported from the anode compartment along with Na^+ ions through the membrane and is about 3-5 moles per 1 mole of Na^+ transferred depending on the operating parameters. Since the cation exchange membrane rejects Cl^- ions, the catholyte contains a very small amount of salt, and caustic evaporation is not required if 15-30 wt % caustic can be used directly. Thus, the membrane cell plant consumes less process energy than a diaphragm cell plant.

The process energy consumption in a membrane cell is small compared to diaphragm and mercury cell operations and the membrane cell caustic is of the same quality as mercury cell caustic. Hence, the membrane cell technology is recognized as the most economical and preferred method for producing chlorine and caustic (see Section 6 for additional details regarding membrane cells).

4. Mercury Cell Process

The typical layout of a mercury cell plant for producing Cl_2 and NaOH is presented in Fig. 6. The brine purification procedure is

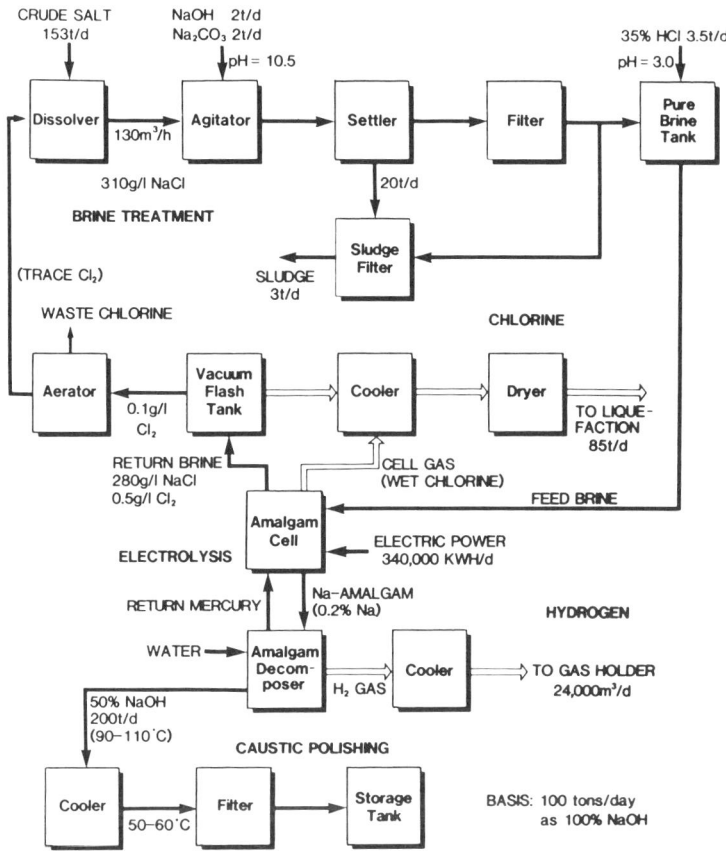

Figure 6. Flowsheet and material balance in the amalgam cell plant.

the same as the one employed in diaphragm-type chlor-alkali operations and the purified brine is acidified prior to electrolysis to minimize the formation of hypochlorite.

The cell gas consisting of wet chlorine is cooled by titanium heat exchanger to condense water vapor and brine mist, and dried using concentrated sulfuric acid prior to liquefaction.

Sodium amalgam containing about 0.2 wt % Na is decomposed with water in the amalgam decomposition tower (see Section 5.1.2 for details related to the design of the amalgam decomposer) of about 1 m high, containing graphite lumps of 1-2 cm diameter to produce 50%-70% caustic. The amalgam decomposition tower is operated at about 100°C and, hence, the caustic is cooled to ~50°C prior to filtration (using activated carbon powder filter aids) to remove trace quantities of mercury and other solid impurities such as graphite particulates and iron oxides. This caustic still contains mercury in the range of 0.05-0.10 mg/liter, which is unacceptable for consumers such as the food industry. Hine and Yagishita designed a caustic filtration system composed of conical filter leaves covered with filter cloth,[34] the precoat layer being activated carbon powder. However, careful operation was required since the precoat layer of activated carbon is fragile contributing to leakage. Plant data depicted in Fig. 7 show that the average mercury level in caustic is reduced to 0.02 ppm. This can be further lowered to 0.005-0.01 ppm by using a small check filter in series.

Hydrogen from the cells is cooled by heat exchangers located at the top of the amalgam decomposer to recover mercury and caustic mist, and returned to the gas holder. The gas is cooled further, if necessary, prior to removal of trace mercury by adsorption using activated carbon and/or ion-exchange resin columns.

Dechlorination of depleted brine is an important step in the operation of mercury cells since the dissolved chlorine affected the

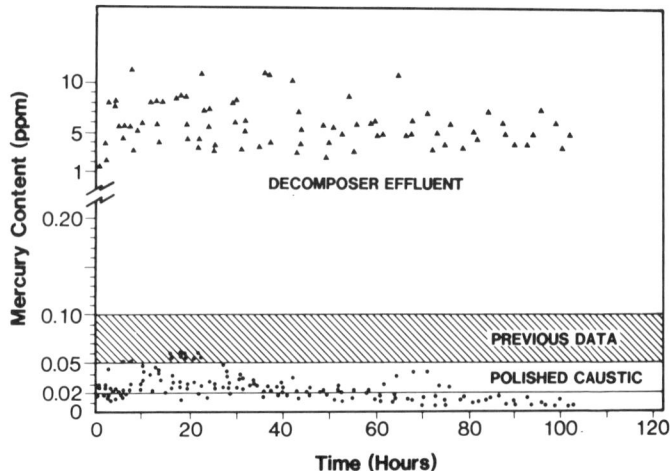

Figure 7. Mercury content in caustic soda solutions.

formation and sedimentation of $CaCO_3$ and $Mg(OH)_2$, and restricted the choice of materials of construction of the brine system. Dissolved chlorine in brine can be removed by Na_2SO_3 as described by reaction (21) or by carbon as illustrated by reaction sequence (22):

$$Cl_2 + Na_2SO_3 + H_2O \rightarrow 2HCl + Na_2SO_4 \quad (21)$$

$$Cl_2 + C + H_2O \rightarrow 2HCl + CO + (\text{or } CO_2) \quad (22)$$

In the second scheme, brine is fed to a rubber-lined steel tank packed with graphite along with caustic to ensure conversion of chlorine to HOCl and/or OCl^-. Alternately, aeration in a perforated plate tower may be employed to dechlorinate the depleted brine as shown by Yokota.[35,36]

IV. ELECTRODE MATERIALS AND ELECTRODE PROCESSES

1. Anodes

Historically, graphite was used as an anode for chlorine production. These anodes have inherent problems which include high chlorine overpotential and dimensional instability caused by gradual oxidation of C to CO_2 which results in widening of the anode–cathode gap and, hence, increased energy consumption for producing chlorine. However, the advent of noble metal oxide coatings on titanium substrates in the late 1960s has revolutionized the industry, the most widely used anodes being $RuO_2 + TiO_2$-coated titanium, which exhibit low chlorine overpotential and excellent dimensional stability. Escalating power costs in the mid-1970s accelerated the transition of the chlor-alkali industry from graphite to metal anodes, and presently all but a few installations have fully converted to the exclusive use of dimensionally stable anodes.[37]

(i) Graphite Anodes

Graphite anodes have a limited life of 6–24 months, owing in part to electrochemical oxidation and physical wear. Other factors promoting graphite degradation include high operating temperature, high anolyte pH, presence of impurities such as SO_4^{2-} in the brine, and concentration of active chlorine.[38]

The anodic polarization behavior of graphite is shown in Fig. 8 along with the dependence of chlorine current efficiency on current density (c.d.). Thus, at current densities beyond $10 \, A/dm^2$, the potential-log c.d. variation is nonlinear, and the chlorine current efficiency decreases with increase in current density. The deviation of the Tafel behavior at high current densities is attributed to the oxide layer on the anode and to the simultaneous discharge of oxygen.[39] Comparison of the kinetic parameters evaluated from the experimental data (see Table 1A) with the theoretical values presented in Table 1B for various reaction pathways suggests the slow-step to be electrochemical desorption in the low current density region:

$$Cl^- \rightleftharpoons Cl_{ad} + e \quad \text{(fast)} \quad (23)$$

$$Cl_{ad} + Cl^- \rightarrow Cl_2 + e \quad \text{(slow)} \quad (24)$$

It should, however, be pointed out here that unequivocal conclusions on mechanisms and values of Tafel parameters are difficult to reach unless the surface state of the electrode is well characterized. For example, studies by Hine were performed on a rough graphite electrode.

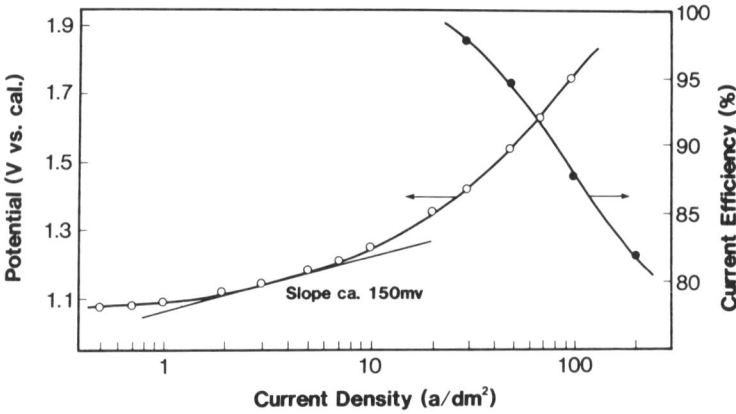

Figure 8. Polarization curve and current efficiency for chlorine evolution reaction with graphite anode having rough or corroded surface in saturated NaCl (pH = 2.3) at 50°C.

Table 1
Kinetic Parameters for the Chlorine Electrode Process

A. Experimental

	Stoichiometric number, ν	Tafel slope (mV/decade)	Reaction order z_{Cl^-}	Reaction order z_{Cl_2}	Exchange current density, i_0 (A/dm^2)
Anodic	1 (ca. 1.2)	ca. 40 at low C.D.[a] 120 ~ 130 at high C.D.[b]	—	—	(i_0)an. = (i_0)ca. = 0.12 In sat'd NaCl (pH 0.5), p_{Cl_2} = 1 atm at 50°C
Cathodic	1 (ca. 1.2)	ca. −120	1 (ca. 0.6) 0	0 1	

B. Theoretical[c]

Mechanism		Rate-determining step	ν	Tafel slope[d] $\theta_{Cl} \to 0$	Tafel slope[d] $\theta_{Cl} \to 1$	z_{Cl^-} $\theta_{Cl} \to 0$	z_{Cl^-} $\theta_{Cl} \to 1$	z_{Cl_2} $\theta_{Cl} \to 0$	z_{Cl_2} $\theta_{Cl} \to 1$
Anodic	I	$Cl^- = Cl_{ad} + e$	2	$2RT/F$	$2RT/F$	1	1	0	0
		$2Cl_{ad} = Cl_2$	1	RT/F	∞	2	0	0	0
	II	$Cl^- = Cl_{ad} + e$	1	$2RT/F$	—	1	—	0	—
		$Cl_{ad} + Cl^- = Cl_2 + e$	1	$2RT/3F$	$2RT/F$	2	1	0	0
Cathodic	I	$Cl_2 = 2Cl_{ad}$	1	$-\infty$	—	0	—	1	—
		$Cl_{ad} + e = Cl^-$	2	$-2RT/F$	$-2RT/F$	0	0	$\tfrac{1}{2}$	0
	II	$Cl_2 + e = Cl_{ad} + Cl^-$	1	$-2RT/F$	—	0	—	1	—
		$Cl_{ad} + e = Cl^-$	1	$-2RT/3F$	$-2RT/F$	−1	0	1	0

[a] The backward reaction is negligible because $p_{Cl_2} = 0$.
[b] $E <$ ca. 1.25V vs. calomel. The Tafel slope increased very much at $E >$ ca. 1.25 V.
[c] Langmuir-type isotherm was adopted for adsorption of Cl_{ad}.
[d] When $\alpha = 0.5$.

(ii) Metal Oxide-Based Anodes

Besides the graphite anode, metal oxides such as magnetite and lead dioxide were examined as chlorine-evolving anodes. However, these were found to be unstable when anodically polarized in chloride media.

During the 1960s, major research efforts were devoted to the oxides of platinum group metal exhibiting high metallic conductivity, and in 1967, Beer invented RuO_2 coatings on Ti substrates for application as anode in chlor-alkali cells.[40,41] These dimensionally stable anodes (DSA®) have been further developed and commercialized by V. deNora.[42]

The $RuO_2 + TiO_2$ solid solution coating is formed by thermal decomposition of Ru and Ti salts applied to Ti mesh substrates and firing at ca. 500°C in air to obtain the mixed oxides of Ru and Ti. While the commercial DSA coating is proprietary, the original Beer-type anode is believed[5,11] to contain a Ti/Ru mole ratio of 2/1, whereas a three-component coating developed by K. J. O'Leary[43] consists of a metal mole ratio of 3 Ru : 2 Sn : 11 Ti with a $RuO_2 + SnO_2$ loading of ~1.6 mg/cm².

$RuO_2 + TiO_2$ coatings exhibit low chlorine overvoltage[44] as shown in Fig. 9 and long life (>5 yr) in commercial cells. However, their performace characteristics depend on the brine concentration and quality, and current density in addition to composition of the coating and the method of preparation. The latter is important since one of the failure mechanisms of these coatings is via the formation of a nonconducting TiO_2 layer at the Ti-catalytic coating interface even though the noble metal loading is high.

In an attempt to understand the excellent electrocatalytic characteristics of RuO_2-based coatings several mechanistics studies were carried out in recent years. While it is beyond the scope of this article to thoroughly review these investigations (see, however, Refs. 11-13 for details), a brief summary of these studies is outlined here. The reaction pathways during anodic Cl_2 evolution are analogous to those in cathodic H_2 evolution. The following are the obvious steps and mechanisms at the metal surface or surface oxide sites M that were proposed by various authors:

Mechanism I:

$$M + Cl^- \rightleftharpoons MCl_{ad} + e \quad \text{(discharge)} \quad (25)$$

$$2MCl_{ad} \rightarrow 2M + Cl_2 \quad \text{(recombination)} \quad (26)$$

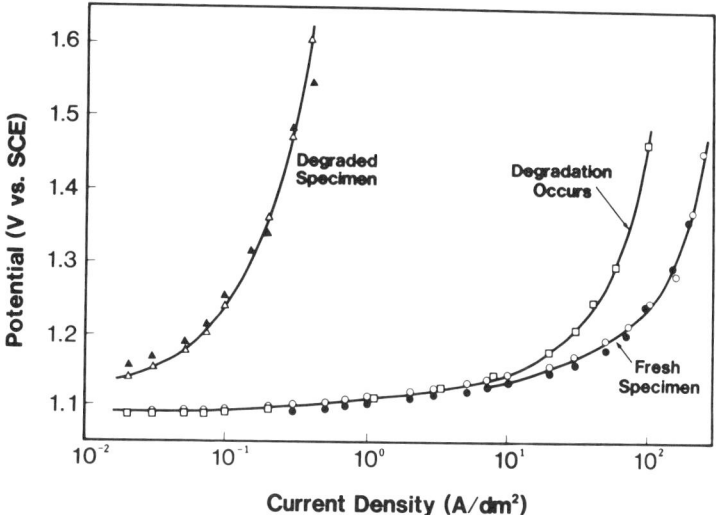

Figure 9. Polarization curves of a RuO_2-TiO_2-coated Ti anode in saturated NaCl (pH = 8) at 40°C.

Mechanism II:

$$M + Cl^- \rightleftharpoons MCl_{ad} + e \quad \text{(discharge)} \quad (27)$$

$$MCl_{ad} + Cl^- \rightarrow M + Cl_2 + e \quad \text{(electrochemical desorption)} \quad (28)$$

Since the experimentally observed Tafel slopes varied in the range of 30–40 mV, recombination and electrochemical desorption have been invoked as the rate-determining step during Cl^- ion discharge. However, recently Erenberg et al., proposed[45,46,46a] another scheme involving chloronium ion, Cl^+, as shown below:

$$M + Cl^- \rightleftharpoons MCl_{ad} + e \quad (29)$$

$$MCl_{ad} \rightarrow MCl_{ad}^+ + e \quad (30)$$

$$MCl_{ad}^+ + Cl^- \rightleftharpoons M + Cl_2 \quad (31)$$

This mechanism, with Eq. (30) as the slow-step, can be supported by reaction order, stoichiometric number, and Tafel slope data. However, the mechanistic data reported in the literature cannot be subjected to a rational comparison since these values strongly depend on the method of preparation of the coating and its compositional and surface state characteristics. Thus, the mechanism of Cl_2 evolution is unresolved and still controversial.

There are several other coatings based on platinum group metals which are commercially available. These include Pt-Ir coatings (by IMI Marston Ltd.), palladium oxides (by TDK Ltd.) and platinum "bronzes" of general formula $M_{0.5}Pt_3O_4$ (by C. Condratty, Nürnberg), where M can be Li, Na, Cu, Ag, Ti, or Sr. Dow Chemical Company has developed a nonnoble metal oxide coating comprised of $M_xCo_{3-x}O_4 \cdot yZrO_2$, where $x \geq 1$, $y \leq 1$, and M is Mn, Cu, Mg, or Zn, which is claimed[5,47,48] to having performance characteristics similar to the $RuO_2 + TiO_2$ coatings.

2. Cathodes

(i) Low-Overvoltage Cathodes

Low carbon steel (in mesh or perforated sheet form) is presently used as a cathode in diaphragm and membrane cells because of its low cost, long life (>20 yr), and good electrocatalytic characteristics for the hydrogen evolution reaction. Thus, the hydrogen overvoltage on steel is typically about 400 mV at 20 A/dm^2 in 2.5 N NaOH at 90°C, the exact value being dependent on the surface state of steel.

Motivated by the escalating power costs in the mid-1970s, various approaches were adopted to realize low cathodic overvoltage by surface modification of the existing steel cathodes. These methods include enhancing the true surface area of the cathode or coating with high surface area materials with better electrocatalytic properties than steel. Composites chosen for coating steel are mainly Ni- and Co-based, and are deposited on the cathode by thermal and/or electrolytic routes along with a second component such as Al or Zn, which is leached out in NaOH solutions to achieve high active area. Several compositions were mentioned in the literature, some of which include Ni-S,[49-52] Ni-Zn,[53,53a] Ni-Al,[53,53a,54] Ni-Mo,[54a] and Co-W-P[55] coatings on steel.

Carlin et al.[55] prepared Co-W(12%-15%)-P(3%-5%) composite which showed low overpotential under the operating conditions of a diaphragm cell. Eltech Systems claims high-surface-area Ni-Zn and Ni-Al coatings[53] exhibiting low overvoltages. Tseung developed a novel cathode material consisting of $NiCo_2S_4$ bonded with Teflon on a nickel screen which showed an overpotential of 40 mV at 25 A/dm^2 in 15% NaOH at 85°C.[56]

While extensive testing of catalytic cathodes is currently ongoing in the chlor-alkali industry, commercialization in large-scale diaphragm plants is still awaited. One of the problems with transition-metal based catalytic cathodes is the corrosion (and, hence, the loss of catalytic activity) of the high-surface-area coatings caused by hypochlorite ions during shutdowns.[56a] This problem does not arise during cell operation since hypochlorite is reduced to chloride at the cathode under limiting current density conditions. Thus, the technical difficulties facing the use of catalytic cathodes in diaphragm cells include (1) selection of a coating technique that could be used to coat the complex cathode assembly without adversely influencing the structural tolerances involved in the fabrication of cathodes, and (2) developing shutdown procedures which would eliminate hypochlorite as quickly as possible to preserve the catalytic activity of the coatings. These problems are not encountered in membrane cells in view of the simplicity of the cathode structure and the anion rejection properties of the membrane.

(ii) Air-Depolarized Cathodes

The operating voltage of a chlor-alkali cell can be lowered considerably by changing the cathode reaction from the discharge of H_2O [Eq. (32)],

$$2H_2O + 2e \rightarrow 2OH^- + H_2 \quad (E_{rev} = -0.827 \text{ V}) \quad (32)$$

to the reduction of O_2 to form OH^-:

$$2H_2O + O_2 + 4e \rightarrow 4OH^- \quad (E_{rev} = +0.401 \text{ V}) \quad (33)$$

Thus, the oxygen reduction reaction offers the possibility of saving 1.23 V based on the difference in the reversible electrode potentials. This voltage saving corresponds to an energy reduction of >900 kW h/s ton of Cl_2.[57,58]

Chlor-alkali producers have been actively pursuing the development of oxygen electrodes for some time and it is unlikely that the full 1.23 V saving can be achieved at current densities typically used in chlor-alkali operations. Many problems remain to be solved before the oxygen cathode is ready for commercial cells. The problems include electrode stability to peroxide, which is in an intermediate formed during oxygen reduction, limited solubility of oxygen in caustic, which may cause mass-transfer limitations, and fouling of the air-cathode by CO_2. However, the huge potential savings offered by the air-cathodes provide sufficient incentive for continued research.

It is worth mentioning that Eltech Systems and Case Western Reserve University (CWRU) were awarded a $2.5MM grant by the U.S. Department of Energy in 1977 to develop low-cost oxygen cathodes, the performance target being the demonstration of >0.8 V savings in an ion-exchange membrane cell at a current density of 30 A/dm^2 and a caustic strength of 30% at 85°C with a life of >10,000 h.

Basic aspects related to electrocatalysis of oxygen reduction on materials such as transition metal complexes are being investigated at CWRU, while Eltech Systems Corporation is involved in scaling up and testing in pilot cells. The air-cathode developed by Eltech operates at a potential of 0.6 V versus RHE at a current density of 31 A/dm^2, which corresponds to a voltage savings of 0.85 V compared to a hydrogen evolving cathode based chlor-alkali cell. Commercialization of oxygen cathode technology in membrane cells was tentatively scheduled for 1984–1985.[59,60] However, there have been no announcements of commercialization as of February 1986.

V. ENGINEERING ASPECTS IN CHLOR-ALKALI OPERATIONS

1. Chemical Engineering Aspects of Amalgam Decomposition

(i) Electrochemistry of Sodium Amalgam

Mercury is a unique liquid metal at room temperature and forms various compounds or alloys, termed amalgams, with many metals which may be classified into three types[61]:

1. "Gallium-type" amalgams with Fe, Ni, Co, Mn, Cr, Al, and Ga which are amalgamated with difficulty, and

insoluble in mercury exhibiting half-wave potentials negative to the M/M^+ standard electrode potentials and are useful for retort processes.
2. "Zinc-type" amalgams with Zn, Cd, Pb, Tl, Sn, and Bi are easily amalgamated and readily soluble in mercury. They exhibit the same potential as the pure metal and are suited for phase separation.
3. "Sodium-type" amalgams with Na, K, Rb, Cs, Ca, Si, and Ba react completely with low solubility in mercury. They can be formed electrolytically and exhibit half-wave potentials positive to the M/M^+ standard electrode potential.

Solubilities and the physicochemical data related to these amalgams are readily available in the literature.[62-65] Of these, sodium amalgam is of importance from the operational viewpoint of mercury cells. The phase diagram of sodium amalgam has been well established (see Fig. 10), and there are various compounds of

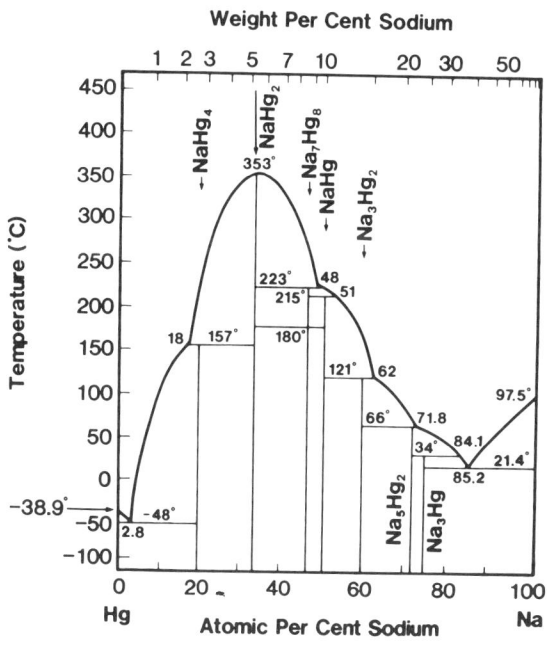

Figure 10. Phase diagram of the Hg-Na system.

the type $NaHg_x$. At room temperature, the solubility of sodium in mercury is small (0.6 wt % or 5.3 at. %). The electrical conductivity of sodium amalgam varies slightly with an increase in the concentration of sodium, the conductivity reaching a minimum value of 0.9852×10^4 mho/cm at 0.268% Na compared to 1.0342×10^4 mho/cm for mercury.[65]

The standard potential of Na/Na^+ at 25°C is -2.7142 V and the potential of the Na/Na amalgam cell is -0.75852 V when the mole fraction of sodium in the amalgam is unity. Hence, the standard electrode potential for the reaction

$$NaHg_x \rightleftharpoons Na^+ + xHg + e \qquad (34)$$

is -1.95568 V versus SHE. However, when the activity of Na^+ (a_{Na^+}) is 1 and the activity of sodium amalgam (a_{NaHg}) is 0.012 (for 0.1% Na amalgam, see Fig. 11), the Nernst equation for reaction (34) can be written as

$$E = \bar{E} + \frac{RT}{F}\left[\ln(a_{Na^+}) - \ln\left(\frac{a_{NaHg}}{0.012}\right)\right] \qquad (35)$$

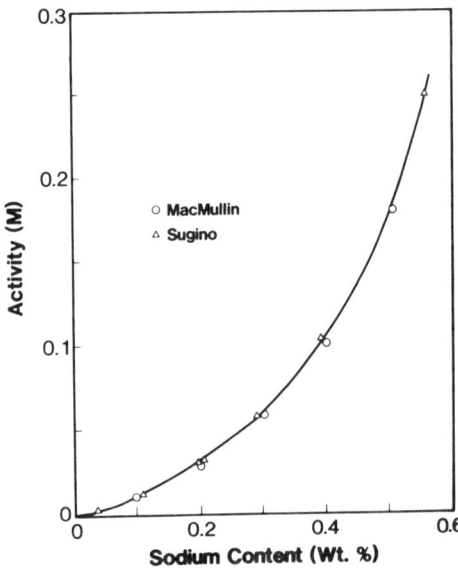

Figure 11. Activity versus concentration curve for Na in sodium amalgam at 25°C.

where $\bar{E}$ refers to the potential of 0.1% sodium amalgam in the solution and has a value of -1.834 V.[66,67]

Figure 12, depicting the polarization behavior of the sodium amalgam electrode at 25°C, shows the potential (uncorrected for ohmic drop) versus current density curve (A) to be linear, the solution ohmic contribution being of the same magnitude as the uncorrected anode potential. Hence, the overvoltage of amalgam electrode is very small even at high current densities (4 mV at 25 A/dm^2). However, with dilute amalgams, diffusion-controlled limiting currents are observed as noted in Fig. 12.[68]

An estimated profile of the diffusion layers across both sides of the amalgam-solution interface is shown in Fig. 13. Thus, the anodic current density (i) is given by

$$\frac{i}{F} = \frac{D_M(a_{Na} - a_{Na}^S)}{\delta_M} = \frac{D_S(a_{Na^+}^S - a_{Na^+})}{\delta_S} \tag{36}$$

where F refers to Faraday (96,500 A s/g eq.), D to the diffusion coefficient (cm^2/sec), a to the activity (mol/cm^3), a^S to the activity at the interface (mol/cm^3), and δ to the thickness of the diffusion layer (cm). The subscripts M and S refer to the metal and solution side, respectively.

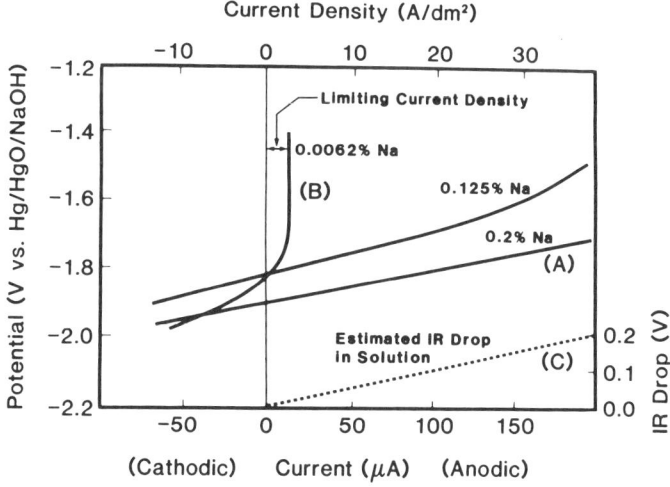

Figure 12. Polarization curves of sodium amalgam electrode in 50% NaOH.

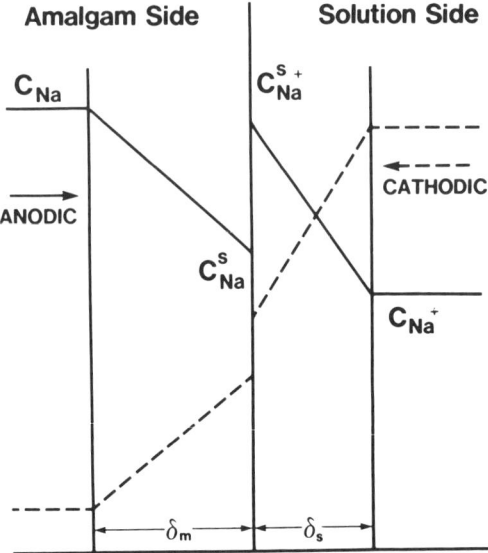

Figure 13. Profile of the amalgam–solution interface.

Since the activation overvoltage is small, any change in the electrode potential during cell operation is only due to change in the concentration of sodium at the amalgam surface. Since the solubility of Na in sodium amalgam is small, the process becomes mass-transfer limited at high current densities, the anodic limiting current $i_L(A)$ being defined as

$$i_L(A) = \frac{FD_M a_{Na}}{\delta_M} \quad (37)$$

The $i_L(A)$ of 0.0062% sodium amalgam ($a_{NaHg} = 6 \times 10^7$ mole/cm^3) is 2.3×10^{-2} A/cm^2 as noted in Fig. 12. Hence, δ_M can be calculated to be 1.86×10^{-3} cm assuming a D_M value of 7.41×10^{-4} cm^2/s. Correspondingly, δ_S estimated from the cathodic limiting current

$$i_L(C) = \frac{FD_S a_{Na^+}}{\delta_S} \quad (38)$$

would be of the same magnitude as δ_M. For example, at $i_L(C) = 10$ A/cm^2, δ_S can be estimated to be 6.27×10^{-4} cm, assuming $D_{Na^+} = 1.3 \times 10^{-5}$ cm^2/s and $a_{Na^+} = 5 \times 10^{-3}$ mol/cm^3.

To avoid getting into diffusion-limited regimes arising from the solubility limitations, the amalgam should be stirred sufficiently to reduce the thickness of the diffusion layer. However, impurities such as Mg^{2+} in the brine can deposit on the amalgam cathode as $Mg(OH)_2$ retarding the convection of the amalgam. Hence, brine purity becomes an important issue in mercury cell operations.

(ii) Design of the Amalgam Decomposition Tower[69]

Sodium amalgam reacts with water to form NaOH and H_2. However, the rate of this reaction is very slow and is enhanced by coupling with graphite in a galvanic cell mode where the following component electrochemical reactions occur:

$$H_2O + e \rightarrow OH^- + \tfrac{1}{2} H_2 \qquad \text{(at the graphite surface)} \qquad (39)$$

$$NaHg_x \rightarrow Na^+ + xHg + e \qquad \text{(at the amalgam surface)} \qquad (40)$$

The driving force of this reaction is the electromotive force of the cell

$$NaHg_x(Y)/NaOH\ aq\ (X); H_2O(W)/H_2\text{---graphite} \qquad (41)$$

given by

$$E = E^0 + \frac{RT}{F} \ln \frac{YW}{X} \qquad (42)$$

where X refers to the activity of NaOH in the solution, Y to the activity of Na in the amalgam, and W to the water activity in NaOH.

Voltage losses in this galvanic cell arise from the anodic and cathodic overvoltages associated with reactions described by Eqs. (40) and (39), respectively, and the ohmic drop between the amalgam and the graphite. There is also an additional ohmic drop in the metallic part of the circuit which was found to be negligible when the cell is short-circuited. Thus, the rate of amalgam decomposition, which is proportional to the cell voltage E, is largely controlled by the cathodic overvoltage and the ohmic resistance in NaOH, the anodic overvoltage being small.

While the hydrogen overvoltage variations with current density may be described by the Tafel equation in general, the experimental results (see Fig. 14) show the overvoltage (η_h) dependence on

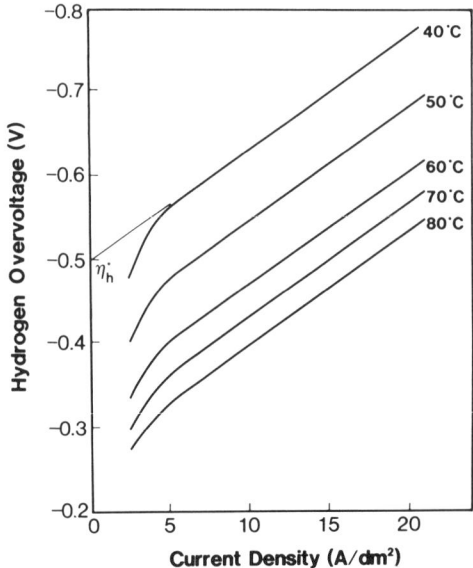

Figure 14. Hydrogen overvoltage of graphite in 40% NaOH solution.

current density to be linear above 5 A/dm². η_h on a graphite cathode in caustic solutions has been shown to be independent of the concentration of the solution. Hence, the relationship between η_h and total current of the system (J) is

$$J = k_h(\eta_h - \eta_h^0) \tag{43}$$

where k_h is a constant which is temperature independent and η_h^0 refers to the potential at zero current which is a function of the temperature.

Thus, the ohmic drop between the amalgam and the graphite is equal to $E - \eta_a - \eta_h$ and, hence,

$$J = k_f(E - \eta_a - \eta_h) \approx k_f(E - \eta_h) \tag{44}$$

where k_f is a constant.

The design used in the industry is the tower-type decomposer wherein graphite lumps (~10 mm diameter) are packed in the tower and sodium amalgam is flashed from the top of the tower. The denuded mercury is recycled to the amalgam cell and caustic soda and hydrogen are recovered from the tower. Advantages associated with this system include savings in floor space and increased reaction rates compared to horizontal-type amalgam decomposer.

A schematic diagram of an amalgam decomposer is shown in Fig. 15. Sodium amalgam flows down the tower at a rate of M (kg moles/h) with a concentration of Y_0 (kg moles Na/Kg moles Hg) and water exits the tower at a rate of L_1 (kg moles/h) with a concentration of X_1 (kg moles of NaOH/kg mole water). Obviously water is consumed in the electrochemical reaction resulting in H_2 production at a rate of $\frac{1}{2}(L_0 - L_1)$. Thus, the mass transfer, W, at a distance h may be represented as

$$W = M(Y_0 - Y) = LX - L_0 X_0 = L_0 - L \tag{45}$$

where the three terms describe the mass transfer rate of amalgam, caustic, and water, respectively.

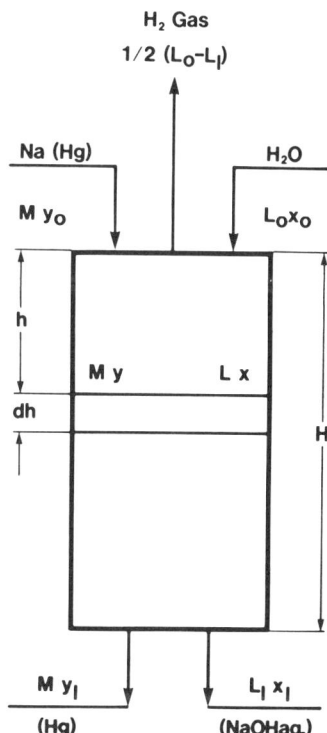

Figure 15. Material balance in the amalgam decomposer.

On the other hand, the electrochemical reaction must follow Faraday's law

$$W = J/F = Ih/F \qquad (46)$$

where $I\ (=J/h)$ refers to the current per unit length of the tower. From Eqs. (43), (44), and (46) it can be shown that

$$I = K_F F S \chi (E - \eta_h) \qquad \text{(ohmic resistance)} \qquad (47)$$

$$= K_H F S (\eta_h - \eta_h^0) \qquad \text{(cathodic reaction)} \qquad (48)$$

K_F in Eq. (47) refers to the coefficient of mass transfer on the amalgam side and is related to k_f in Eq. (44) by the relationship $K_f = k_f / FSh$. Similarly, the mass transfer coefficient at the graphite surface, K_H, is related to k_h in Eq. (43) as $K_H = k_h / FSh$. χ in Eq. (47) refers to the specific conductivity of NaOH and S is the cross-sectional area of the column. Thus,

$$\frac{I}{FS} = \frac{E - \eta_h}{(1/K_F \chi)} = \frac{\eta_h - \eta_h^0}{(1/K_H)} \qquad (49)$$

Defining $I/FS = E_h K$, where $E_h = (E - \eta_h^0)$, it can easily be shown using the addition theorem (or Componendo-Dividendo method) that

$$\frac{1}{K} = \frac{1}{K_F \chi} + \frac{1}{K_H} \qquad (50)$$

The mass transfer across a section of column of thickness dh is equal to, from Eqs. (45), (46), and (50),

$$d(Lx) = KE_h S\, dh \qquad (51)$$

$$-M\, dy = KE_h S\, dh \qquad (52)$$

The integrals of these equations being

$$SH = \int_{L_1 X_0}^{L_1 X_1} \frac{d(LX)}{KE_h} \qquad (53)$$

$$\frac{SH}{M} = \int_{Y_0}^{Y_1} \frac{dY}{KE_h} = -\frac{1}{K} \int_{Y_0}^{Y_1} \frac{dY}{E_h} \qquad (54)$$

where H is the required height of the tower.

These equations permit estimation of the required volume of the amalgam decomposer. However, the important factors involved in the design of a decomposer column are the height and the diameter of the column and not its volume. Height, H, can be calculated, following Chilton and Colburn's concept for packed columns,[70] as

$$H = (\text{H.T.U}) \cdot (\text{N.T.U}) \tag{55}$$

where the first term refers to the height of transfer unit and the second term to the number of transfer units. For electrochemical systems, H.T.U. and N.T.U. will have different dimensions for the amalgam decomposer since the mechanism of mass transfer involves charge transfer reactions. Thus, from Eq. (54), it can be shown that

$$\text{N.T.U.} = \int_{Y_0}^{Y_1} \frac{dY}{E_h} \tag{56}$$

and

$$\text{H.T.U.} = M/KS \tag{57}$$

where the subscripts 0 and 1 in Eqs. (53), (54), and (56) refer to the top and the bottom of the tower, respectively.

The design of the amalgam decomposition tower is thus achieved by combining the electrochemical principles with the chemical engineering concepts involved in describing packed columns.

2. Chemical Engineering Aspects of Porous Diaphragms

(i) Transport Characteristics through Porous Media

Brine flow through a porous mat such as the vacuum-deposited asbestos diaphragm is akin to the filtration operation which is well described in many chemical engineering textbooks.[71] Thus, the flow rate V in (m^3/s) through the porous structure is related to its characteristics as

$$\frac{\Delta P g_c}{L} = K_1 \frac{(1-\varepsilon)^2}{\varepsilon^3} \cdot S^2 \cdot V\mu \tag{58}$$

where K_1 is a dimensionless constant and ΔP refers to the pressure drop in N/m², L to the thickness of porous modium (m), ε to the void fraction or porosity, S to the total surface area per unit bed volume, and μ to viscosity in units of Kg/m s. Equation (58) is valid under laminar flow conditions where the viscous term dominates the inertial term. In typical diaphragm-type chlor-alkali cells, the brine flow rate is $\sim 10^{-3}$ cm/s and, hence, laminar flow is assumed to prevail.

Void fraction or porosity, ε, is given by the ratio of the superficial velocity v or V/A (where A is the sectional area in m²) to the effective flow velocity, v_e, (in units of m/s) in the pores as

$$\varepsilon = \frac{v}{v_e} \tag{59}$$

It is useful to describe a parameter m called the hydraulic radius of the bed, which is defined by

$$m = \frac{\text{Cross section of a conduit}}{\text{Wetted perimeter}}$$

$$= \frac{\text{Volume of voids in the packet bed}}{\text{Total area of the solid}}$$

$$= \frac{AL\varepsilon}{ALS(1-\varepsilon)} = \frac{\varepsilon}{S(1-\varepsilon)} \tag{60}$$

Another term, which finds extensive use in the description of porous beds, is permeability, P (in units of m²), described by the equation

$$P = \frac{\mu v L}{\Delta P} \tag{61}$$

From Eqs. (58) and (61), it can be shown that

$$P = \frac{\varepsilon^3}{K_1 S^2 (1-\varepsilon)^2} \tag{62}$$

where K_1 varies in the range of 5-5.5. Also it can be shown that

$$\frac{m^2}{P} = \frac{K_1}{\varepsilon} \tag{63}$$

from Eqs. (60) and (62).

MacMullin et al.[72] describe the porosity of a packed bed in terms of resistance of the bed with and without the packing. If R is the electrical resistance of the porous bed filled with electrolyte and R_0 is its electrical resistance without any packing, then

$$\frac{R}{R_0} = \frac{\text{Tortuosity}}{\text{Porosity}} = \frac{\tau}{\varepsilon} \qquad (64)$$

Combining Eqs. (63) and (64) yields

$$\frac{m^2}{P} = \frac{K_1}{\tau} \cdot \frac{R}{R_0} \qquad (65)$$

Detailed investigations by MacMullin et al.[72] show that the value of K_1/τ for many porous beds is 3.666 ± 0.098. Since the tortuosity of a porous asbestos diaphragm is ~1.5 and K_1 varies in the range of 5-5.5, the model describing the porous beds is applicable for the separators used in diaphragm chlor-alkali cells.

The flow velocity, v, is given by

$$v = \frac{P}{\mu L} \cdot \Delta P \qquad (66)$$

and if the filter medium is compressible, the permeability, P, depends on the pressure drop.[73] Hence,

$$\log\left(\frac{v}{v^0}\right) = (1 - s) \log(\Delta P) \qquad (67)$$

where s refers to the compressibility factor and v^0 to the flow velocity at $\Delta P = 1$. Filtration studies through the deposited asbestos mats at constant pressure show s to be about 0.2 based on log-log plots v versus ΔP.[73]

(ii) Complications during Cell Operation

Although the model proposed for porous beds is applicable to porous asbestos diaphragms under "static" conditions, complications arise during electrolysis. Asbestos fiber consists of ~40% SiO_2 and ~40% MgO. This fiber dissolves readily in acid media while it is resistant to caustic[73] at low temperatures. During electrolysis, the magnesium layer is believed to dissolve preferentially on the anolyte side and reprecipitate within the diaphragm. Thus, the diaphragm consists of a silica-rich layer on the anolyte side of the diaphragm and a Mg-rich layer on the catholyte side.[74-78]

It is well known[79] that the flow rate through an asbestos diaphragm is significantly reduced when the current is switched on. MacMullin attributed this effect to electrokinetic phenomena.[80] However, experimental results[81] did not support this hypothesis since the brine strength is near its saturation limit. Hine et al.[73] showed that the reduction in flow rate is caused by the penetration of hydrogen bubbles into the diaphragm. Thus, when the asbestos separator is placed away from the cathode, the brine flow rate is unaffected when the current is turned on.

Regular asbestos diaphragm swells and in extreme cases touches the anode, resulting in high cell voltages. This can be prevented by thermally setting the asbestos mat with small amounts of PTFE or Halar powder (see Ref. 10 for detailed patent literature) without any adverse effects on the flow characteristics. Thus, the structure of the mat is unaffected by heat treatment while the polymer contributes to set the fiber.

As stated previously, asbestos fiber is chemically attacked in acid media. Repeated current interruptions cause the diaphragm to "tighten" by the dissolving Mg layer on the anolyte side which gets precipitated in the separator. The diaphragm structure is thus altered by either repeated shutdowns or load changes. It is in this sense that the chemical stability of asbestos diaphragms is inferior to that of membranes or PTFE-based separators. Several chloralkali manufacturers are attempting to substitute microporous PTFE diaphragms for asbestos (see Ref. 10 for detailed patent literature) to overcome the stability problems associated with asbestos and its carcinogenic effects.[37]

(iii) Mass Transfer through the Diaphragm

Mass transfer through separators in chlor-alkali cells has been studied by several authors and the general one-dimensional model (see Fig. 16) proposed by Mukaibo[82] is discussed in this section. Anolyte flows from the anode compartment through the diaphragm toward the cathode side under a hydrostatic head with a velocity of v. Hydroxyl ions generated in the cathode compartment move

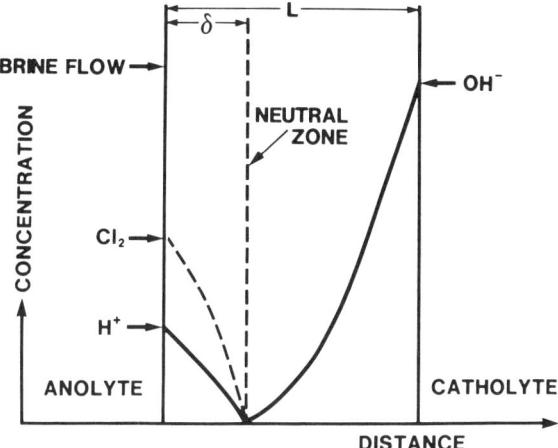

Figure 16. Mass transfer through the diaphragm and the concentration profile of H^+, OH^-, and Cl_2 in a diaphragm-type chlor-alkali cell.

into the anolyte through the diaphragm by diffusion and migration. On the other hand, H^+ and dissolved Cl_2 (Cl_2 + HOCl) move into the cathode compartment. Thus, a neutral zone is established in the diaphragm at a distance of δ from the anolyte/diaphragm interface. Thus, the flux, J, expressed in kg mole/m² s, is composed of contributions from diffusion, J_d, migration J_m, and convection J_h as

$$J = J_d + J_m + J_h \tag{68}$$

J_d in Eq. (69) is given by

$$J_d = -D_i \frac{dCi}{dx} \tag{69}$$

where D is the diffusion coefficient through the porous diaphragm having a porosity ε, and C is the concentration of species i. D can be evaluated from the diffusion coefficient in the electrolyte, D^0, as

$$F_f = \frac{R}{R_0} = \frac{D^0}{D} \tag{70}$$

where F_f is called the MacMullin number or formation factor.[83]

The migration and convection terms are given by

$$J_m = euC\frac{d\phi}{dx} \tag{71}$$

$$J_h = C_i v \tag{72}$$

where e refers to the valency of the ion, u to ionic mobility, C to the concentration in kg mole/m^3, and $d\phi/dx$ to the potential gradient in units of V/m.

The differential equations governing mass transfer through a porous diaphragm can now be described by Eqs. (73)-(76) under steady-state conditions[82]:

$$j_H = -D_H \frac{dC_H}{dx} + k_H C_H i + C_H v \tag{73}$$

$$j_{Cl} = -D_{Cl} \frac{dC_{Cl}}{dx} + C_{Cl} v \tag{74}$$

$$j_{OH} = -D_{OH} \frac{dC_{OH}}{dx} - k_{OH} C_{OH} i + C_{OH} v \tag{75}$$

$$j_H + j_{Cl} + j_{OH} = 0 \quad \text{(within the neutral zone)} \tag{76}$$

where the subscripts H, Cl, and OH refer to H$^+$, Cl$_2$ dissolved, and OH$^-$, respectively; i refers to the current density in A/cm^2; and k with subscripts H and OH refers to the contribution from migration (see Ref. 73 for details). Note that D's in Eqs. (73)-(75) refer to the diffusion coefficient of species i in the porous matrix evaluated from the bulk diffusion coefficient values using Eq. (70).

Assuming $d\phi/dx$ to be constant, these equations can be solved to obtain the caustic current efficiency (CE), j_{OH} and the caustic concentration in the catholyte, C_{OH}^0:

$$CE = \frac{i - (Fj_{OH})}{i} \times 100(\%) \tag{77}$$

$$-j_{OH} = \frac{(k_{OH} i - v) C_{OH}^0}{1 - \exp\left[-\left(\dfrac{k_{OH} i - v}{D_{OH}}\right)(L - \delta)\right]} \tag{78}$$

$$C_{OH}^0 = \left\{ v + \frac{k_{OH}i - v}{1 - \exp\left[-\left(\frac{k_{OH}i - v}{D_{OH}}\right)(L - \delta)\right]} \right\}^{-1} \frac{i}{F} \quad (79)$$

Experimental data obtained by Hine et al.[73] agree well with Mukaibo's model.

Several detailed modeling investigations[84-91] were made in recent years taking into account the diaphragm characteristics such as τ and ε. However, all these models would degenerate into the simple description proposed by Stender et al.[84] if the significance of D in Eq. (75) is properly acknowledged as noted in Eq. (70). Thus, the flux of OH^- ions is described as

$$j_{OH} = -D_{OH} \frac{dC_{OH}}{dx} - UC_{OH} \quad (80)$$

and

$$U = \frac{F}{RT} \frac{D_{OH}}{\chi} i - v \quad (81)$$

with the boundary conditions

$$C_{OH} = C_{OH}^0 \quad \text{at } x = L$$

and

$$C_{OH} = 0 \quad \text{at } x = 0$$

the following equations can be obtained:

$$-j_{OH} = \frac{C_{OH}^0 U}{1 - \exp(-uL/D_{OH})} \quad (82)$$

and

$$C_{OH} = C_{OH}^0 \left\{ \frac{1 - \exp(-UX/D_{OH})}{1 - \exp[-(U/D_{OH})L]} \right\} \quad (83)$$

U in Eq. (81) represents the difference between the migration term and the convective term. Hence, with thick separators, U is greater than zero since the convective flow is very small and the C_{OH} versus x behavior is convex. When $U < 0$, the C_{OH} versus x curve is concave as shown in Fig. 17, which is the pattern observed in

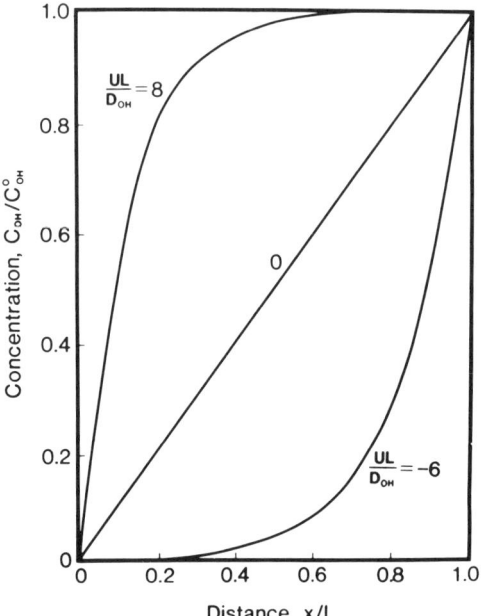

Figure 17. Profile of OH^- ion concentration in the diaphragm. Note that C_{OH}^0 is a function of UL/D_{OH}.

commercial cells. It may be noted that if δ in Mukaibo's model is assumed to be zero, it can be readily shown that Eqs. (78) and (82) are identical with $k_{OH} = (F/RT)(D_{OH}/X)$ and $U = (k_{OH}i - v)$. In practice, δ is 10^{-3} cm and, hence, it is not surprising that both models are substantiated with the experimental results (see Refs. 73 and 83).

From Eqs. (82) and (83), the caustic current efficiency can be represented as follows:

$$CE = \frac{1}{1 + (U/v)\{1 - \exp(UL/D_{OH})\}^{-1}} \times 100(\%) \qquad (84)$$

when the diaphragm is relatively thin, UL/D_{OH} is $\ll 1$ and, hence, Eq. (84) can be simplified to

$$CE \approx \frac{1}{1 + D_{OH}/vL} \times 100(\%) \qquad (85)$$

Assuming $D_{OH} = 10^{-5}$ cm^2/s, $L = 0.5$ cm, $v = 10^{-3}$ cm/s, $X =$

0.5 mho/cm, $T = 373$ K, and $i = 0.2$ A/cm^2, $UL/D_{OH} = -7$ and, hence, CE is 98%. Thus, the NaOH concentration versus distance curve in practical cells is concave as shown in Fig. 17 and the caustic current efficiency depends on the loss of OH$^-$ ions by diffusion and back migration.

All the proposed models are valid, in principle, for polymeric diaphragms as well. If CE, i, and v are fixed, which fixes C_{OH}^0 automatically, one can calculate the thickness of the separator for a chosen τ/ε. The minimum ohmic drop across the separator (ΔV) is then given by

$$\Delta V = \frac{i}{\chi} \cdot \frac{\tau}{\varepsilon} \cdot L \tag{86}$$

VI. ION-EXCHANGE MEMBRANES AND MEMBRANE TECHNOLOGY

Ion-exchange membranes were first developed for electrodialysis applications such as desalination of sea water or brackish water. The electrodialyzer consists of an anode and a cathode assembly between which a stack of anion and cation exchange membranes is placed. During electrolysis, the field across the membranes permits transfer of anions and cations producing dilute and concentrated NaCl streams. Thus, the primary role of the electrodes is one of providing electric field for the transport of ions across the membranes. On the other hand, in ion-exchange membrane electrolytic cells, products are generated at the electrodes and the membrane serves the purpose of an "efficient separator" for producing pure products in the anolyte and catholyte compartments.

1. General Requirements of Membranes for Chlor-Alkali Production

Ion-exchange membranes for chlorine-caustic production should satisfy the following requirements,[92] some of which tend to be conflicting with each other as will be elaborated in Section 6.2:
- Chemical stability
- Physical stability
- Uniform strength and flexibility

- High current efficiency
- Low electric resistance
- Low electrolyte diffusion

These requirements are met by a judicious choice and combination of the following properties of the membranes:

- Water content
- Type and distribution of ion-exchange groups
- Polymer structure and composition
- Membrane thickness

Membranes used in chlor-alkali operations are typically hydrolyzed copolymers of tetrafluoroethylene and perfluoro vinyl ether monomer containing an ion-exchange group, represented by the following general formula[93]:

$$(CF_2CF_2)_n-CF_2CF-\underset{\underset{CF_3}{|}}{(OCF_2CF-)_m}OCF_2CF_2-X$$

where X is an ion-exchange group or its precursor such as SOOF, SOOR, COOR. In chlor-alkali cells, the membranes consisting of sulfonic acid groups and/or the carboxylic acid groups find extensive use. Commercially available membranes used in the industry[93a] are manufactured by DuPont (Nafion®), Asahi Glass (Flemion®), Tokuyama Soda (Neosepta-F®), and Asahi Chemical Industry; they are structurally similar although there are minor compositional variations. Mechanical strength of these membranes is generally achieved by reinforcing with Teflon cloth or mesh (e.g., Nafion 324).

Dow Chemical Company recently announced new high-performance membranes employing only sulfonic acid groups. With shorter suspended groups, it is claimed[93b] that the water of hydration is significantly reduced and the concentration of active sites on the membrane can be increased without creating channels for back migration of caustic.

2. Properties of Membranes and Their Performance Characteristics

Significant advances have been made in recent years to develop high-performance membranes for chlor-alkali cells by a judicious selection of the relevant properties of the membranes. An attempt is made here to provide an appreciation of the basic properties that led to the development of these membranes.

The wettability or the water content of a membrane depends on the equivalent weight (EW) or the ion-exchange capacity (IEC), pretreatment, solution composition, operating temperature, etc. The equivalent weight is defined as the weight of the polymer which is neutralized by one equivalent of alkali and is the reciprocal of IEC as

$$IEC \cdot \frac{(m.\ eq)}{g\ dry\ resin} = \frac{1000}{EW\left(\frac{g\ dry\ resin}{equivalent}\right)} \quad (87)$$

The water content of a perfluoro sulfonic acid membrane and a perfluorocarboxylic acid membrane illustrated in Fig. 18 clearly

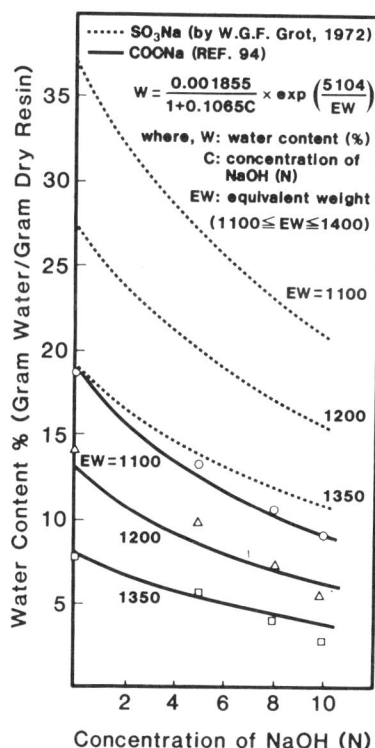

Figure 18. Water content, equivalent weight, and concentration of NaOH of perfluorocarboxylic and perfluorosulfonic acid membranes (reproduced with permission from the authors).

shows that the water content of the sulfonic acid group is more than that for the carboxylic acid group for the same EW of the membrane.[94] On the other hand, the electrical conductance of a membrane generally increases with increase in water content, and decreases with increase in equivalent weight. As a result, the sulfonic acid form of the membrane exhibits[95] lower electric resistance as shown in Figs. 19 and 20. However, the caustic current efficiency of the sulfonic acid membrane is low (~80%) because of its high water content while the current efficiency achieved with carboxylic acid membrane is quite high (~96%). Thus, most of the high-efficiency membranes are bilayer membranes, with the sulfonic acid side facing the anolyte and the carboxylic acid side of the membrane facing the catholyte. The thickness of the carboxylic acid part of the membrane need be only 5-7 μm to achieve high current efficiencies without incurring much of the voltage penalty.

Multilayer membranes with perfluorocarboxylic and perfluorosulfonic acid groups can be prepared by lamination of two different films containing the respective polymers[96-98] or by chemical conversion of perfluorosulfonic acid to perfluorocarboxylic acid.[99-102] The latter method is preferred to achieve the desired thickness of 5-10 μm of the carboxylic acid pendant group.

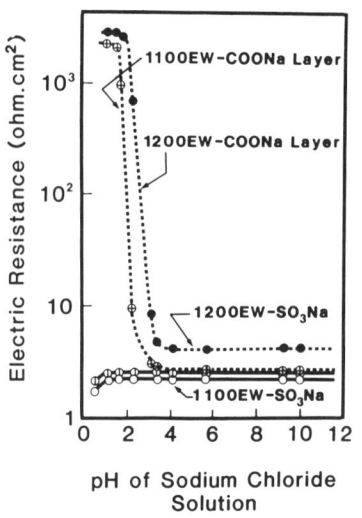

Figure 19. Variation of electric resistance of membranes with pH of NaCl solution at 25°C (reproduced with permission from the authors and Society of Chemical Industry, London).

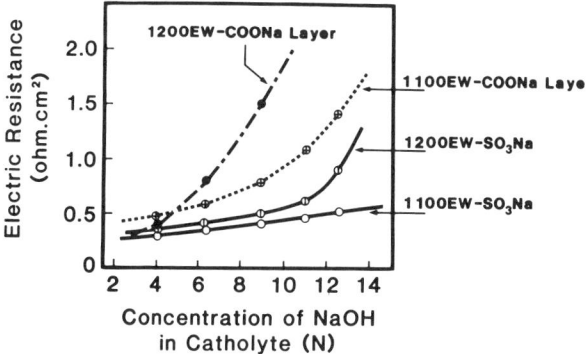

Figure 20. Variation of electric resistance of membranes with caustic concentration at 80°C (reproduced with permission from the authors and Society of Chemical Industry, London).

There is yet another reason for the choice of multilayer membrane as it provides excellent chemical stability. During operation of membrane cells, HCl is added to the anolyte to maintain the pH as low as possible to minimize the accumulation of $NaClO_3$ in the anode compartment. The chemical stability of the perfluorosulfonic acid group is excellent in acidic media since the pK_a value of sulfonic acid group is less than 1. The pK_a of the carboxylic acid group is 2-3 and, hence, its stability is quite acceptable during electrolysis. However, HCl addition for neutralizing the back-migrating caustic is plausible only with the sulfonic acid membranes facing the anolyte because of pK_a restrictions and consequent resistance increase of carboxylic acid membranes (see Fig. 19).

Many of the membrane producers treat the membrane surface so that gas bubbles, especially H_2, do not adhere to the surface thus causing an increase in the cell voltage. Asahi Chemical Industry claims[103] minimization of the voltage increase even with decreased anode-cathode distances, by roughening the membrane surface facing the catholyte. Surface modification with porous materials such as "fluorocarbon polymer and inorganic materials" is also claimed[95] to improve bubble release from the catholyte side of the membrane. Thus, the surface treatment of the membrane has permitted designing zero-gap cells such as the DN system[95] developed by Tokuyama Soda and the AZEC system[104] by Asahi Glass Company.

Figure 21 depicts trends in the variation of cell voltage with interelectrode distance (obtained with Flemion membranes).

There have been several theoretical modeling investigations aimed at understanding the structural and transport characteristics of ion-exchange membranes. The reader is referred to Refs. 105 and 106 for details.

3. Engineering Design Aspects of Ion-Exchange Membrane Cell Technology

(i) Ion Transport across Membranes

The NaCl concentration profile across a membrane during electrolysis is shown in Fig. 22. The material balance for the depleted layer near the anolyte side of the membrane can be expressed as[107]:

$$C - C_0 = \frac{\delta}{D} \cdot \frac{i}{F}(t_M - t_{Na^+}) \tag{88}$$

where C and C_0 refers to NaCl concentration in the bulk and at the surface of the membrane, respectively, i to current density, F to Faraday's constant, t_M to the transport number of Na^+ in the membrane, t_{Na^+} to the transport number in the anolyte, D to the diffusion coefficient, and δ to the thickness of the boundary layer.

The current density at which C_0 approaches zero is defined as the limiting current density, i_L which is given by

$$i_L = \frac{C}{\delta} \frac{DF}{t_M - t_{Na^+}} \tag{89}$$

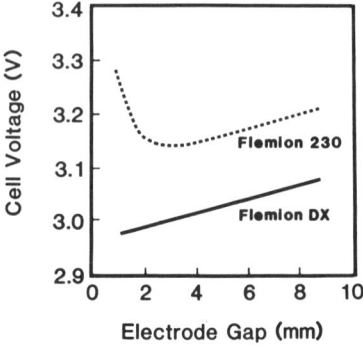

Figure 21. Dependence of cell voltage on anode–cathode gap at 90°C and at a current density of 20 A/dm² (NaOH concentration: 35 wt.%; Anolyte concentration: 3.5 N; reproduced with permission from the authors and Society of Chemical Industry, London).

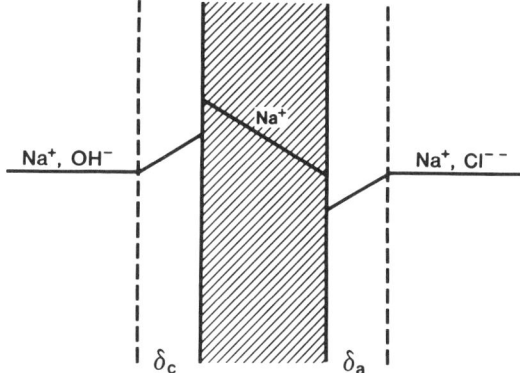

Figure 22. Concentration profile across the membrane and the diffusion layers (reproduced with permission from the author and *J. Electrochem. Soc.*).

Thus, at current densities above the limiting current, NaCl concentration at the membrane surface will be zero and discharge of water (or OH^- migration) would occur, which results in decreased current efficiency. Hence, it is important to minimize δ in order to operate the cell at as high a current density as possible, high current density operation being desirable in view of the high cost of the membranes.

(ii) Current Density Distribution

Bipolar cells are claimed to have superior current density distribution compared to monopolar cells.[103,108] However, properly designed monopolar cells should also have uniform current distribution if several contact points are provided between the anode and current collector as well as the cathode and current collector. In the case of bipolar cells, the metal frame is the current collector which results in negligible structural ohmic drop and uniform current distribution.

A majority of membrane cells operate with the membrane resting on the anode. It is conceivable that the structure of the anode is an important design parameter. Asahi Chemical Industry claims to have optimized the structure of the anode by the use of a perforated plate. However, it is theoretically possible to design an expanded mesh anode which is as good as the perforated plate anode and is cost effective.

Another important factor which affects the current distribution is the void fraction within the electrolyte. It is obvious that the design variables involved to minimize the void fraction would be (1) electrode height, (2) liquid flow rate through the anode and cathode compartments, and (3) anode structure. Asahi Chemical operate at high pressures to compress the bubble volume and hence, the void fraction as practiced (see Ref. 14 for details) in some water electrolyzers. Recent studies[109,110] on the effect of electrode structure on void fraction show that expanded mesh electrodes generally give low void fraction.

Thus, uniform current density is achieved by proper electrode/current collector design, optimal anode/cathode structure (e.g., geomerty, height), and maximal flow rate through the electrolyzer.

(iii) Water Transport across Membrane

An ideal membrane should transport as much water as possible without incurring losses in current efficiency. This would result in (1) increased NaCl concentration in the anolyte (hence, reduced brine resaturation requirements of the anolyte stream), and (2) decreased process soft water requirements on the catholyte side (for a desired caustic strength). Different water transport characteristics of individual membranes can offer the above process advantages.

(iv) Effect of Ca^{2+} and Mg^{2+} in Feed Brine and Soft Water

Brine and soft water of ultrapure quality is essential for smooth and efficient operation of membrane cell process since Ca^{2+} and Mg^{2+} ions can harm the performance of ion-exchange membranes in the following ways. The precipitated $Ca(OH)_2$ and $Mg(OH)_2$ offer increased electrical resistance across the membrane thereby increasing the cell voltage. Furthermore, the anolyte diffusion layer characteristics are affected which would alter the optimum current density. More seriously, the membrane performance is affected irreversibly via the formation of micropores within the membrane which would lower the caustic current efficiency.

High hardness levels in soft water result in the deposition of hydroxide films on the cathode surface thereby increasing the cathodic overpotential. With porous cathode matrices, plugging may occur leading to maldistribution of flow within the porous structure. Other impurities that are adverse to membrane performance include sulfates, chlorates, and silica.

(v) HCl Addition, O_2 Evolution, and Anode Life

A majority of membrane cells operate with the membranes resting on the anode.[9,102-104,108] Hence, it is important that the anode have inherent stability in a high-pH medium, as one of the failure modes of $RuO_2 + TiO_2$-based anodes (anode passivation phenomenon) is believed to be caused by an O_2 evolution reaction favored in alkaline media. Thus, addition of HCl to the anolyte not only suppresses the oxygen evolution reaction thereby increasing the anode life but also lowers the chlorate content in the anolyte. Enhanced anode life may also be achieved by using $RuO_2 + TiO_2$ solid-solution-based coating with dopants such as Ir, Hf, and Sn which suppress[11] the oxygen discharge reaction.

4. Advantages Afforded by Membrane Technology

The advent of membrane cells is largely dictated by environmental regulations related to the hazardous nature of mercury effluents (from mercury cell process) and asbestos (used in diaphragm cell technology).[37] However, there are several advantages offered by the membrane cell technology as noted below.

(a) High-Purity Caustic. Caustic concentrations of 28–33 wt % containing less than 50 ppm NaCl and 20 ppm $NaClO_3$ can be directly produced.

(b) Low Steam Costs. Since the caustic strength is quite high compared to diaphragm cell liquor, steam economies accrue. In addition, the evaporator system can be operated easily compared to a diaphragm cell effluent because of the lower salt content.

(c) High Current Density Operation. The brine gap in membrane cells can be made very small unlike in asbestos diaphragm

cells, and the voltage drop across the membrane is lower compared to asbestos. These facts weighted against the relatively high cost of membranes force one to operate the membrane cells at relatively high current densities to achieve high space-time yields (or economy of floor space).

(d) Interrupted Operations. It is very difficult to operate asbestos diaphragm cells when there are too many power interruptions and load sheddings since they create high anolyte levels. These problems are nonexistent in membrane cells, and, hence, these cells are uniquely suited to handle current interruptions and load changes[111] resulting in effective utilization of the cheap off-peak power.

(e) Pressurized Operations. Since membranes are impervious to gases, pressurized operation is possible and is claimed[103] to reduce membrane vibrations and, hence, the consequent damage to the membrane. Furthermore, drying for Cl_2 and H_2 gases would be somewhat simplified.

(f) Suitability for Bipolar Cells. Membranes are ideally suitable for bipolar cell design with negligible hardware voltage drop. For example, the structural ohmic drop is estimated to be only 3 mV even at a current density as high as 50 A/dm^2.[108]

(g) Suitability with Nonnoble-Metal-Based Activated Cathode. Most of the nonnoble metal activated cathode coatings are prone to corrosion by hypochlorite during shutdowns in diaphragm cells. Membranes prevent the anolyte and catholyte from readily mixing and provide sufficient time to remove hypochlorite from the anode compartment thereby preventing the anodic dissolution of the catalytic cathode coatings. Thus, nonnoble metal based cathode coatings find extensive use in membrane cells to achieve reduction in energy consumption.

5. State of the Art of Membrane Cell Technology

Since the first membrane cell installation at the Nobeoka plant by Asahi Chemical Industry in 1975, several membrane cell plants have been constructed, especially in Japan, as a pollution-free chlor-alkali process. By the end of 1982, the total capacity of the membrane cell process in the world was estimated to be about 600,000 tons of NaOH per year.[112]

The mercury cell requires 2700–3000 kW h/M.T NaOH, depending on the current density, producing 50% caustic directly by the amalgam decomposer without any additional energy. Although the diaphragm cell consumes[5] relatively small amounts of energy, 2200–2500 kW h/M.T NaOH, the steam requirement to concentrate the cell liquor is about 2–4 tons/M.T of NaOH. The energy consumption of modern membrane cells is claimed to be less than 2100 kW h/M.T NaOH as noted below:

- Tokuyama Soda[95]: 2080 dc kW h/M.T NaOH at 30 A/dm^2 and 28% NaOH using activated cathodes.
- Asahi Chemical[103]: 2100 dc kW h/M.T NaOH with activated cathode, and 2200 dc kW h/M.T NaOH with perforated steel cathode at 25% NaOH and 30 A/dm^2.
- Asahi Glass[104]: 2080 dc kW h/M.T NaOH at 30 A/dm^2.
- Eltech Systems[113]: 2100 dc kW h/M.T NaOh at 30 A/dm^2 using activated cathode.

There have been significant activities in membrane retrofit technology for mercury cell or diaphragm cell plants. Diaphragm retrofit technology has been proven in Japan by several companies including Kanegafuchi, Mitsui Toatsu, Tokuyama Soda, and Toyo Soda.[114,115] Conversion of mercury cell circuits to membrane cell technology was recently addressed by Noli.[116] Diaphragm retrofit technology using membranes should pose a serious challenge to microporous PTFE separators as a replacement for asbestos.

Currently, major research and developmental emphasis is toward achieving an energy consumption of less than 2000 kW h/M.T NaOH. The membrane technology is so advanced that a reduction of ~100 kW h/M.T NaOH will reach the practical minimum value. Thus, the membrane cell process promises to be the main technology for chlor-alkali production in the near future.

NOTATION†

a area of electrolyte film per m^3 of electrode (m^2/m^3)
C total molar concentration (mol/m^3)

† All current densities are defined to be current/projected area of the electrode.

C_1^0	solubility of oxygen in the electrolyte at a partial pressure of 1 atmosphere (mol/Nm)
$C_1(r_a, Z)$	concentration of oxygen at the film–agglomerate interface (mol/m^3)
C_T	double layer capacity of the active layer (farad)
C_C	double layer capacity of the current collector (farad)
D	diffusion coefficient of oxygen in the liquid phase (m^2/sec)
$D_{i,j}^{\text{eff}}$	effective binary diffusion coefficient (m^2/s)
$D_{1,2}^0$	binary diffusion of oxygen in nitrogen (m^2/s)
$E(t)$	potential drop at the active layer at $Z_{\max}$; see Fig. A.1 (V)
$E_1(t)$	potential drop at the current collector; see Fig. A.1 (V)
$E_2(t)$	potential drop at the contact resistance; see Fig. A.1 (V)
E_{meas}	potential drop between the current collector and the luggin tip
$E(Z)$	potential difference between the electrode and the solution (V)
$\Delta E(t)$	change in the measured potential between the current collector and the SCE reference electrode due to a current pulse (V)
$E^*(Z^*)$	dimensionless potential
E_a	porosity–tortuosity factor for the agglomerates (dimensionless)
$E_e(Z)$	potential in the electrode (V)
$E_i(Z)$	potential in the electrolyte (V)
E_k	kth experimental potential difference between the electrode and solution (V)
E_n	fraction of the electrode consisting of agglomerates (dimensionless)
E_P	porosity–tortuosity factor for gas diffusion in porous backing (dimensionless)
E_r	reference potential (V)
E_T	porosity–tortuosity factor for gas diffusion in active layer (dimensionless)
F	Faraday's constant (96487 C/mol of electrons)

REFERENCES

[1] J. W. F. Hardie, *Electrolytic Manufacture of Chemicals from Salt*, The Chlorine Institute, Atlanta, Georgia, 1975.

[2] J. E. Currey and G. G. Pumplin, *Encyclopedia of Chemical Processing and Design,* Ed. by J. J. McKetta and W. A. Cunningham, Vol. 7, p. 305, Marcel Dekker, New York, 1978.

[3] *Diaphragm Cells for Chlorine Production,* Society of Chemical Industry, London, 1977.

[4] Y. C. Yen, Process economics program report No. 61 (1970), 61A (1974), 61B (1978), 61C (1982); Stanford Research Institute, Menlo Park, California.

[5] D. L. Caldwell, *Comprehensive Treatise of Electrochemistry,* Vol. 2, Chap. 2, Ed. by J. O'M. Bockris, B. E. Conway, E. A. Yeager, and R. E. White, Plenum Press, New York, 1981.

[6] A. T. Kuhn, *Industrial Electrochemical Processes,* Elsevier, New York, 1971.

[7] Proceedings of the Oronzio DeNora symposium on chlorine technology, Ed. by O. DeNora, Impiante Elettrochimica S.P.A., Italy, 1979.

[8] Final report on improvements in energy efficiency of industrial electrochemical processes, Ed. by T. R. Beck, ANL/OEPM-77-2, 1977.

[9] *Modern Chlor-Alkali Technology,* (A) Ed. by M. O. Coulter, 1980; (B) Vol. 2, Ed. by C. Jackson, 1983, Society of Chemical Industry, London.

[10] Chlorine production processes: Recent and energy saving developments, Ed. by J. S. Robinson, Noyes Data Corporation, New Jersey, 1981.

[11] D. M. Novak, B. V. Tilak, and B. E. Conway, *Modern Aspects of Electrochemistry,* No. 14, Chap. 4, Ed. by J. O'M. Bockris, B. E. Conway, and R. E. White, Plenum Press, New York, 1982.

[12] *Electrodes of Conductive Metallic Oxides,* Part A and Part B, Ed. by S. Trasatti, Elsevier, New York, 1981.

[13] S. Trasatti and W. E. O'Grady, *Advances in Electrochemistry and Electrochemical Engineering,* Vol. 12, p. 177, Ed. by H. Gerischer and C. W. Tobias, Wiley, New York, 1981.

[14] B. V. Tilak, P. W. T. Lu, J. E. Colman, and S. Srinivasan, *Comprehensive Treatise of Electrochemistry,* Vol. 2, Chap. 1, Ed. by J. O'M. Bockris, B. E. Conway, E. Yeager, and R. E. White, Plenum Press, New York, 1981.

[15] A. J. Appleby, H. Kita, M. Chemla, and G. Bronöel, *Encyclopedia of Electrochemistry of the Elements,* Vol. IX, Part A, Chap. IXa-3, Ed. by A. J. Bard, Marcel Dekker, New York, 1982.

[16] M. S. Kircher, H. R. Engle, B. H. Ritter, and A. H. Bartlett, *J. Electrochem. Soc.* **100** (1953) 448.

[17] F. Hine, M. Yasuda, and T. Tanaka, *Electrochem. Acta.* **22** (1977) 429.

[18] F. Hine, M. Yasuda, and K. Fujita, *J. Electrochem. Soc.* **128** (1981) 2314.

[19] C. J. Molnar and M. M. Dorio, paper presented at the Electrochem. Soc. Meeting, Atlanta, Georgia, October 1977.

[20] G. Angel and T. Lunden, *J. Electrochem. Soc.* **99** (1952) 435, 442; **100** (1953) 39; **102** (1955) 124, 243; and **104** (1957) 167.

[21] W. E. Cowley, B. Lott, and J. H. Enteisle, *Trans. Inst. Chem. Eng.* **41** (1963) 372.

[22] K. Hass, *Electrochem. Technol.* **5** (1967) 246.

[23] F. Hine, S. Matsuura, and S. Yoshizawa, *Electrochem. Technol.* **5** (1967) 251.

[24] F. Hine, M. Yasuda, F. Wang, and K. Yamakawa, *Electrochem. Acta* **16** (1971) 1519.

[25] R. B. MacMullin, *Chlorine,* ACE monograph No. 154, Ed. by J. S. Sconce, p. 151, Reinhold, New York, 1962.

[26] F. Hine, S. Yoshizawa, and S. Okada, *Kogyo Kagaku Zashi (J. Chem. Soc. Jpn. Ind. Chem. Sec.)* **62** (1959) 769, 773, 778.

[27] F. Hine, T. Sugimori, S. Yoshizawa, and S. Okada, *Electrochem. Technol.* **62** (1959) 955.

[28] F. Hine, S. Yoshizawa, S. Okada, N. Yokota, T. Kadota, and J. Kushiro, *Electrochem. Technol.* **62** (1959) 961.
[29] H. A. Sommers, *Chem. Eng. Prog.* **61**(3) (1965) 94.
[30] F. Hine and M. Yasuda, *J. Electrochem. Soc.* **119** (1972) 1057.
[31] F. Hine, M. Yasuda, and M. Higuchi, Chlorine bicentennial symposium, p. 278, Electrochem. Soc., 1974.
[32] Y. Sekine and C. Motohashi, *Soda to Enso (Soda and Chlorine)* **33** (1982) 66.
[33] J. J. Wolff and R. E. Anderson, paper presented at the AIChE Meeting, Orlando, Florida, May, 1982.
[34] F. Hine and Y. Yagishita, *Soda to Enso (Soda and Chlorine)* **24** (1973) 131.
[35] N. Yokota, *Soda to Enso (Soda and Chlorine)* **9** (1958) 495.
[36] N. Yokota, *Kagaku Kogaku (J. Soc. Chem. Eng. Jpn.)* **23** (1959) 438; **25** (1961) 170.
[37] P. J. Kienholz, paper presented at the Chemical Marketing Research Association, New York, May 1983.
[38] F. Hine, M. Yasuda, I. Sugiura, and T. Noda, *J. Electrochem. Soc.* **121** (1974) 220.
[39] F. Hine and M. Yasuda, *J. Electrochem. Soc.* **121** (1974) 1289.
[40] British patent 6490/67, 1967.
[41] H. B. Beer, *Diaphragm Cells for Chlorine Production*, p. 11, Society of Chemical Industry, London, 1977.
[42] O. DeNora, *Chem. Ing. Tech.* **42** (1970) 222; **43** (1971) 182.
[43] K. J. O'Leary, U.S. Patent No. 3,776,834, 1973.
[44] F. Hine, M. Yasuda, T. Noda, T. Yoshida, and J. Okuda, *J. Electrochem. Soc.* **126** (1979) 1439.
[45] R. G. Erenburg, L. I. Khristalik, and V. I. Bystrov, *Electrokhimiya* **8** (1972) 1740.
[46] R. G. Erenburg, L. I. Khristalik, and I. P. Yaroshevskaya, *Electrokhimiya* **11** (1975) 1068, 1072, 1236.
[46a] V. E. Kazarinov, L. I. Khristalik, and Yu. V. Pleskov, *Trans. SAEST (India)* **12** (1977) 287.
[47] British patent 877,901 (1961); U.S. Patents Nos. 3,977,958, 4,142,005, and 4,061,549.
[48] M. J. Hazelrigg and D. L. Caldwell, Abstract No. 457, Electrochem. Soc. Meeting, Seattle, Washington, May 1978; Chap. 10 in Ref. 9A.
[49] S. Matsuura, Y. Ozaki, K. Motani, Y. Ohashi, and Y. Onoue, Abstract No. 417, Electrochem. Soc. Meeting, St. Louis, Missouri, May 1980.
[50] Japanese patent Sho 25-2305.
[51] K. Kanzaki and K. Fukatsu, *Denki Kagaku (J. Electrochem. Soc. Jpn.)* **19** (1951) 255; **23** (1955) 169.
[52] F. Hine, M. Yasuda and M. Watanabe, *Denki Kagaku (J. Electrochem. Soc. Jpn.)* **47** (1979) 401.
[53] I. Malkin and J. R. Brannan, Abstract No. 262, Electrochem. Soc. Meeting, Boston, May 1979.
[53a] A. M. Greenberg, *New Materials and New Processes*, Ed. by M. Nagayama *et al.*, p. 238, J.E.S. Press, Cleveland, Ohio, 1981.
[54] D. W. Carnell and C. R. S. Needes, Abstract No. 260, Electrochem. Soc. Meeting, Boston, May 1979.
[54a] D. E. Brown, F. O. Fogarty, M. N. Mahmood, and A. K. Turner, Chap. 14 in Ref. 9B.
[55] W. W. Carlin and W. B. Darlington, Abstract No. 261, Electrochem. Soc. Meeting, Boston, May 1979; W. B. Darlington, p. 30 in Ref. 7.
[56a] U.S. Patent No. 4,358,353 (1982); K. Viswanathan and B. V. Tilak, *J. Electrochem. Soc.* **131** (1984) 1551.

[56] M. C. M. Man and A. L. C. Tseung, Abstract No. 263, Electrochem. Soc. Meeting, Boston, May 1979; British patent 1,556,452.
[57] E. Yeager, *Electrochemistry in Industry—New Directions*, Ed. by U. Landau, E. Yeager, and D. Kortan, p. 29, Plenum Press, New York, 1982.
[57a] E. Yeager, Proc. Renewable Fuels and Advanced Power Sources for Transportation on Workshop, June 1982, Ed. by H. L. Chum and S. Srinivasan, p. 27, Report No. SERI/CP-234-1707; DE 83011988.
[58] V. H. Thomas, Paper presented at the 5th Symposium on Soda Industry, Kyoto, November 1981.
[59] L. J. Gestaut, J. A. Moomaw, and V. H. Thomas, paper presented at the Chlorine Institute's 25th Plant Operations Seminar, Atlanta, Georgia, February 1982.
[60] L. J. Gestaut, T. M. Clere, C. E. Graham and W. R. Bennett, Abstract No. 393; L. J. Gestaut, T. M. Clere, A. J. Niksa and C. E. Graham, Abstract No. 124, Electrochem. Soc. Meeting, Washington, D.C., October 1983.
[61] R. B. MacMullin, *Chem. Eng. Prog.* **46** (1950) 440.
[62] *Constitution of Binary Alloys*, Ed. by M. Hansen, McGraw-Hill, New York, 1958.
[63] *Constitution of Binary Alloys*, 1st Suppl., Ed. by R. P. Elliott, McGraw-Hill, New York, 1965.
[64] *Constitution of Binary Alloys*, 2nd Suppl., Ed. by F. A. Shunk, McGraw-Hill, New York, 1969.
[65] *Supplement to Mellor's Comprehensive Treatise on Inorganic and Theoretical Chemistry*, Vol. 2, Suppl. 2, p. 581, Longmans, London, 164.
[66] J. Balej, *Electrochim. Acta.* **21** (1976) 953.
[67] F. Hine, *Muki Kogyo Kagaku (Inorganic Industrial Chemistry)*, p. 21, Asakura, Tokyo, 1967.
[68] F. Hine, M. Okada, and S. Yoshizawa, *Denki Kagaku* **27** (1959) 419.
[69] F. Hine, *Electrochem. Technol.* **2** (1964) 79.
[70] T. H. Chilton and A. P. Colburn, *Ind. Eng. Chem.* **27** (1935) 255.
[71] W. L. McCabe and J. C. Smith, *Unit Operations of Chemical Engineering*, Second Edition, McGraw-Hill, New York, 1967.
[72] R. B. MacMullin and G. A. Muccini, *AIChE J.* **2** (1956) 393.
[73] F. Hine, M. Yasuda, and T. Tanaka, *Electrochim. Acta* **22** (1977) 429.
[74] B. Nagy and T. F. Bates, *Am. Mineralogist* **37** (1952) 1055.
[75] G. J. Young and F. H. Healy, *J. Phys. Chem.* **58** (1954) 881.
[76] F. L. Pundsack, *J. Phys. Chem.* **59** (1955) 892.
[77] S. C. Clark and P. F. Holt, *Nature* **4708** (1960) 237.
[78] I. Choi and R. W. Smith, *J. Colloid Interface Sci.* **40** (1972) 253.
[79] M. S. Kircher, *Chlorine*, A.C.S. Monograph 154, Ed. by J. S. Sconce, p. 105, Reinhold, New York, 1962.
[80] R. B. MacMullin, private communication.
[81] F. Hine and M. Yasuda, *J. Electrochem. Soc.* **118** (1971) 166.
[82] T. Mukaibo, *Denki Kagaku* **20** (1952) 482.
[83] R. E. White, J. S. Beckerdite, and J. Van Zee, Abstract No. 560, Electrochem. Soc. Meeting, San Francisco, May 1983; J. Van Zee and R. E. White, *J. Electrochem. Soc.* **132** (1985) 818.
[84] V. V. Stender, O. S. Ksenzhek, and V. N. Lazarev, *Zh. Prikl. Khim.* **40** (1967) 1293.
[85] O. S. Ksenzhek and V. M. Serbrit-Skii, *Sov. Electrochem.* **4** (1968) 1294.
[86] V. M. Serebrit-Skii and O. S. Ksenzhek, *J. Appl. Chem. U.S.S.R.* **43** (1970) 69; *Sov. Electrochem.* **7** (1971) 1592.
[87] V. L. Kubasov, *Sov. Electrochem.* **12** (1976) 72.
[88] A. W. Bryson, P. G. Cook, and M. G. Lawrence, *Chem. Eng. Commun.* **3** (1979) 53.

[89] W. H. Koh, *AIChE Symp. Ser.* **77**(204) (1981) 213.
[90] D. L. Caldwell, P. A. Poush, J. W. Van Zee, and R. E. White, *Proc. Symp. Electrochem. Process and Plant Design*, p. 216, Ed. by R. C. Alkire, T. R. Beck, and R. D. Varjian, Proceedings Vol. 83-6, The Electrochem. Soc., New Jersey, 1983.
[91] K. Viswanathan, Abstract No. 415, Electrochem. Soc. Meeting, Minneapolis, Minnesota, May 1981.
[92] M. Seko, S. Ogawa, and K. Kimoto, paper presented at the Am. Chem. Soc. Centennial Meeting, New York, April 1976.
[93] D. J. Vaughan, *DuPont Innovation* **4**(3) (1973) 10.
[93a] D. Bergner, *J. Appl. Electrochem.* **12** (1982) 631.
[93b] *Chemical Week*, Nov. 17 (1982) 35.
[94] M. Seko, paper presented at the ACS Polymer Division Workshop on Perfluorinated Ionomer Membranes, Lake Buena Vista, Florida, February 1982.
[95] T. Sata, K. Motani, and Y. Ohashi, *Ion Exchange Membranes*, Ed. by D. S. Flett, Chap. 9, p. 137, Society of Chemical Industry, London, 1983.
[96] M. Seko, U.S. Patent No. 4,178,218, 1979.
[97] C. J. Molnar, E. H. Price, and P. R. Resnick, U.S. Patent No. 4,176,215, 1979.
[98] T. Asawa, Y. Oda, and T. Gunjima, Japanese Patent. Appl. No. 52-36589, 1977.
[99] M. Seko, Y. Yamakoshi, H. Miyauchi, M. Fukomoto, K. Kimoto, I. Watanabe, T. Hane, and S. Tsushima, U.S. Patent No. 4,151,053, 1979.
[100] W. G. Grot, G. J. Molnar, and P. R. Resnick, Belgian Patent No. 866122, 1978.
[101] T. Sata, A. Nakahara, and J. Ito, Japanese Pat. Appl. No. 53-137888, 1978.
[102] Wm. D. Morrison, paper presented at the Chlorine Institute's 26th Chlorine Plant Operations Seminar, New Orleans, Louisiana, February 1983.
[103] M. Seko, A. Yomiyama, and S. Ogawa, *Ion Exchange Membranes*, Ed. by D. S. Flett, Chap. 8, p. 121, Society of Chemical Industry, London, 1983.
[104] H. Ukihashi, T. Asawa, and H. Miyake, *Ion Exchange Membranes*, Ed. by D. S. Flett, Chap. 11, p. 165, Society of Chemical Industry, London, 1983.
[105] K. A. Mauritz and A. J. Hopfinger, *Modern Aspects of Electrochemistry*, No. 14, Chap. 6, Ed. by J. O'M. Bockris, B. E. Conway, and R. E. White, Plenum Press, New York, 1982.
[106] D. N. Bennion, *Proc. Symp. on Membranes and Ionic and Electronic Conducting Polymers*, p. 78, Ed. by E. B. Yeager, B. Schuum, K. Mauritz, K. Abbey, D. Blankenship, and J. Akridge, proceedings Vol. 83-3, The Electrochem. Soc., 1983.
[107] J. Jorne, *J. Electrochem. Soc.* **129** (1982) 722.
[108] M. Seko, S. Ogawa, M. Yoshida, and H. Shiroki, paper presented at the International Society of Electrochemistry, 34th Meeting, Erlangen, Germany, September 1983.
[109] F. Hine, M. Yasuda, Y. Ogata, and K. Hara, *J. Electrochem. Soc.* **131** (1984) 83.
[110] G. H. Sedahmed and L. W. Shemilt, *J. Appl. Electrochem.* **14** (1984) 123.
[111] S. M. Ibrahim, E. H. Price, and R. A. Smith, p. 53, Ref. 9B.
[112] S. Ogawa, paper presented at Indian Inst. Chem. Eng., New Delhi, March 1980.
[113] Brochure entitled MGC—Monopolar Membrane Electrolyzer—Data sheet, ECL-MGC-1 by Eltech Systems Corporation, 1983.
[114] M. Esayian and J. H. Austin, paper presented at the 27th Chlorine Plant Operations Seminar, Washington, D.C., February 1984.
[115] T. Tozuka, H. Aikawa, T. Onishi, K. Yamaguchi, I. Kumagai, and T. Ichisaka, paper presented at the 27th Chlorine Plant Operations Seminar, Washington, D.C., February 1984.
[116] B. Noli, paper presented at the 27th Chlorine Plant Operations Seminar, Washington, D.C., February 1984.

6

Phenomena and Effects of Electrolytic Gas Evolution

Paul J. Sides

Department of Chemical Engineering, Carnegie-Mellon University, Pittsburgh, Pennsylvania 15213

I. INTRODUCTION

Electrolytic gas evolution is a significant and complicated phenomenon in most electrochemical processes and devices. In the Hall process for aluminum production, for example, bubbles evolved on the downward-facing carbon anodes stir the bath and resist the current, both of which directly affect the heat balance and the cell voltage. Bubbles appear as a result of primary electrode reactions in chlorine and water electrolysis, and as the result of side reactions in the charging of lead–acid batteries and some metal electrowinning. Stirring of the electrolyte by gas evolution is an important phenomenon in chlorate production. Electrolytically evolved bubbles have also been used in mineral flotation. Relatively few major electrochemical processes do not evolve gas.

Electrolytic gas evolution is a dynamic phenomenon affected by interactions among all the process variables. The interaction of the potential, electrode, and electrolyte not only determines the rate at which gas is evolved, but also affects the contact angles of the bubbles that determine, in conjunction with the electrolyte surface tension, the fundamental forces binding the bubbles to the electrode. Since the process occurs at a surface, small quantities of impurities may have a large effect. The dynamics of bubble evolution depend on current density, electrode morphology, and pretreatment. Bubble evolution is generally rapid and special techniques such as high-speed cinematography and microscopy have been used as well as electrochemical measurements to study it.

The physical process of gas evolution can be divided into three stages: nucleation, growth, and detachment. Bubbles nucleate at electrode surfaces from solutions highly supersaturated with product gas and grow by diffusion of dissolved gas to the bubble surface or by coalescence with other bubbles. They detach from the electrode when buoyancy or liquid shearing forces pulling the bubbles away overcome the surface forces binding them.

The effects of gas bubbles include their obstruction of electric current and the stirring of electrolyte within a cell. Bubbles decrease the effective conductivity of the electrolyte and hence increase ohmic losses in the cell. Mixing the electrolyte in the crucial region near the surface, bubbles improve heat transfer away from the electrode to the walls or mass transfer of diffusion-controlled species to the electrode.

Electrolytic gas evolution can be discussed on two scales of length. The macroscopic or process scale is important to the overall design of equipment and includes modeling the overall distribution of gas in the reactor and the effects of gas bubbles on the gross electrolyte flow pattern. The microscopic scale is where the details of bubble events and their consequences are found. In this review, I concentrate on the latter, microscopic scale.

II. NUCLEATION, GROWTH, AND DETACHMENT OF BUBBLES

1. Nucleation

Nucleation theory has been advanced for vaporization of pure substances[1] and for nucleation of bubbles from solutions containing dissolved gas.[2] Bubbles of a critical radius and larger grow while bubbles having radii less than this dimension tend to decay. The result of nucleation theory is the prediction of the maximum attainable limit of supersaturation. Two equations are sufficient for this calculation:

$$J = Z \exp\left[-\frac{16\pi\sigma^3}{3kT(P''-P')^2}\right] \quad (1)$$

$$P'' = \frac{P'C'}{\nu_2 C_0} + \frac{P_0}{\nu_1}\left\{\exp\left[\frac{v_1(P'-P_0)}{kT}\right]\exp(-C')\right\} \quad (2)$$

where P_0 is the vapor pressure of the pure solvent, P'' is the pressure of dissolved gas inside the bubble, P' is the external pressure in the liquid, C' is the concentration of dissolved gas in solution, C_0 is the saturation concentration of dissolved gas, σ is the surface tension of the liquid, Z is the preexponential frequency factor, J is the nucleation rate, ν_1, ν_2 are activity coefficients of solvent and solute, v_1 is the specific volume of the pure solvent, and k is Boltzmann's constant. On the right-hand side of the first equation is the exponential function that governs the rate of nucleation while the second equation relates the pressure inside the critical bubble to the concentration of dissolved gas in the surrounding liquid. Solving these equations for hydrogen gas dissolved in 1 N sulfuric acid, one calculates a supersaturation of a thousandfold as the limit.[3] Supersaturation in the vicinity of gas evolving electrodes does not reach this magnitude, however, as demonstrated by several investigators who found supersaturations of $O(10^2)$ in aqueous solutions, because these theories do not describe nucleation at solid surfaces that inevitably have imperfections acting as nucleation "sites" such as those discussed for boiling nucleation.[4,5] Westerheide and Westwater[6] reported that nucleation on their microelectrode occurred at preferred sites such as pits and scratches, and Janssen and Hoogland[7] observed that bubbles nucleated on a rotating platinum wire at specific sites that depended on pretreatment as well as current density. Dapkus and Sides[3] investigated hydrogen evolution on mercury in sulfuric acid in order to study nucleation at an ideally smooth electrode. In their cell design, the mercury flowed in a way that continuously renewed the crucial area for nucleation. They found nucleation at supersaturations much less than predicted by theory, which indicated a possible effect of the intense electric field in the electric double layer at the surface.

2. Growth

(i) Growth

Bubbles grow initially from the critical radius by expansion due to the high internal pressure and by mass transport of dissolved gas to the gas/liquid interface. The equations governing this stage of growth are continuity, motion, and convective diffusion. The continuity equation can be combined with the equation of motion for a radially symmetric system in a viscous liquid supersaturated with a sparingly soluble gas to give, after one integration,[8]

$$\frac{P_0 + P'' - P' - 2\sigma/R}{\varepsilon \rho_L} = R\ddot{R} + \frac{3}{2}\dot{R}^2 + 4\nu \frac{\dot{R}}{2} \quad (3)$$

where R is the bubble radius, ν is the kinematic viscosity, ρ_L is the density of the liquid, and ε is the ratio of the density difference between liquid and gas to the liquid density. This equation, a force balance between the driving pressures on the left-hand side and the inertial and viscous forces on the right, governs the expansion of the cavity in response to the forces at the gas/liquid interface. The equation of convective diffusion in these circumstances,

$$\frac{\partial C}{\partial t} = D\left(\frac{\partial^2 C}{\partial r^2} + \frac{\partial}{r}\frac{\partial C}{\partial r}\right) - \frac{\varepsilon R^2 \dot{R}}{r^2}\frac{\partial C}{\partial r} \quad (4)$$

describes the transport of gas dissolved in the electrolyte to the bubble. These two equations govern the growth of a spherical cavity from the critical radius to the point where the bubble coalesces with other bubbles or detaches. Contributions to knowledge in this area can be understood as simplifications of this pair. Rayleigh[9] derived an expression with neglect of viscous and surface tension forces. Since his case did not involve any concentration of thermal gradients, the driving pressure was constant. Epstein and Plesset[10] recognized the requirement of a surface tension term in the equation of motion and solved the diffusion equation with neglect of the convective term in Eq. (4). Plesset and Zwick[11] solved the problem of the initial stages of phase growth with neglect of viscous forces and temperature gradients. In the same contribution, they solved approximately the diffusion equation with convection included. Forster and Zuber[12] also contributed an approximate solution to the problem including convection. Scriven,[8] using a similarity transform, presented a general analysis of the diffusion controlled growth of a bubble. The growth rate is given by

$$R = 2\beta(Dt)^{1/2} \tag{5}$$

where R is the radius, D the diffusivity, t the time, and β is a coefficient characteristic of the degree of supersaturation. Cheh and Tobias[13] provided additional solutions for the case of nonuniform supersaturation of the electrolyte.

An overall mechanism for the initial development of a bubble emerges from this work. The rate of growth of a critical-size bubble is zero because it is in chemical and mechanical equilibrium with the surrounding liquid. A positive disturbance in the radius of the bubble must initiate phase growth by giving the internal pressure of the bubble a slight favor over the force due to curvature of the surface and background pressures. Although growth is relatively slow at first, the expansion by the pressure inside the bubble quickly accelerates until the supply of new molecules for the bubble limits growth. Mass or heat transfer then governs the final stage of this process. To state that the initial growth from the critical bubble is "slow" is relative, because the time in absolute value is quite small so mass or heat transfer takes control of the process at an early stage. The error introduced by assuming that the entire process is governed by mass or heat transport is small.

Westerheide and Westwater,[6] using a microelectrode and a high-speed camera, photographed individual electrolytic hydrogen bubbles and quantitatively compared their growth data to the diffusive square-root-of-time growth dependence as shown in Fig. 1. They found agreement for a single bubble but multiple bubbles interfered with each other's growth. Nevertheless this work experimentally established the importance of mass transfer of dissolved gas to the bubble surface as the asymptotic mechanism by which bubbles grow before and between coalescence. Glas and Westwater[14] extended this work to include oxygen, chlorine, and carbon dioxide evolved on various electrode materials.

(ii) Growth by Coalescence

The mechanism of the initial stage of growth being known, other investigators reported the importance of coalescence in gas

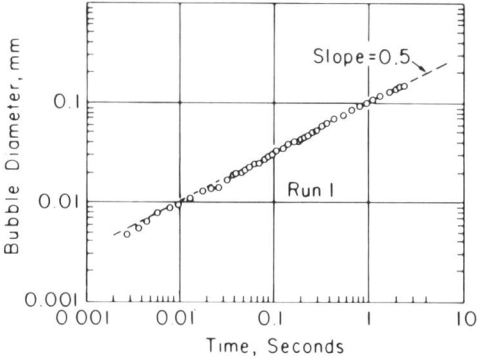

Figure 1. Experimental data on diffusion-controlled growth of an electrolytically evolved hydrogen bubble, from Westerheide and Westwater.[6] Note the slope of 0.5 on the logarithmic plot [see Eq. (5)]. (Reproduced by permission of the American Institute of Chemical Engineers.)

evolution. Janssen and van Stralen[15] observed bubbles of oxygen evolved in aqueous potassium hydroxide on a transparent nickel electrode; photographing at 70 frames per second, they observed frequent coalescence, lateral mobility of bubbles still attached to the surface, consumption of small bubbles by large ones at high current density, and a radial movement of small bubbles toward large ones. Putt[16] reported that hydrogen bubbles produced in acid grew large by a scavenging mechanism in which the bubbles slid along the electrode and consumed other smaller bubbles. One concludes from these studies that coalescence may largely determine the bubble size and that several modes of coalescence may be involved.

Sides and Tobias[17] documented coalescence phenomena with high-speed motion pictures of oxygen evolution taken from the back side of a transparent tin oxide electrode in alkaline electrolyte. They identified a variety of types of coalescence, some of which had been reported previously, and organized them into an overall mechanism for oxygen bubble growth. The first type of coalescence was among small bubbles which touched on the electrode surface during growth by diffusion. The small bubbles appeared to touch and immediately coalesce in much less than 0.0001 s. In the second mode of coalescence, bubbles of a medium size, about 40 μm, established themselves as central collectors and received the smaller bubbles nucleating and growing around them. The effect was called radial specific coalescence because smaller bubbles translated radially across the electrode surface toward stationary central bubbles. A film sequence recording this mode of coalescence appears in Fig. 2. Small bubbles moved specifically in the direction of, and coalesced with, the medium bubble in whose sphere of influence, extending about two bubble diameters from the collector bubble, they were located. The central bubble swelled with the addition of the smaller bubbles but was not otherwise disturbed as

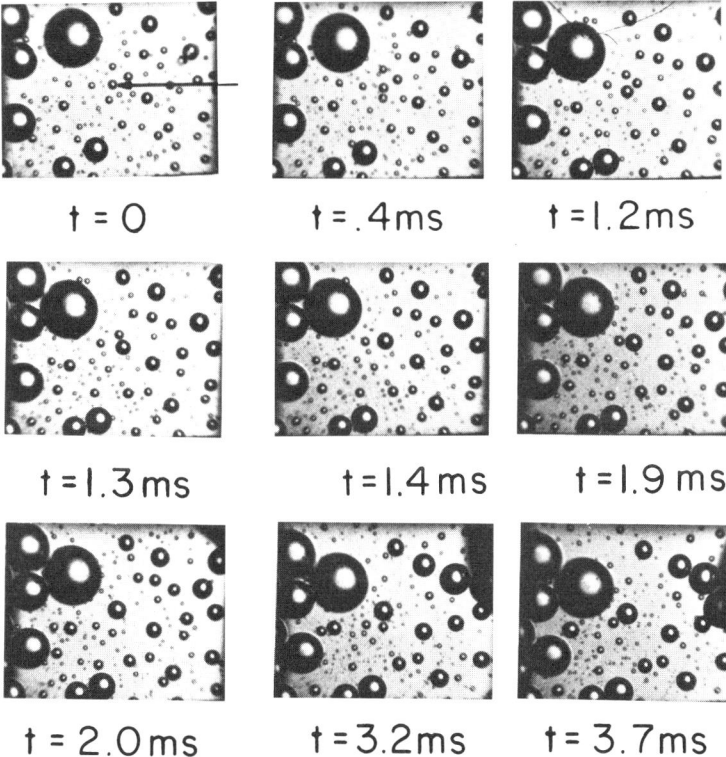

Figure 2. Documentation of specific radial coalescence, from Sides and Tobias.[17] The indicated bubble at time zero receives the four bubbles around it sequentially and visibly grows. One may observe this same effect around other bubbles in the sequence. Conditions: 10,000 frames per second, oxygen evolution, 298 K, no forced convection, 3% (wt) KOH, 500 mA/cm^2. (Reprinted by permission of the publisher, The Electrochemical Society, Inc.)

it coalesced with them. Although there was an overall flow impressed on the system by movement of the buoyant bubbles, the radial movement of bubbles was symmetric 360° around the collector bubble, which remained anchored to the electrode; hence Sides and Tobias concluded that the radial motion was a local phenomenon. The movement may have been a result of local flows established by continual coalescence that entrains other bubbles toward the collector. Requiring a transparent electrode to be revealed, this phenomenon was also observed only by Janssen and van Stralen[15] on their transparent nickel electrode.

After the medium size bubbles attained a size between 50 and 100 μm, large bubbles still attached to the surface and sliding from below the field of view scavenged them or coalesced between themselves and moved out of view. (The bubbles' flattened areas of attachment were visible from behind the transparent electrode.) Scavenging of medium bubbles by large ones was the third mode of coalescence observable on the transparent electrode and was the mechanism by which bubbles attained their final size. A film clip showing this appears in Fig. 3. After a group of large bubbles passed, a new group of bubbles nucleated and grew to a size of 10-15 μm in diameter. These small bubbles coalesced to form medium size bubbles that then grew by the radical coalescence phenomenon. Large bubbles (200 μm), attached to and moving along the surface, scavenged the small and medium size bubbles in their path.

On the basis of these films, Sides and Tobias[17] proposed a cyclic mechanism of oxygen bubble growth in basic medium; it organizes previously observed phenomena into a process consisting of nucleation followed by growth by diffusion, coalescence of small bubbles, radial movement of many small bubbles to coalesce with stationary medium size bubbles, and scavenging coalescence of the medium bubbles by large ones moving along the electrode. This is the process by which large oxygen bubbles are built in basic medium. They termed this sequence of events (nucleation, growth by diffusion, coalescence of small bubbles, coalescence by radial motion, and scavenging coalescence) the cyclic process of bubble growth because it was repeated continually on the electrode and is the general way bubbles were "built". The period of the cycle, at a current density of 100 mA/cm^2, was approximately 0.1 s. These

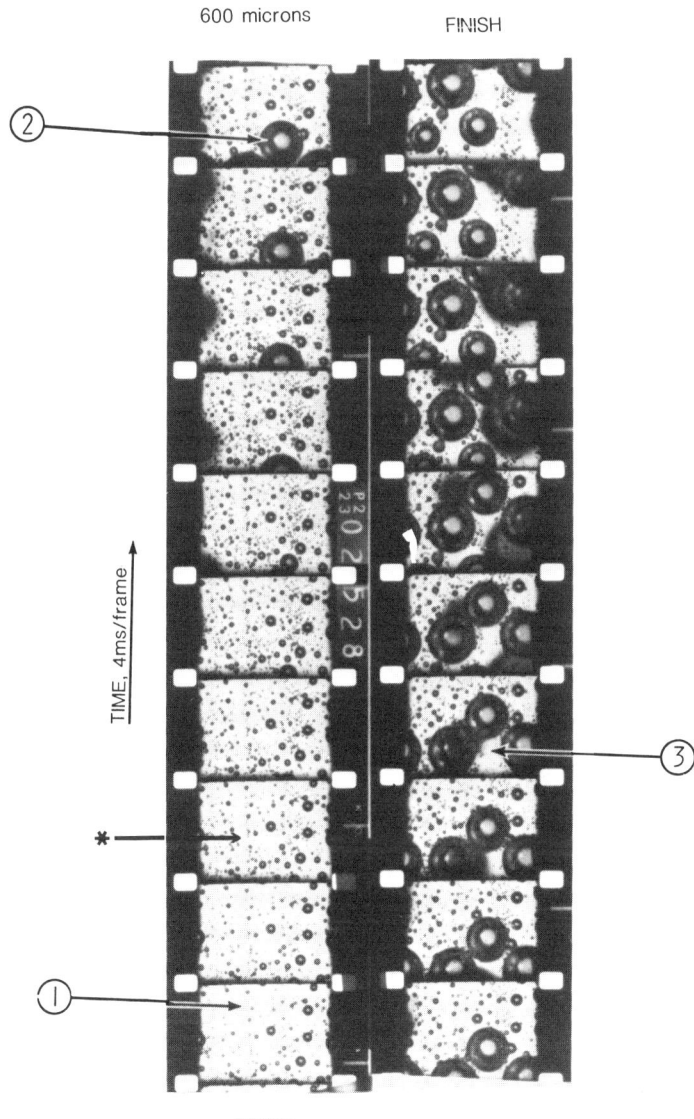

Figure 3. Scavenging coalescence, from Sides and Tobias.[17] (1) Dispersion of small bubbles which grew on the electrode after a group of large bubbles swept through. (2) Several large bubbles travel along the surface and scavenge the small and medium size bubbles before them. (3) Empty area behind a large bubble showing its path. *, Trail left by an earlier scavenging bubble. (Reprinted by permission of the publisher, The Electrochemical Society, Inc.)

phenomena may or may not occur on different electrodes evolving other gases. In fact, as discussed subsequently, coalescence seems unimportant in hydrogen evolution from basic solution because the bubbles evolved there are quite small. The foregoing discussion demonstrates the variety of growth phenomena, but does not provide a general mechanism for all types of gases and electrodes.

The cyclic process of bubble growth has been seen on a much larger scale by Fortin et al.[18] in a physical model designed to investigate the behavior of bubbles underneath the carbon anodes of the Hall process for aluminum production. When the electrode was tilted a few degrees from horizontal, bubbles forming uniformly under the simulated electrodes coalesced to produce a large bubble "front" that moved across the surface and scavenged other bubbles in its path. The process was repeated at a frequency of 1-3 Hz when the gas evolution rate was equivalent to 1 A/cm^2.

3. Detachment

The third stage in the physics of gas evolution, detachment, also has been theoretically and experimentally investigated. In equilibrium measurements, Frumkin and Kabanov[19] found that the buoyant gas bubbles detach when surface adhesive forces, related to bubble contact angles, can no longer restrain them. Westerheide and Westwater,[6] observing the dynamics of gas evolution, noted that the bubble resulting from the coalescence of two large bubbles jumped off the electrode and sometimes even returned. They concluded that the expanding boundaries of the new bubble mechanically forced it off the electrode and they speculated that bubbles' movement toward the electrode could be influenced by electrostatic forces operating on a moving bubble or by surface forces varying with concentration. Other investigators[15,16] have also noted that coalescence often precedes detachment. In an unusual and as yet unexplained mode of detachment, bubbles sometimes were ejected from the electrode in what Glas and Westwater[14] termed "rapid fire emission." Janssen and Barendrecht[20] observed that hydrogen bubbles formed trains of noncoalesced small bubbles which left the electrode.

Detachment of a bubble from the transparent electrode was also observed in work with the transparent tin oxide electrode.[17]

The sequence shown in Fig. 4 indicates that coalescence of two bubbles to form a new bubble, which being compressed against the electrode pushes away from it, is an important mechanism by which bubbles depart from vertical electrodes. This confirms the observation of Janssen and van Stralen.[15] As mentioned by other investigators, especially by Glas and Westwater,[14] the bubble, once detached, may return, which occurred on the tin oxide electrode also. One can see the contact area under the bubbles in Fig. 4 as a slightly darkened spot on the left side of the light central area. When the bubbles coalesced and the new bubble had ceased its

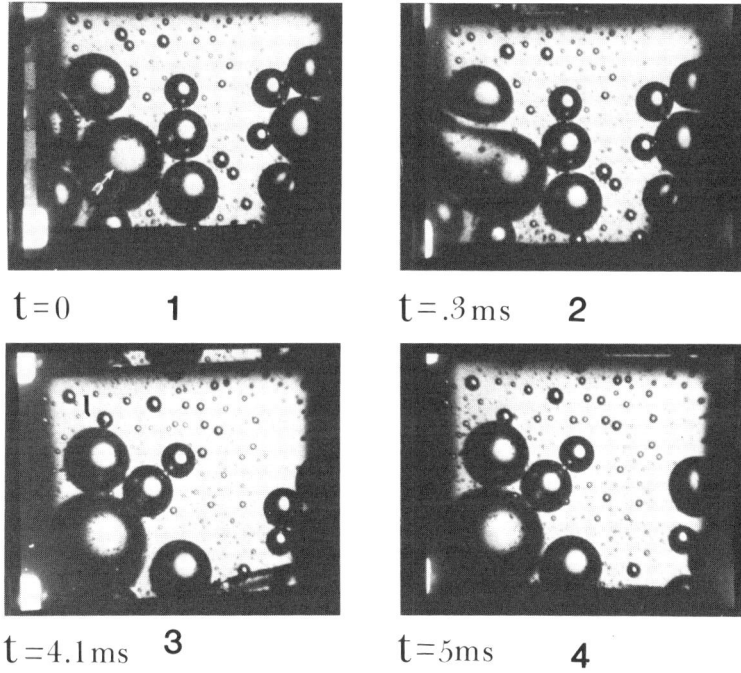

Figure 4. Detachment and return of a bubble and after coalescing with another bubble, from Sides and Tobias.[17] (1) A bubble on the electrode touches another off the frame to the left. (2) Two bubbles coalesce. (3) The new bubble is off the surface. There are small bubbles on the surface between the large bubble and the surface. (4) The bubble has reattached. Conditions: 10,000 frames per second, oxygen evolution, 298 K, no forced convection, 28% (wt) KOH, 100 mA/cm². (Reprinted by permission of the publisher, The Electrochemical Society, Inc.)

violent vibrations and had established its spherical boundary, the contact spot vanished but it subsequently reappeared. Coehn and Neumann[21,22] would argue that the return to the electrode was caused by the attraction of a charged bubble to the electrode surface; however the charge on a bubble, which might exist as a result of differential adsorption of ionic species, is probably too small to have an effect. Glas and Westwater[14] speculated that a surface tension gradient caused the bubbles' return.

Characterizing the detachment of a bubble from a vertical electrode is difficult because the buoyancy force does not act perpendicularly to the electrode; the only force continuously pushing the bubble away from the surface (at equilibrium) is exerted by the internal pressure of the bubble against the flattened bubble base. Although the forces holding a bubble on the electrode at equilibrium and the role of coalescence in breaking bubbles away are established, the criteria or kinetics by which one may judge whether a given coalescence will lead to detachment or not are unclear; furthermore, the return of a detached bubble to the electrode surface and the rapid emission of bubbles await explanation.

Related to the detachment of bubbles is their mobility on electrode surfaces. Dussan and Chow[23] explored the forces holding a drop on an inclined plane and concluded that contact angle hysteresis, the difference between bubbles' advancing and retreating contact angles, is responsible for the sticking of drops to surfaces inclined from the horizontal. The leading cause of hysteresis is roughness of the surface. Their work and a more recent contribution[24] that can also be applied to the case of gas bubbles established criteria for judging whether or not a bubble should move on an inclined surface. Figure 5, taken from Dussan,[24] illustrates the relation between the hysteresis angle, the contact angle, and the volume of the largest bubble that sticks to a surface facing downward and inclined by γ degrees to the horizontal. Consider an example in which the average contact angle is 10°, the electrode is vertical, the surface tension of the electrolyte is 73 dyn/cm (water), and the hysteresis in the contact angle is 10°. The largest bubble which would be immobile in these circumstances is about 25 μm in radius. The largest bubble that would be immobile on such a surface at the same inclination and with the same hysteresis would have a contact angle of 72° and a radius of about 250 μm.

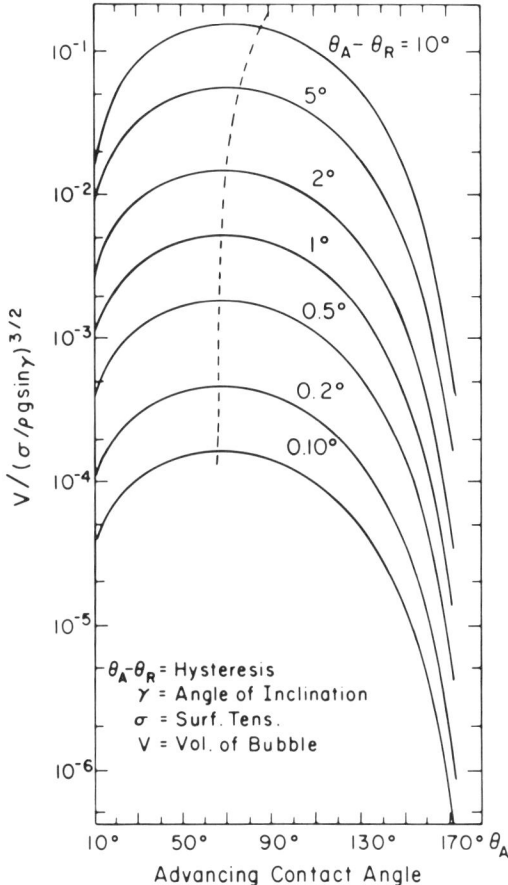

Figure 5. Dimensionless volume of largest bubble or drop that adheres to an inclined surface; graph from Dussan.[24] (Reprinted from the *Journal of Fluid Mechanics* by permission of Cambridge University Press.)

These values bracket the sizes of electrolytically generated bubbles in aqueous solution.

4. Effect of Additives and Operating Parameters

The electrode substrate, current density, and additives influence the process of bubble growth and hence the ultimate bubble size.

Venczel[25] photographed bubbles from the front of graphite, iron, copper, and platinum electrodes and from the back side of glass plates on which thin layers of platinum, chromium, nickel, and gold were vacuum deposited. Large bubbles formed on platinum because the bubbles grew uniformly and coalesced, but small ones formed on copper and iron because the bubbles detached from the electrode before reaching a size for coalescence. Ibl and Venczel[26] observed that bubbles evolved on platinum were much larger than those evolved on copper.

Current density has also been shown to affect bubble size, but there is some contradiction in the literature about its effect. Janssen and Hoogland[7] attributed variations in bubble size with current density to bubbles' coalescing at higher current densities. In other work, Janssen and Hoogland[27] evolved gas on horizontal and vertical platinum disks and determined that as the current density was increased past $10 \, \text{mA/cm}^2$, all bubbles but hydrogen evolved in base increased in size; the writers attributed this to increased coalescence. The results of their experiments appear in Fig. 6. Note that hydrogen bubbles evolved in alkaline medium are the smallest and their size is not affected by current density; oxygen bubbles evolved in base, however, quickly increase in size at current densities higher than $30 \, \text{mA/cm}^2$ owing to increased coalescence. Ibl,[28] reviewing the physics of gas evolution, pointed out the discrepancy between the work of Venczel,[25] who said that bubble size decreased with current density, and the work of Janssen and Hoogland,[27] who found that bubbles increased in size with current density. Landolt, Acosta, Muller, and Tobias[29] photographed hydrogen evolution from the side of a transparent cell which simulated the high electrolyte flow rates characteristic of electrochemical machining. Agreeing with Janssen and Hoogland,[27] but contrary to Venczel,[25] and contrary to the theory of Frumkin and Kabanov,[19] bubbles increased in size with current density. Thus some disagreement remains in the literature about the effect of current density on bubble size.

In addition to the influence of the electrode material and current density, the effect of additives on the bubble behavior and size has been investigated. Venczel[25] added gelatin, glycerine, and beta-naphthochinolin to his electrolyte; in most cases the bubble size decreased and in some cases a frothy mixture resulted. The

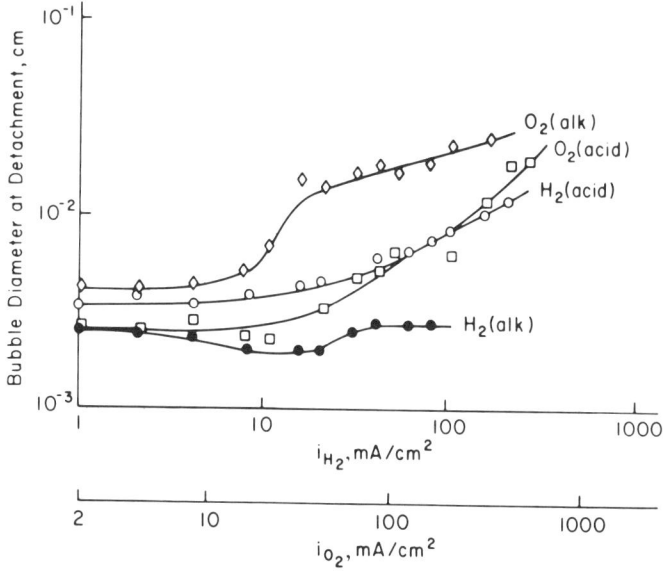

Figure 6. Bubble breakoff diameters as a function of current density for a Pt gas-evolving electrode (horizontal) in various solutions, from Janssen and Hoogland.[27] (Reprinted with permission from *Electrochimica Acta* **18**, L. J. J. Janssen and J. Hoogland, The effect of electrolytically evolved gas bubbles on the thickness of the diffusion layer—II, Copyright, 1973, Pergamon Press.)

additives reduced the ratio of the bubble diameter to the contact diameter by half; Venczel[25] suggested that the increased wettability of the electrode in the presence of inhibitors led to a thick film of electrolyte between the gas and the electrode and he asserted that this film was less adhesive than thin films; however, the force holding the bubble to the electrode was probably weakened proportionally with the reduction in both the perimeter of the contact area and the contact angle; furthermore, the inhibitors might have stabilized the bubble interfaces and prevented coalescence.

In addition to current density, additives, and the electrode itself, other process characteristics affect bubble size and behavior. Landolt *et al.*[29] found that the bubble size decreased with flow rate in the interelectrode area; in fact, they achieved current densities much higher than those which in stagnant electrolyte would produce anode effect. Electrode orientation and configuration must also affect bubble size.[30]

III. ELECTRICAL EFFECTS OF GAS EVOLUTION

Electrolytically evolved gas bubbles affect three components of the cell voltage and change the macro- and microscopic current distributions in electrolyzers. Dispersed in the bulk electrolyte, they increase ohmic losses in the cell and, if nonuniformly distributed in the direction parallel to the electrode, they deflect current from regions where they are more concentrated to regions of lower void fraction. Bubbles attached to or located very near the electrodes likewise present ohmic resistance, and also, by making the microscopic current distribution nonuniform, increase the effective current density on the electrode, which adds to the electrode kinetic polarization. Evolution of gas bubbles stirs the electrolyte and thus reduces the supersaturation of product gas at the electrode, thereby lowering the concentration polarization of the electrode. Thus electrolytically evolved gas bubbles affect the electrolyte conductivity, electrode current distribution, and concentration overpotential and the effects depend on the location of the bubbles in the cell. Discussed in this section are the conductivity of bulk dispersions and the electrical effects of bubbles attached to or very near the electrode. Readers interested in the effect of bubbles dispersed in the bulk on the macroscopic current distribution in electrolyzers should see a recent review of Vogt.[31]

1. Conductivity of Bulk Dispersions

The conductivity of a heterogeneous medium depends on three characteristics: the ratio of the conductivities of the dispersed and continuous phases, the volume fraction occupied by the dispersed phase, and its state of aggregation. If the conductivity of the medium is greater or less than that of the surrounding medium, the dispersed phase can enhance or retard transport, respectively. Volume fraction, characterizing the amount of dispersed phase present, is likewise significant and often is the only parameter, other than the aforementioned ratio, that appears in equations predicting the effect of a given dispersed phase on the overall conductivity. By state of aggregation I mean both the shape of the dispersed phase, which can range from spheres to cylinders to plates, and its distribution in the system's container. For example, the effect of a small amount of finely distributed gas bubbles in an electrolyzer might be insignificant, but if the same amount of gas blankets an electrode, the effect might be quite large. Meredith and Tobias[32] have reviewed the roles these three characteristics play in the conductivity of heterogeneous media.

Gas bubbles dispersed in the bulk electrolyte, common in industrial electrolysis, are essentially randomly distributed spheres having zero conductivity. There are a number of different approaches to describing the effect of such dispersions on the overall conductivity. Simplification of the problem is possible when the dispersed phase is dilute or when a limited range of void fraction is considered. Some writers discuss media in which the dispersed phase occupies well-defined lattice positions while others treat random dispersions. One may also consider spheres of equal size or dispersions containing a distribution of sizes. I classify the approaches by the type of dispersion they aim to describe and compare the theory for these classes to appropriate experimental data.

(i) Randomly Distributed Spheres of Equal Size

Maxwell,[33] deriving a fundamental result in the theory of heterogeneous conductivity for randomly distributed spheres, considered a spherical surface having unequal resistivities on the inside and outside. By expanding the potential in both regions in spherical harmonics and by requiring equivalence of currents and potentials at the spherical surface, he evaluated the disturbance wrought by the sphere in the overall impressed linear field. Maxwell then placed n spheres, which by superposition disturb the potential n times the amount due a single sphere, into a larger sphere. He equated the disturbance obtained by superposition of the n spheres to the disturbance due to the one larger sphere and thereby obtained the conductivity of the dispersion:

$$K_m = (1-f)/(1+f/2) \qquad (6)$$

where K_m is the ratio of the conductance with the dispersed phase present to the conductance in the absence of the dispersed phase. This result, appropriate for dilute random dispersions, may also be derived as the first approximation to the conductivity of a cubic array, as shown by Rayleigh,[34] which attests to the fact that it is a fundamental result in heterogeneous conductivity.

Theoreticians rederive and experimenters[35] continue to advocate Maxwell's result even though it has been shown to fail at moderate to high void fractions.[36] Hashin[37] derived Maxwell's relation by considering a sphere (representing the dispersed phase) embedded concentrically in another sphere (representing the continuous phase), which is in turn embedded in a third statistically homogeneous phase consisting of dispersed and continuous phases. The total current in the concentric spheres is the sum of the two partial currents and the average current density in the spheres is the same as that in the outer medium; Hashin combined this fact with a similar argument about the electric field and electric potential relationships similar to Maxwell's. In fact, if one considers Maxwell's n spheres to be a single sphere having an equivalent volume, the two methods are identical. Neale and Nader[38] used Hashin's model as the basis for their derivation; however, they evaluated the disturbances in the concentric continuous phase instead of those in the dispersed phase. The various derivations of the same result indicate that Maxwell's result is a fundamental relation of heterogeneous conductivity and they illustrate some of the techniques one can use to theoretically investigate such systems.

Buyevich[39] modeled the spheres surrounding a central sphere as point dipoles, but one calculates conductivities from his equation which are higher than those of Maxwell, a physical absurdity pointed out by Turner[40] and noticed by Buyevich himself. He concluded that the dipole model cannot account for the constriction of current between spheres. Extending Maxwell's result, Jeffery[41] calculated the resistivity of a suspension of random spheres to order $O(f^2)$ by writing a general formulation for the flux of current through such a suspension and then using the interaction of two spheres as a model for the interactions occurring throughout the suspension. He found the coefficient of f^2 explicitly for the case where all possible configurations of pairs of spheres are equally probable:

$$K_m = 1 - 1.5f + 0.588f^2 \qquad (7)$$

O'Brien[42] also calculated the conductivity function to $O(f^2)$ by relating the dipole strength of a sphere to integrals over the surrounding particles and then averaging the resulting expression for the dipole strength. Avoiding Rayleigh's convergence problems by properly evaluating the macroscopic boundary integral he improved on Maxwell's solution at void fractions less than 0.5. Higuchi[43] tried to improve Clausius and Mossotti's classical result for the dielectric constant analogous to Maxwell's result by deducing a correction to the average field which surrounds the particle of interest. He calculated a reaction field proportional to an unknown constant and due to dipole interactions between adjacent spheres; he summed this to the conventionally assumed average field and evaluated the constant by comparing his equation with a compilation of data by Pearce.[44] Prager[45] applied the principle of minimum entropy to obtain bounds on the diffusion coefficient of a solute in a suspension of solid particles. Since the ratio of the diffusion can be considered analogous to the conductivity ratio, one may use the result to estimate the conductivity of heterogeneous electrolyte. His result, exact for a suspension of spheres, agrees with Maxwell's result at low void fractions and significantly improves on it over medium and concentrated ranges of void fraction:

$$K_m = 1 - 1.5f + 0.5f^2 \qquad (8)$$

Chiew and Glandt[46] modeled the structure of a dispersion of identical spheres as an equilibrium hard sphere fluid. They used pair correlation functions to compute the contributions of pairs of spheres to the effective thermal conductivity of the dispersions. Their results, derived for arbitrary values of the conductivity of the two phases, reduce to the following equation for the case of gas bubbles in electrolyte:

$$K_m = (1 - f - 0.162f^2 - 0.241f^3)/(1 + f/2) \qquad (9)$$

Data for the conductivity of random dispersions of spheres of equal sizes was obtained by DeLaRue and Meredith. DeLaRue[36,47] experimented with uniform dispersions of glass spheres suspended by gyrating his cell during the measurements until a steady state value for the conductance was obtained. He used alternating current and a nearly saturated solution of aqueous zinc bromide as the continuous phase to match the density of the glass spheres. Tabulations of data[48] and graphs published in 1961[49] on the conductivities of emulsions are the results of Meredith's careful experiments using alternating current.

The equations of Maxwell,[33] Jeffery,[41] Prager,[45] and Chiew and Glandt[46] are compared to the experimental data in Fig. 7. Maxwell's result, accurate below a void fraction of 0.1, yields conductivities too large above that point although it follows the trend of the data rather well and never grossly deviates from it. Jeffery's result improves on Maxwell's at void fractions between 0.1 and 0.5. Prager's result and the equation of Chiew and Glandt bracket the data. These equations are more accurate at higher void fractions than the others even though they seem to underestimate the conductivity. Study of DeLaRue's data reveals that the fine structure in the data above a void fraction of 0.3 may be due to the use of different sizes of spheres in the three points at each void fraction. The smallest spheres used gave the lowest conductivity while the largest spheres yielded the largest conductivity at a given void fraction. Thus the larger the ratio of the characteristic length

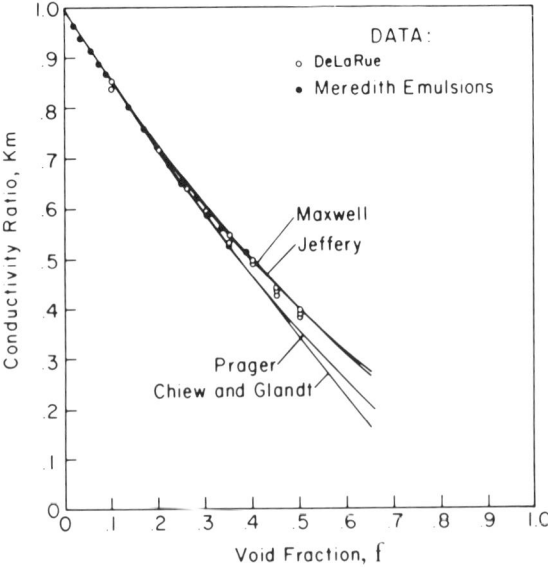

Figure 7. Comparison of equations predicting the reduced conductivity of random dispersions of monosized spheres with data. Maxwell Eq. (6), Jeffery Eq. (7), Prager Eq. (8), Chiew and Glandt Eq. (9).

of the cell to the average diameter of the dispersed phase, the closer the results approach the theoretical predictions, as expected.

(ii) Random Arrangements of Spheres of Unequal Sizes

There are two theoretical models based on size distributions of dielectric particles and one equation said to be applicable to dispersions of spheres of unequal sizes. Bruggeman[50] treated a "pseudocontinuous" distribution of sizes by accumulating the contributions of a range of bubble sizes. It is "pseudocontinuous" because each size fraction of bubbles must be very different from each of the other sizes included in the integration. Adding infinitesimal size fractions, he treated the mixture already present as continuous with a bulk conductivity. One can write this process in differential form,[48] from which integration gives

$$K_m = (1-f)^{1.5} \tag{10}$$

Meredith and Tobias[49] noted that Bruggeman's equation overcorrects in the concentrated ranges and devised another approach called the Distribution Model by considering only two size fractions. As in the Bruggeman equation, the smaller size fraction is added first and then is considered as part of a continuous medium having its own bulk conductivity when the larger size fraction of bubbles is added:

$$K_m = 8(1-f)(2-f)/(4+f)(4-f) \tag{11}$$

Experimental data for random dispersions of spheres of unequal sizes come from DeLaRue[36,47] who, using the same techniques discussed in the previous section, experimented with multisized dispersions of glass spheres. The data used in this comparison also include data previously discussed and other data for which the size distribution was not specified. Clark[51] experimented with foams and thus obtained data at very high void fractions. He used alternating current at 1 kHz in the experiments. Slawinski[52] used alternating current techniques to measure the conductivity of dispersions of castor oil in a mixture of gum arabic and 0.05 N KCl. Because his data deviate from the other data and do not follow Maxwell's equation at low void fractions, they must be suspect. Neale and Nader[38] investigated conduction through packed spheres at void fractions of 60%-70%. They used a tubular plastic cell, copper electrodes, copper sulfate, and alternating current.

The equations derived for distributions of spheres of unequal sizes are compared to these data in Fig. 8 along with other equations that converge to zero conductivity at a void fraction of 1.0. Since multiple sizes of spheres may fill a volume completely, the equations are compared with the entire range. Bruggeman's result agrees with Maxwell's at low void fractions and with DeLaRue's data for multiple sizes of spheres in dilute and intermediate ranges. Intermediate between Maxwell's and Bruggeman's equations, the Distribution Model is as accurate as Bruggeman's at low void fraction, seems not quite as accurate in the intermediate range, and brackets Clark's data on foams. The results of Meredith and Tobias[49] and

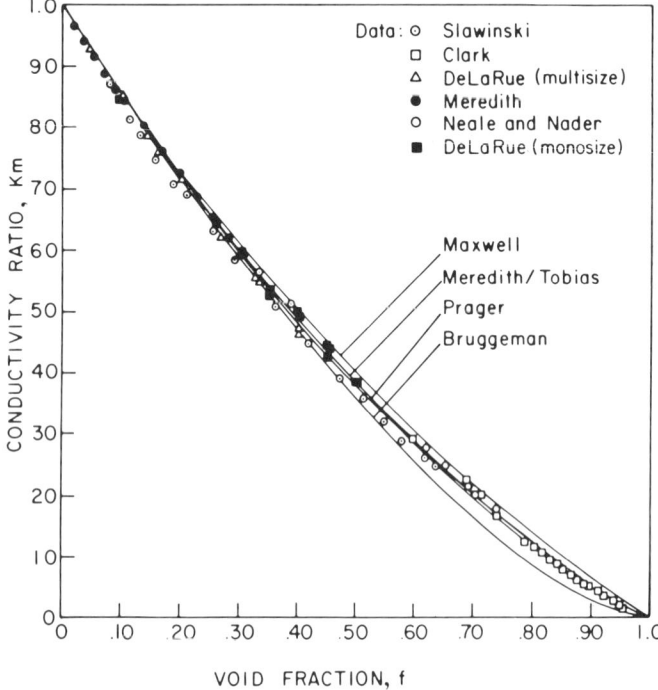

Figure 8. Comparison of equations predicting the reduced conductivity dispersions of spheres of unequal sizes with data. Bruggemann Eq. (10), Meredith/Tobias Distribution Model Eq. (11). Also included are Maxwell Eq. (6) and Prager Eq. (8) for comparison.

DeLaRue[47] follow the same line at low void fractions but, beginning between void fractions of 0.1 and 0.2, DeLaRue's conductivities of dispersions containing spheres of various sizes are consistently lower than both Meredith's conductivities and DeLaRue's own conductivities of dispersions of monosized spheres. For example, note the difference at a void fraction of 0.4 where the conductivities of DeLaRue's dispersion of spheres of various sizes are lower than the conductivities of dispersions of monosize spheres.

One may understand these differences with the help of the equations represented in the graphs. Maxwell's equation (6) accurately predicts conductivities at void fractions below 0.1 and is not sensitive to the dispersion's size distribution (indeed all of the equations agree with Maxwell's result in the limit of low void fraction), but it does not account for the concentration effects shown in DeLaRue's monosized-sphere data which fall below the predictions of Maxwell's equation. Bruggeman's equation (10), derived for dispersions of spheres of various sizes, accurately predicts the conductivities of such dispersions at least to the limit of DeLaRue's multisized-sphere data; unfortunately, no data are available for comparison to the markedly low conductivities predicted by Bruggeman's equation for dispersions containing high volume fractions of multisized spheres. Prager's equation (8) and the Distribution Model (11) predict similar accurate conductivities for the packed columns of Neale and Nader and the foams of Clark at high void fractions.

(iii) Conductivities of Ordered Arrangements

Rayleigh[34] calculated conductivity ratios of cubic arrangements of spheres of equal size, but his results contained an error in one of the coefficients. Runge[53] corrected the constant by a factor of π in 1925, and in 1952 De Vries[54] obtained the correct coefficient of the $f^{10/3}$ term, as Churchill[55] pointed out in an extensive review of heterogeneous thermal conductivities. McPhedran and McKenzie[56] also noted the correction in 1977. The corrected Rayleigh equation appears below:

$$K_m = 1 + \frac{3f}{\dfrac{\alpha + 2}{\alpha - 1} - f - (1.31)\left(\dfrac{\alpha - 1}{\alpha + 4/3}\right)f^{10/3}} \quad (12)$$

The parameter α is the ratio of the conductivity of the dispersed phase to that of the continuous phase. Meredith and Tobias[57] extended this result to higher-order terms. Zuzovsky and Brenner[58] used a multipole expansion technique to calculate the effective conductivity of simple cubic, body-centered cubic, and face-centered cubic arrays of spheres. Their technique allowed for fourfold symmetry in the arrays, while those of previous authors did not. McPhedran and McKenzie[59] and McKenzie, McPhedran, and Derrick[60] extended Rayleigh's[34] method for calculating the conductivities of lattices of spheres. Their method includes the effects of multipoles of arbitrarily high order; specifically, their equation gives the numerical value of the f^6-order term referred to by Zuzovsky and Brenner.[58] Sangani and Acrivos[61] also used a fourfold potential to calculate effective conductivities of simple cubic, body-centered cubic, and face-centered cubic lattices to $O(f^9)$. They corrected a "numerical slip" in the work of Zuzovsky and Brenner. Their equation is

$$K_m = 1 - \left[3f \bigg/ -\frac{1}{L_1} + f + aL_2 f^{10/3} \left(\frac{1 + bL_3 f^{11/3}}{1 - CL_2 f^{7/3}} \right) \right.$$
$$\left. + dL_3 f^{14/3} + eL_4 f^6 + gL_5 f^{22/3} \right] \qquad (13)$$

where

$$L_n = \frac{\alpha - 1}{\alpha + \dfrac{2n}{(2n - 1)}}$$

the coefficients of which appear in Table 1.

Predictions of this equation and those of the corrected Rayleigh and Meredith and Tobias equations appear for the two extreme cases—infinitely insulating and infinitely conductive dispersed phases—in Table 2. The improvement over the corrected version of Rayleigh's original equation for the case of gas bubbles in electrolyte is slight. For the case of infinitely conducting spheres, the improvement is significant, 6%, at the maximum packing for a simple cubic array. For the purposes of this work I present results for the value of the ratio being zero, suitable for bubbles in an electrolyte. The equation of Sangani and Acrivos is compared to Meredith's data on simple cubic arrays[57] in Fig. 9. The agreement

Table 1
Coefficients of the Equation of Sangani and Acrivos, Eq. (13)[a]

Coefficient	SC array	BCC array	FCC Array
a	1.3047	0.129	0.07529
b	0.2305	−0.41286	0.69657
c	0.4054	0.76421	−0.74100
d	0.07231	0.2569	0.04195
e	0.1526	0.0113	0.0231
g	0.0105	0.00562	$9.14(10)^{-7}$

[a] Reference 61.

Table 2
Predictions of the Heterogeneous Conductivity of Simple Cubic Arrays, by Rayleigh's Corrected Equation (12),[a] Meredith/Tobias,[b] and the Equation of Sangani and Acrivos (13)[c]

Void fraction	Rayleigh, corrected	Meredith/Tobias	Sangani and Acrivos
Conductivity of the dispersed phase = 0			
0.1	0.8571	0.8571	0.8571
0.2	0.7267	0.7267	0.7267
0.3	0.6057	0.6057	0.6057
0.4	0.4902	0.4904	0.4903
0.5	0.3757	0.3769	0.3764
0.5236	0.3482	0.3498	0.3492
Conductivity of the dispersed phase = infinity			
0.1	1.334	1.334	1.334
0.2	1.756	1.756	1.756
0.3	2.331	2.332	2.333
0.4	3.230	3.245	3.250
0.5	5.054	5.197	5.269
0.5236	5.836	6.094	6.231

[a] Reference 34.
[b] Reference 57.
[c] Reference 61.

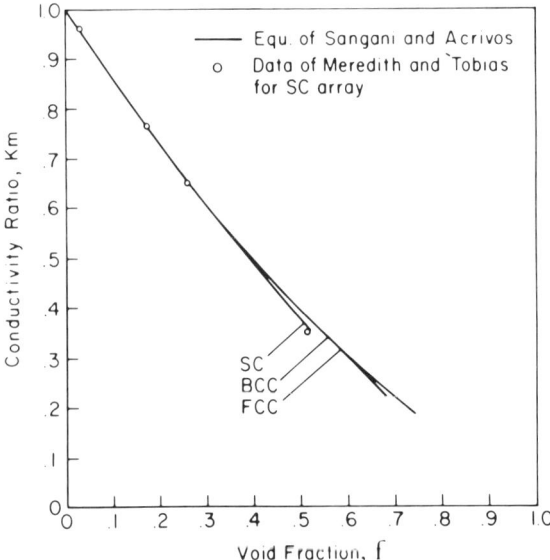

Figure 9. Comparison of equations predicting the reduced conductivity of a dispersion in the form of a simple cubic array. Sangani and Acrivos Eq. (13).

of the equation with the data is within 2% at the highest void fraction possible for a simple cubic array.

(iv) Experiments on Operating Systems

Apart from these theoretical investigations and experiments on static models, investigators have studied dynamic gas evolving electrodes. Hine and co-workers,[62] in early work on bubble resistance in chlor-alkali electrolysis, demonstrated the effect of electrode orientation on ohmic losses. Hine et al.[63,64,65,66,67] reported the results of experiments on large-scale vertical electrolyzers and concluded that Bruggeman's equation (10) described the results when properly applied, that perforated electrodes help minimize ohmic losses due to bubbles, and that flow conditions have a significant effect on the cell voltage of electrolyzers. Sigrist et al.[35] also experimented with electrodes evolving bubbles but concluded that Maxwell's equation fit their data best. Janssen et al.[68] investigated ohmic losses in water electrolysis due to the presence of bubbles as a function of electrode shape and height, electrolyte flow velocity, and current density. They derived a correlation from their results that incorporates the effects of these parameters on the ohmic resistance due to bubbles:

$$\Delta R_{wm} = K_1 \text{Re}^n (1 + K_2 h^m + K_3 d^p) \qquad (14)$$

where ΔR_{wm} is the resistance between the working electrode and the cell diaphragm divided by the resistance which would exist were there no bubbles present. h is the reduced height in the electrolytic cell and d is normalized distance. The remaining parameters are empirical constants.

(v) Conclusions

For dilute dispersions ($0 < f < 0.1$), the effective conductivity obeys Maxwell's equation and is independent of the size distribution and order. In dispersions where $0.1 < f < 0.6$, the difference between the conductivities of monosized and multisized dispersions is perceptible and accurate prediction of conductivity depends on allowing for the size distribution of the spheres; however, any of the equations, with the exception of Maxwell's, predicts values within ten percent. For dispersions having void fractions greater than 0.6, one must distinguish between the monosized and multisized distributions. The Distribution Model of Meredith/Tobias, Eq. (11), and Prager's equation (8) predict accurate values for Clark's data on the conductivity of foams, but no information on the size distribution in these experiments is available. Bruggeman's equation for the multisized dispersions predicts conductivities in the concentrated range that differ from the predictions of the other models by 50% or more, though this has not been adequately tested in the concentrated region ($f > 0.6$). There may be substantial effects of the size distribution at these void fractions but more experiments are required to elucidate them.

Investigators have reported that Maxwell's equation (6) represents their data even though it was derived for dilute solutions. Neale and Nader[38] endorsed Maxwell's equation even at void fractions greater than those at which spheres touch in some packings. Sigrist et al.[35] reported agreement with Maxwell's equation at void fractions which have been shown[36,47] to reduce the conductivity below Maxwell's prediction. Turner[40] found that the conductivity at low void fractions exceeded Maxwell's predictions, but he blamed this discrepancy on experimental error or inhomogeneities in the fluidized dispersion; despite these discrepancies at low void fractions where Maxwell's result should be accurate, he accepted the agreement between his results and Maxwell's equation at high void fraction where it should not be accurate. Perhaps the durability of Maxwell's equation can be explained. Since it represents the conductivity of a random dispersion of spheres whose size is infinitely small compared to the overall size of the vessel, inhomogeneities or finite spheres in a dispersion must give a different conductivity. The more the spheres are concentrated in the direction of overall current flow, the limiting case of which is a homogeneous column of gas parallel to the overall current, the higher the apparent conductivity, but the more the spheres are concentrated in a direction perpendicular to the overall current, the lower the apparent conductivity for the same void fraction. The absolute highest conductivity for a volume having a rectangular geometry, the limit of the former case, is given by $(1 - f)$ for any void fraction while the lowest, corresponding to the latter, is zero. For experiments in which the electrodes are oriented parallel to gravity and the densities of the two phases are different, the measured conductivities may therefore exceed Maxwell's predictions at low void fraction and agree with it at high void fractions where the increased disturbance in the current flow caused by closer packing compensates for the increase in conductivity because of inhomogeneities for large size with respect to the system's dimensions. For experiments in which the electrodes are oriented horizontally and the densities are not matched, the conductivity will be less than Maxwell's prediction. In experiments where attention is paid to matching densities of the dispersed and continuous phases and to using other techniques to maintain a random dispersion such as used by DeLaRue,[47] the conductivity of concentrated dispersions will be less than predicted by Maxwell's equation and will be accurately represented by Chiew and Glandt's, Meredith/Tobias' or Prager's equations.

2. Electrical Effects of Bubbles on Electrodes

Just as bubbles dispersed in the bulk electrolyte increase ohmic losses and alter the macroscopic current distribution within electrolyzers, bubbles attached to or located very near the electrode resist current and alter the microscopic current distribution. Furthermore, bubbles at the electrode affect the supersaturation of electrolyte with product gas and thereby influence the concentration overpotential of the electrode. To distinguish among these effects, I analyze the potential measured between a hypothetical gas-evolving electrode and a reference electrode of the same kind placed just outside the sheath of bubbles covering the electrode:

$$\Delta\Phi_T = V_w - \Phi_r \qquad (15)$$

or, as a voltage component statement,

$$\Delta\Phi_T = \Delta\Phi_{ohm} + \eta_s + \eta_c \qquad (16)$$

The first term on the right-hand side of Eq. (16) is the ohmic potential difference between any point on the electrode and the reference electrode, the second term is the local surface overpotential for the reaction, and the third term is the local concentration overpotential. These three terms may vary from point to point on the electrode, but their sum must always be $\Delta\Phi_T$. Distinguishing among the three components of the experimentally measured voltage of the electrode is a complicated problem, but necessary if one wishes, for example, to obtain fundamentally meaningful values for the surface overpotential as a function of current density for a gas-evolving electrode. Consider the following cases.

(i) Conductivity of the Bubble Layer

If the electrode reaction is reversible and product gas does not supersaturate the electrolyte near the electrode, Eq. (16) simplifies to

$$\Delta\Phi_T = \Delta\Phi_{ohm} \qquad (17)$$

The potential drop between the working and reference electrodes is thus entirely ohmic. As in bulk dispersions, bubbles on the electrode force the current to take longer paths and flow through constricted areas. Calculation of the potential drop requires knowledge of the electrical conductivity of the bubble layer either from theory or experiment. Theoretical treatments of the bubble layer are few because solving Laplace's equation in the complicated asymmetric environment is difficult; nevertheless, solution of the problem for a layer of spherical bubbles tangent to an infinite planar electrode is possible if the bubbles are spaced sufficiently far apart so the electric fields around them do not interact.[69] While idealizing many aspects of the problem, this model for the bubble layer contains one essential feature, the asymmetric environment of the bubble that is attached to the electrode on one side and faces an electrolyte of infinite extent on the other. Laplace's equation with homogeneous boundary conditions such as zero potential at the electrode surface, zero flux at the bubble surface, and a disturbance vanishing at infinity due to the bubble governs the primary current distribution on the plane. Solution of the problem[69] yields the current distribution around the sphere and the ohmic resistance presented by it.

The current distribution around a single bubble attached to an electrode appears in Fig. 10. The abscissa is normalized distance from the contact point while the ordinate is normalized current

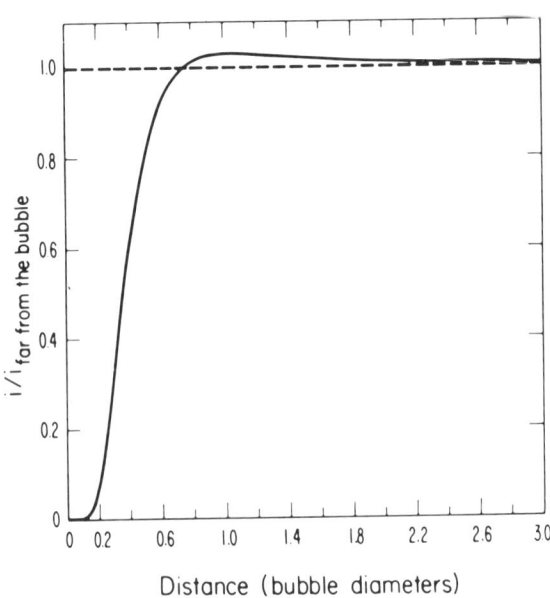

Figure 10. Current distribution around a dielectric sphere tangent to an electrode, from Sides and Tobias.[69] (Reprinted by permission of the publisher, The Electrochemical Society, Inc.)

density. Zero at the point of tangency, the current rises to 80% of its undisturbed value at the outermost circle on the plane shadowed by the bubble and reaches a maximum 2% greater that the undisturbed value at a distance of one diameter from the contact point. At a distance of three diameters, the current density is not substantially different from the undisturbed value. The resistance presented by such a bubble, evaluated by integrating the disturbance due to the bubble over a plane so far from the surface and parallel to it, is 10% less than that presented by a comparable bubble in the bulk because the equipotential electrode terminates the disturbance in the potential. Translating this result into the resistance presented by an array of bubbles on an electrode, however, one obtains a value too low for void fractions in the bubble layer greater than 0.1 because this calculation cannot account for the interactions of fields around closely spaced bubbles.

Closely packed arrays of bubbles attached to the electrode present significant resistance, but analyzing the electrical interactions between bubbles on electrodes is a complicated three-dimensional problem that Sides and Tobias[70] avoided by experimenting with a large-scale analog of a hexagonal array of bubbles. Carefully machined spheres were placed one at a time in the bottom of the cell to simulate varying degrees of surface coverage. The resistance was translated into an equivalent conductivity for the bubble layer, which appears in Fig. 11. Shown in the diagram are data for single spheres and for the spheres with wedges placed in the corners to simulate layers of smaller bubbles under the larger ones. Also provided are various theoretical predictions for comparison. Note the data for single spheres given by (+). Predicted by the tangent sphere calculation, the experimental conductivity values at small void fractions are greater than those predicted by the Maxwell relation. At close packing the conductivity of the bubble layer is reduced by a factor of nearly 5. In accord with the evidence from the tangent sphere calculation that bubbles on the surface present less resistance than corresponding bubbles in the bulk, the data for the single spheres remain above the prediction of the various models for bulk dispersions up to a void fraction of 0.4. A noticeable effect is the downward turn of the results for the bubble layer at void fractions greter than 0.5, which reflects the validity of a simple calculation of the effect of constrict-

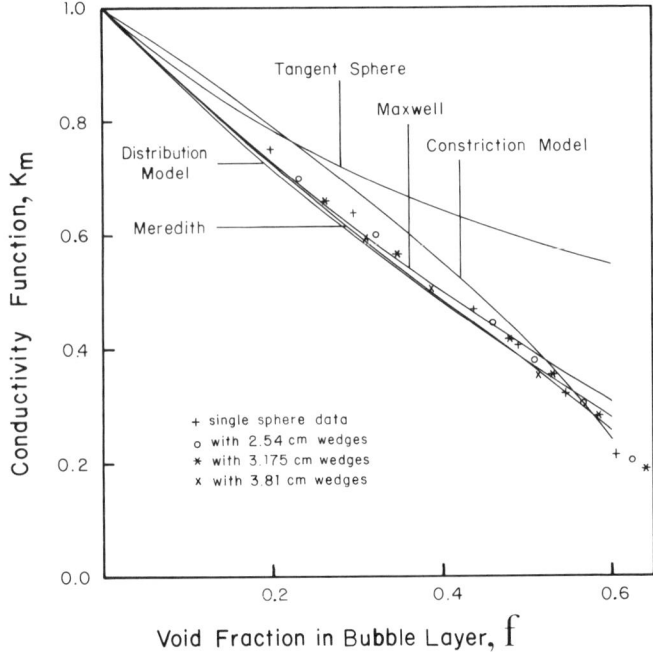

Figure 11. Reduced conductivity of a close-packed array of dielectric spheres resting on an electrode, from Sides and Tobias.[70] (Reprinted by permission of the publisher, The Electrochemical Society, Inc.)

ing the current to flow in the small interstices between closely spaced bubbles.[70] Called the "construction model," it was extended by Lanzi and Savinell[71] to cover high void fractions in the bubble layer.

Investigators of dynamic gas-evolving systems have been interested in the ohmic effects of bubbles attached to electrodes for at least 30 years. Early writers[72,73] noted bubble effects on measurements of overpotential. The importance of considering bubble layer effects has been demonstrated by experiments on operating cells. Takata et al.[74] experimentally investigated the resistance of a chlorine gas bubble layer on electrodes by varying the distance between the reference electrode probe and the gas-evolving electrode for various orientations of the electrode. Kubasov and

Volkov[75] studied the voltage increase caused by a bubble layer on graphite anodes which face downward during the chlorine evolution. They found voltage losses as high as 3 V. Hine *et al.*[64] distinguished between a region near the electrode containing a higher void fraction of gas than the bulk electrolyte. Haupin,[76] using a scanning reference electrode to investigate the potential profile in an operating Hall cell, observed voltages attributable to bubble polarization. Kuhn[77] provided both a review of previous work on potential measurements at gas-evolving electrodes and new data on the resistance of bubble layers at current densities less than 250 mA/cm^2. An anomalous result in his work, that the resistance before the appearance of oxygen bubbles was greater than the resistance during gas evolution, indicates that the effects are complicated and desired values may be masked. He concluded that the thickness of the bubble layer may be smaller than previously thought, less than 0.05 cm as opposed to 0.3 cm. Janssen and Barendrecht[78] used impedance techniques to investigate the electrolytic resistance of solution layers at hydrogen and oxygen evolving electrodes as a function of electrode area, material, and position, temperature, and pressure. The resistance correlated with the logarithm of the current density. The equations derived for bulk electrolytes worked reasonably well. Ngoya,[79] experimenting with bubble layers on electrodes facing downward such as found in mercury-type chlorine cells and in the Hall cell for aluminum production, found that even small angles of inclination away from the horizontal ($<3°$) made a large difference in the resistance presented by such layers.

The appreciable reduction in conductivity of the bubble layer has been identified in theory and experiment. One should determine whether ohmic losses in the layer are significant in any gas-evolving process under investigation. Furthermore, efforts to decrease the interelectrode gap in processes where there are significant bubble layers may not yield energy savings as large as anticipated because a significant resistance at the electrode surface remains unchanged, or perhaps increased. For many practical purposes, the equations developed for bulk dispersions of gas bubbles give a reasonable estimate of the bubble layer conductivity if correct account of the actual void fraction in the layer is taken.

(ii) Microscopic Current Distribution and Hyperpolarization of the Electrode

If the hypothetical electrode reaction is not reversible and gas still does not supersaturate, one must add the second component to the right-hand side of Eq. (17):

$$\Delta \Phi_T = \overline{\Delta \Phi}_{\text{ohm}} + \overline{\eta_s} \tag{18}$$

where $\overline{\Delta \Phi}_{\text{ohm}}$ is the ohmic drop corresponding to the primary current distribution at the given current level as measured by an interruption technique[80] and $\overline{\eta_s}$ represents an average over the distribution of overpotentials accompanying the nonuniform current distribution. $\overline{\eta_s}$ must itself be the sum of the surface overpotential accompanying the electrode reaction if no bubbles were present (i.e., the electrode operates at a uniform superficial current density) plus an amount of voltage due to the effectively higher current density which must exist because the bubbles cause a nonuniform current distribution on the electrode:

$$\overline{\eta_s} = \eta_{so} + \eta_h \tag{19}$$

The first term on the right-hand side of Eq. (19) is the overpotential which would exist were no bubbles present and the second term on the right-hand side in Eq. (19) stands for the portion of the surface overpotential related to the nonuniform current density, hyperpolarization of the electrode.

If the gas-evolving reaction of this discussion can be characterized by Tafel constants a and b on a given smooth electrode in the absence of gas bubbles, then

$$\eta_{so} = a + b \log(I/A) \tag{20}$$

where I is the total current to the gas evolving electrode and A is the actual electrode area. Hine et al.,[56] discussing the anode shift potential caused by the local current density's exceeding the superficial current density because the bubbles screen the electrode surface, interpreted their experimental results in terms of a reduced area available for current. In the spirit of this idea, one may assume that with bubbles present, there must be an effective average area available for current flow to the electrode, A_h, which is smaller than A, the superficial electrode area. The total surface overpotential must now be written as

$$\overline{\eta_s} = a + b \log(I/A_h) \tag{21}$$

Subtracting (20) from (21), one obtains

$$\eta_h = b \log(A/A_h) \tag{22}$$

Thus bubbles on an electrode introduce an additional current-dependent term in the kinetic equation. In addition, A_h must be a function of the exchange current density, temperature, kinetic transfer coefficients, the average bubble diameter, and the void fraction, all of which determine the secondary current distribution on the electrode.

Knowledge of the secondary current distribution around bubbles on electrodes would be of some interest, for example in electroplating where gas pits in the plated metal are a problem. Calculating the secondary current distribution around bubbles on an electrode, however, requires a numerical solution. Nevertheless some conclusions can be drawn from dimensional analysis of the problem. Newman[81] has shown that two parameters characterize nonuniform secondary current distributions:

$$J = (\alpha_a + \alpha_c) i_0 aF/\kappa RT \tag{23}$$

$$\delta = J(i_{\text{avg}}/i_0) \tag{24}$$

These parameters arise from the boundary conditions on Laplace's equation when finite kinetic rates are included. The first is a ratio between the exchange current density and ohmic parameters such as length and conductivity. The second parameter is a dimensionless current level. Values of either or both parameters much greater than unity indicate that ohmic effects dominate and the current distribution resembles the primary case. Low values of both parameters indicate that kinetics limit the process and the current distribution is uniform. Even though a low value of J indicates substantial kinetic effects, the current distribution resembles the primary case at high values of δ because the linear ohmic dependence on current density dominates the logarithmic overpotential dependence. These

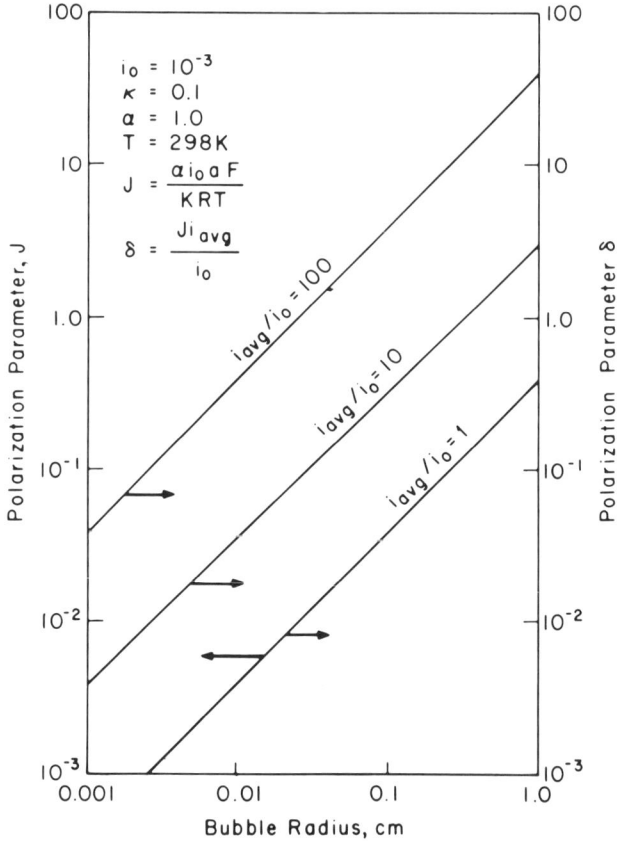

Figure 12. Polarization parameters J and δ for determining the current distribution around bubbles on electrodes.

quantities can readily be applied to bubbles on an electrode if the radius of the bubbles is used as the length parameter. Figure 12 shows values of J and δ as a function of bubble size for physical parameters appropriate for aqueous electrolytic gas evolution. The values of the two parameters tend to be less than unity for the small gas bubbles of these processes, an indication that the actual microscopic current distribution may be much more uniform than the primary current distribution.

(iii) Effect of Gas Bubbles on Concentration Overpotential

In the general case, the potential of a gas-evolving electrode is given by

$$\Delta\Phi_T = \overline{\Delta\Phi}_{ohm} + \overline{\eta_s} + \overline{\eta_c} \qquad (25)$$

Equation (25) includes ohmic, kinetic, and concentration effects. The equation is analogous to Eq. (16) except that it is now written in terms of quantities which can be measured directly or at least isolated. Use of current interruption still yields the correct ohmic drop (if ionic concentration gradients are negligible), but the measured overpotential comprises kinetic and concentration-dependent components.

The concentration of product gas dissolved in the electrolyte in immediate contact with the electrode determines the average concentration overpotential that may be calculated through equations for the chemical potential of the gas. Three investigators have measured these product-gas supersaturations in gas-evolving systems. Breiter *et al.*[82] measured hydrogen concentrations at cathodes. Bon,[83] measuring the concentrations of dissolved gas next to a hydrogen-evolving electrode with hydrogen reference electrodes made of thin Pt foil, found supersaturations of 4-60 atm in the range of current densities 0.2-40 mA/cm^2. Shibata[84,85] measured the supersaturation of hydrogen and oxygen electrodes with a galvanostatic transient technique and found a maximum supersaturation of hydrogen of 160 atm at 200 mA/cm^2 and a maximum supersaturation of oxygen of 70 atm at 100 mA/cm^2. A graph representing the results of Breiter,[82] Bon,[83] and Shibata[84] appears in Fig. 13. Vogt[86] distinguished between the concentration in the bubble layer that is important for bubble growth from the concentration measured by Shibata and Bon that determines the concentration overpotential. Recently, Janssen and Barendrecht,[87] using a novel rotating ring cone electrode, found smaller supersaturations of 14 for oxygen evolution in alkaline medium; however, their results were complicated by loss of oxygen from solution as the supersaturated electrolyte flowed from the working cone electrode to the detecting ring. Supersaturations of hundreds of atmospheres are well established by these studies for aqueous solutions of hydrogen and oxygen.

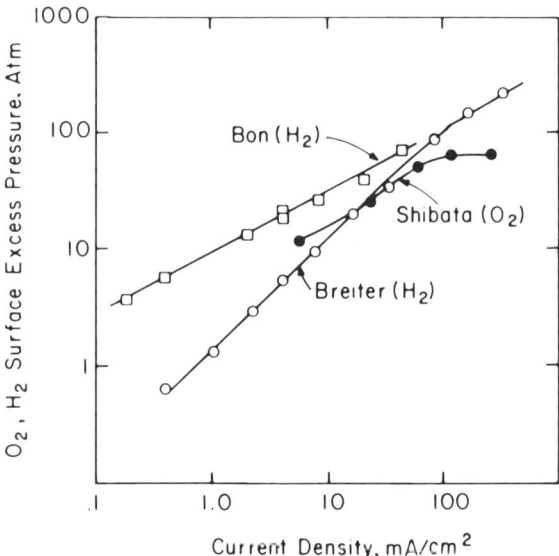

Figure 13. Supersaturation at gas-evolving electrodes from Bon,[83] Shibata,[85] Breiter.[82] The surface excess pressure is the supersaturation in atmospheres at the electrode surface.

The results of these investigations may be used to separate the concentration overpotential from the surface overpotential in Eq. (25). Further isolation of hyperpolarization from the bubble-free kinetic polarization is not possible unless measures are taken to remove the bubbles from the surface. The concentration of dissolved gas also affects the microscopic current distribution around the bubbles and may change the conclusions one might reach from analysis of the kinetic parameters given by Eqs. (23) and (24).

(iv) Summary of Bubble Layer Effects

To conclude this section, I estimate the voltage components of the bubble layer in circumstances roughly appropriate to oxygen evolution in 1 N sulfuric acid and carbon dioxide evolution in Hall-Heroult electrolysis. The values chosen should be taken as approximations designed to show upper bounds of the individual components for the purposes of illustration. Equation (26) represents the ohmic voltage loss associated with the presence of the bubbles on the electrode. Note that the voltage loss that would occur were the bubbles not there has been subtracted:

$$\overline{\Delta\Phi}_{ohm} = \frac{id(1 - K_m)}{\kappa(K_m)} \quad (26)$$

Hyperpolarization of the electrode is calculated from Eq. (22) with estimated values for the effective area available for current. The concentration overpotential is calculated from the standard thermodynamic expression where activity coefficients have been ignored for the purposes of this rough estimate:

$$\overline{\eta}_c = (RT/4)\ln(p^*) \quad (27)$$

where p^* is the supersaturation ratio. The results of these calculations appear in Table 3. In aqueous electrolysis, the three components are roughly comparable and sum to around a tenth of a volt. In the molten salt cell, the high current density and thick bubble layer on the underneath of the carbon anodes drive the ohmic loss to over half a volt, which agrees with Thonstad's estimate[88] but may be higher than the actual loss. Hyperpolarization may be responsible for small but measureable amounts of polarization at gas-evolving electrodes.

Table 3
Gas Evolution Parameters and Voltage Components of the Bubble Layer

Parameter	Oxygen evolution (1 N sulfuric)	Carbon dioxide (Hall bath, cryolite)
Layer thickness (cm)	0.2	0.5
Electrolyte conductance (mho/cm)	0.3	2.5
Void fraction	0.25	0.5
Supersaturation (atm)	150	150?
Temperature (°C)	100	1000
A/A_h	1.6	2
Current density (A/cm^2)	0.2	1
Tafel slope (V)	0.12	0.26
Component (V)		
Extra ohmic loss	0.07	0.36
Hyperpolarization	0.03	0.08
Concentration overpotential	0.04	0.14
Total	0.14	0.58

IV. MASS TRANSFER AT GAS-EVOLVING ELECTRODES

Electrolytically evolved bubbles enhance transport of heat or mass at gas-evolving electrodes because the growing and detaching gas bubbles mix electrolyte near the surface with electrolyte in the bulk. If there is a reacting species in solution and its rate of reaction is mass transfer controlled, bubbling accelerates its transport to the surface. For example, Roald and Beck[89] in 1951 noted that the rate of magnesium dissolution in acid electrolyte was controlled by transport of acid to the electrode surface and that stirring by the evolved hydrogen bubbles increased the rate. Mass transfer enhancement is also important in chlorate production and perhaps zinc electrowinning where the 10% current going to hydrogen evolution at the cathode may increase the deposition rate. Evolving gas bubbles can accelerate mass transfer to rates achieved by only intense mechanical stirring or flow and is thus very effective where circumstances permit its use.

Phenomena of gas evolution discussed at the outset are responsible for mixing at the electrode. The growing bubbles mix the electrolyte as their surfaces expand and as they coalesce with each other. When a bubble detaches, electrolyte flows to fill the vacancy, hence fresh electrolyte is brought to the electrode surface. The effective density of a heterogeneous mixture of gas and electrolyte, being lower than that of the surrounding liquid, stirs the electrolyte by gas lift. These are complicated phenomena and it seems all but impossible to mathematically analyze the flows, thus engineers have constructed theories that explain and correlate the experimental results and patterns. Three schools of thought in Switzerland, the Netherlands, and Germany may be identified from research published over the last 25 years.

1. Penetration Theory

The first school began with Venczel's dissertation in Zurich in 1961[90] on the transport of ferric ion to an electrode evolving hydrogen gas from one molar sulfuric acid. Venczel found that mass transfer increased rapidly with the onset of gas evolution. Ibl and Venczel[26] reported mass transfer at gas-evolving electrodes as Nernst boundary layer thicknesses that are functions of gas evolution rate

(cm^3/cm^2 s) or current density and plotted the relation in log/log coordinates. The definition of the Nernst boundary layer thickness is

$$\delta_N = nFDC/i \qquad (28)$$

where δ_N is the boundary layer thickness, n is the equivalents per mole, D is the diffusivity of mass transfer-controlled species, C is the concentration of mass transfer-controlled species, F is Faraday's constant, and i is the current density; and it is essentially the reciprocal of a mass transfer coefficient divided by the molecular diffusivity. Figure 14 is a graph from Ibl et al.[91] from their work on mass transfer enhancement in acid solutions. The boundary layer thickness decreases as the gas evolution rate increases. At the highest gas evolution rates, corresponding to 10 A/cm^2, the boundary layer thickness is quite thin, on the order of 1 μm. By comparison, a mechanically well-stirred laboratory vessel may have a boundary layer thickness of 10 μm, which corresponds to 100 mA/cm^2 in the graph.

Ibl and Venczel[26] hypothesized that diffusion of the reactant from fresh electrolyte, brought to the surface after a bubble detaches, is the mechanism of mass transfer enhancement. They

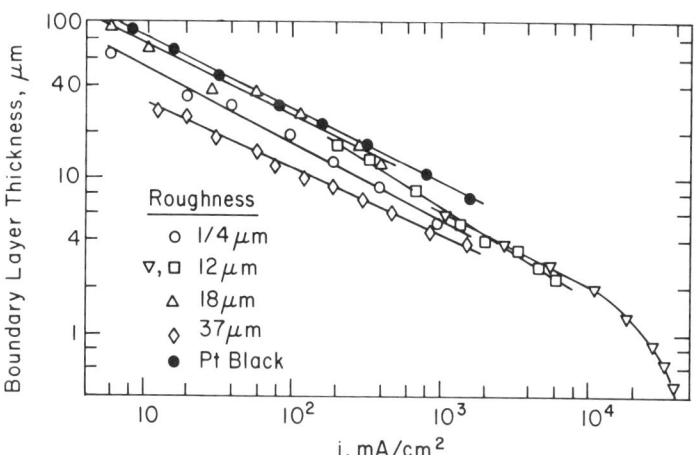

Figure 14. Effect of gas bubble evolution on the thickness of the diffusion layer, from Ibl et al.[91]

used surface renewal or penetration theory as the basis of a derivation which predicted a dependence of the mass transfer rate on the square root of the gas evolution rate:

$$\delta_N = [\pi Dr/6 V_G(1 - \tau)]^{0.5} \qquad (29)$$

where r is the average radius of the bubbles, V_G is a velocity found by dividing the gas volume evolution rate by the area of the electrode, and τ is the fractional surface coverage by bubbles. One must know the bubble size at departure and the fractional area of the electrode shadowed by bubbles on the surface to use this model. Venczel studied the problem of electrode coverage and Vogt[31] correlated the results with

$$\tau = 0.5 Re^{0.18} \qquad (30)$$

where $Re = V_G d/\nu$. Experimentally, Ibl and Venczel found exponents varying from 0.4 to 0.6 in Eq. (29), but a significant number of data were grouped around 0.53, which they took as confirmation of the theory. In later work, Alkire and Lu[92] investigated mass transfer to a vertical electrode in copper deposition with simultaneous hydrogen evolution. They found that above 80 mA/cm^2, the exponent was 0.5 in accord with this model.

2. The Hydrodynamic Model

The second school of thought developed in the Netherlands at Eindhoven. Janssen and several co-workers (Hoogland, Barandrecht, van Stralen) in a series of articles[7,15,20,27,93] investigated mass transfer enhancement in alkaline electrolyte in which hydrogen bubbles are very small (20 μm) while oxygen bubbles can be quite large (200 μm).[20] They found that the mass transfer enhancement caused by hydrogen bubbles obeyed a smaller exponent (0.31) of the gas current density than reported by Ibl et al.[26,28] for low current densities and a large exponent, 0.87, at higher current densities. This change in slope occurred at the same gas evolution rate as the change in bubble size of oxygen bubbles from around 60 μm to the 200 μm mentioned earlier. They related the low exponent to a hydrodynamic model in which free-convective flows, engendered by detaching bubbles, account for the mass transfer. Their equation for this case is

$$\delta_N = C_1(\nu D v_t / Z_d g V_G)^{0.333} \tag{31}$$

where C_1 is a constant, v_t is the terminal rise velocity of a bubble, and Z_d is the drag coefficient on a bubble. Thus Janssen and co-workers disputed the surface renewal model for low current densities and gas evolution where bubble coalescence is not important. Janssen and Hoogland[27] attributed the higher exponent to increased coalescence.

Figure 15 is a graph from Janssen and Hoogland[27] which is the basis of their arguments for the hydrodynamic theory of mass transfer in the absence of coalescence. Note the slope change, around 30 mA/cm², which corresponds to a change in bubble size for these oxygen bubbles evolved in basic solution and is indicative of the onset of coalescence between bubbles on the electrode.

Sides and Tobias[17] observed the coalescence of large bubbles causing fluid motions close to the electrode surface which may be important in the mass transport enhancement due to gas evolution

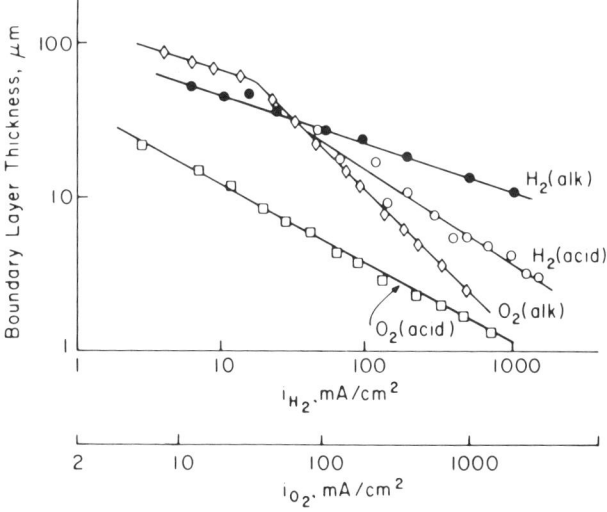

Figure 15. Effect of gas bubble evolution on the thickness of the diffusion layer, from Janssen and Hoogland.[27] (Reprinted with permission from *Electrochimica Acta* **18**, L. J. J. Janssen and J. Hoogland, The effect of electrolytically evolved gas bubbles on the thickness of the diffusion layer—II, Copyright, 1973, Pergamon Press.)

reported by Ibl,[28] Venczel,[90] and Janssen and Hoogland.[27] This mode joins the flow due to gas lift and fluid replacement due to bubble departure as an accelerator of mass transfer. A coalescence sequence between two large bubbles near the electrode and moving along it appears in Fig. 16. The two bubbles appeared to touch for

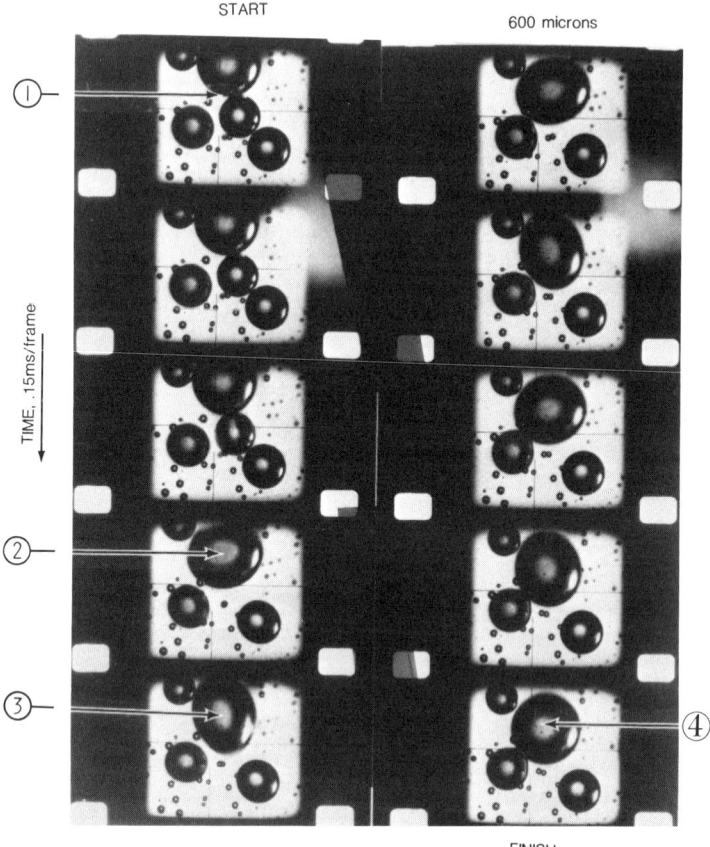

Figure 16. Coalescence and vibration of bubbles, from Sides and Tobias.[17] (1) Two bubbles touch for a number of frames while the film drains from between them. (2) The resulting bubble is compressed against the electrode. (3) The new bubble vibrates as it establishes its equilibrium boundaries. (4) The oscillations end after a half millisecond. (Reprinted by permission of the publisher, The Electrochemical Society, Inc.)

many frames as the film between them thinned and finally ruptured. The bubbles coalesced so quickly that the film rupture and the change from two bubbles to one occurred between two frames, that is, in less than 100 μs. The new bubble was compressed along the axis of coalescence by the fluid rushing into the space behind the coalescing bubbles. The bubble vibrated like this for several frames before becoming spherical again. This is an example of coalescence of two relatively large bubbles, both traveling along the electrode surface.

The Nernst boundary layer thickness is a simple characteristic of the mass transfer but its definition is formal since no boundary layer is in fact stagnant and least of all boundary layers on gas-evolving electrodes; furthermore, the Schmidt number, known to influence mass transfer, is not incorporated in the usual dimensionless form. For this reason, lines representing data from gas evolution in two different solutions can be displaced from one another because of viscosity differences. Nevertheless, the exponent in the equation

$$\delta_N = ai^b \tag{32}$$

is independent of the dimensionless groups and is the main difference between the schools of thought. The exponent b takes on values from -0.25 to -0.87.[91]

3. The Microconvection Model

Stephan and Vogt,[94] representing the third school of thought, correlated the results of others with a model, called the microconvection model, in which the primary enhancement came from the convective flows caused by expansion of the bubble during growth. They adapted mass transfer theory for laminar flow over planes to their system and integrated the resulting equation over the period of bubble growth. Their equation correlates data from other investigators:

$$\text{Sh} = 0.93 \text{Re}^{0.5} \text{Sc}^{0.487} \tag{33}$$

where d is the bubble breakoff diameter, ν is the kinematic viscosity, $\text{Sh} = kd/D$, k is the mass transfer coefficient, and $\text{Sc} = \nu/D$. In addition, Vogt recognized the usefulness of rendering the results of others in dimensionless form. For example, Eq. (29) becomes[31]

$$Sh = (24/\pi)^{0.5}[Re\ Sc(1-\tau)]^{0.5} \qquad (34)$$

and Eq. (31) for the hydrodynamic model becomes

$$Sh = const(Z_d\ Re\ Sc)^{0.333} \qquad (35)$$

Despite the differences among the three main schools of thought on mass transfer at gas-evolving electrodes, the writers agree on some points. Gas evolution is an effective means of enhancing transport. Roughness of the electrode on the order of the boundary layer thickness does not strongly influence the effect. There is a slight dependence on electrode orientation and on the length of a vertically oriented electrode, but these are secondary effects.

4. Microscopic Investigation

Dees,[95] using a novel micromosaic electrode, compared mass transfer enhancement due to a detaching bubble with the mass transfer resulting from coalescence of two bubbles. The electrode, consisting of 100 segments each 100 μm square in a 10×10 array, resolved previously undetected microscopic mass transfer phenomena. He found that coalescing bubbles mixed electrolyte better than disengaging bubbles, and that mixing due to coalescence exceeded the mixing of natural convection by an order of magnitude.

V. SUMMARY

Although much is known about electrolytic gas evolution, it is still a challenging subject for research. Nucleation of bubbles at the electrified interface should be explored. There are subtle phenomena in the growth, coalescence, and detachment of bubbles to be explained. The conductivity of electrolytes containing gas bubbles is theoretically understood, but the modeling of these effects in large-scale systems, particularly on large inverted electrodes such as Hall-cell anodes, is yet to be achieved. Details of the voltage balance at the gas-evolving electrode are yet to be quantified. Many experiments on the enhancement of mass and heat transfer at gas-evolving electrodes have been performed, but the complexity of the problem leaves room for more comprehensive semiempirical analysis.

Electrolytic gas evolution is a complicated and important problem in many industrial processes. The details of bubble formation and the effects of the bubbles presented in this review are but the microscopic aspects of a phenomenon that affects the macroscopic behavior of electrochemical cells. The knowledge summarized here applies in part to the even larger world of boiling liquids in which bubbles also nucleate, grow, coalesce with each other, and detach. In general, the sudden appearance of a phase much less dense than its parent phase will always be an important phenomenon for research because that phase will profoundly affect the process in which it appears.

ACKNOWLEDGMENTS

I wish to acknowledge the help of Professor Charles W. Tobias, whose guidance and counsel influenced much of this work. I also wish to recognize the support of ALCOA, the Aluminum Association, and the American Electroplater's Society, without whose support this work would not have been possible.

NOTATION

a	Tafel intercept (V)
a (13)†	constant given in Table 1
a (23)	bubble radius (cm)
a (32)	empirical constant
A	superficial area of a gas-evolving electrode (cm^2)
A_h	area available for current flow on a gas-evolving electrode (cm^2)
b	Tafel slope (V)
b (13)	constant given in Table 1
b (32)	exponent varying between -0.25 and -0.87

† Numbers in parentheses refer to the equations in which the terms are used.

c (13)	constant given in Table 1
C_1	empirical constant
C'	concentration of dissolved gas (mole/cm^3)
C_0	saturation concentration of dissolved gas (mol/cm^3)
C	concentration of dissolved gas (mol/cm^3)
d	thickness of a bubble layer (cm)
d (13)	constant given in Table 1
d (36)	bubble breakoff diameter (cm)
D	molecular diffusivity (cm^2/s)
e	constant given in Table 1
f	void fraction
F	Faraday's constant (C/equivalent)
g	gravitational constant (cm/s^2)
g (13)	constant given in Table 1
h	reduced height of an electrode (cm)
i	current density (A/cm^2)
i_{avg}	superficial current density (A/cm^2)
i_0	exchange current density (A/cm^2)
I	electric current (A)
J	polarization parameter
J (1)	nucleation rate (s^{-1})
k	mass transfer coefficient (cm/s)
K_m	ratio of the effective conductivity to the conductivity of the continuous medium
K_1	empirical constant
K_2	empirical constant
K_3	empirical constant
L_n	function of conductivity ratio for use in Eq. (13)
m	empirical constant
n	empirical constant
p	empirical constant
P^*	ratio of the equivalent pressure of dissolved gas at the surface of a gas-evolving electrode to the pressure in the bulk
P''	pressure of dissolved gas inside the bubble (atm)
P_0	vapor pressure of the pure solvent (atm)
P'	external pressure in the liquid (atm)
r	radius (cm)
R	radius of growing bubble (cm)

$\dot{R}$	velocity of bubble interface, rate of growth of bubble (cm/s)
Re	Reynolds' number
ΔR_{wm}	resistance between the working electrode and the cell diaphragm divided by the resistance which would exist if there were no bubbles present (Ω)
Sc	Schmidt number
Sh	Sherwood number
t	time (s)
v_1	specific volume of the pure solvent (cm^3/mole)
v_t	terminal rise velocity of a bubble (cm/s)
V	volume of a bubble (cm^3)
V_w	voltage of the working electrode (V)
V_G	effective velocity of gas evolution (cm/s)
Z	pre-exponential frequency factor (cm^{-1})
Z_d	drag coefficient on a bubble
α	ratio of the conductivity of the dispersed phase to the conductivity of the continuous phase
α_a	anodic transfer coefficient
α_c	cathodic transfer coefficient
β	growth coefficient
γ	angle of inclination from the horizontal
δ	dimensionless current level, polarization parameter
δ_N	mass transfer boundary layer thickness (cm)
$\Delta\Phi_{ohm}$	local ohmic voltage drop between the electrode and the reference electrode (V)
$\overline{\Delta\Phi_{ohm}}$	average ohmic voltage drop between the electrode and the reference electrode given by the product of the primary resistance and the total current (V)
$\Delta\Phi_T$	voltage measured between the working electrode and a reference electrode of the same kind (V)
ε	ratio of the density difference between liquid and gas to the liquid density
η_s	surface overpotential at the working electrode (V)
$\overline{\eta_s}$	average surface overpotential at the working electrode
η_c	concentration overpotential at the working electrode (V)
$\overline{\eta_c}$	effective concentration overpotential at the working electrode (V)

η_{so} surface overpotential of the working electrode if no bubbles were present and the electrode operated at the superficial current density (V)
η_h hyperpolarization of the working electrode due to nonuniform current distribution associated with the presence of gas bubbles (V)
Θ_A advancing contact angle
Θ_R receding contact angle
κ intrinsic electrolyte conductivity (mho/cm)
ν kinematic viscosity (cm^2/s)
ν_i activity coefficient of phase i
ρ density (g/cm^3)
ρ_L density of the liquid phase (g/cm^3)
σ surface tension (dyn/cm)
τ fractional surface coverage by bubbles
Φ_r potential of a reference electrode of the same kind

REFERENCES

[1] M. Blander and J. L. Katz, *Am. Inst. Chem. Engrs. J.* **21** (1975) 833.
[2] T. W. Forest and C. A. Ward, *J. Chem. Phys.* **66** (1977) 2322.
[3] K. Dapkus and P. Sides, *J. Coll. Interface Sci.* **111** (1986) 133.
[4] H. B. Clark, P. S. Strenge, and J. W. Westwater, *Chem. Eng. Prog. Symp. Ser. 29*, **55** (1957) 103.
[5] R. Cole, *Adv. Heat Transfer* **10** (1974) 85.
[6] D. E. Westerheide and J. W. Westwater, *Am. Inst. Chem. Engrs. J.* **7** (1961) 357.
[7] L. J. J. Janssen and J. H. Hoogland, *Electrochim. Acta* **15** (1970) 1013.
[8] L. E. Scriven, *Chem. Eng. Sci.* **27** (1959) 1753.
[9] Lord Rayleigh, *Phil. Mag.* **34** (1917) 94.
[10] P. Epstein and M. Plesset, *J. Chem. Phys.* **18** (1950) 1505.
[11] M. S. Plesset and S. A. Zwick, *J. Appl. Phys.* **25** (1954) 493.
[12] H. Forster and N. Zuber, *J. Appl. Phys.* **25** (1954) 474.
[13] H. Cheh and C. Tobias, *Int. J. Heat Mass Transfer* **11** (1968) 709.
[14] J. P. Glas and J. W. Westwater, *Int. J. Heat Mass Transfer* **7** (1964) 1427.
[15] L. J. J. Janssen and S. van Stralen, *Electrochim. Acta* **26** (1981) 1011.
[16] R. Putt, M.S. thesis, University of California, October 1975.
[17] P. Sides and C. Tobias, *J. Electrochem. Soc.* **132** (1985) 583.
[18] S. Fortin, M. Gerhardt, and A. Gesing, *Light Metals* **March** (1983) 721-741.
[19] B. Kabanov and A. Frumkin, *Z. Phys. Chem.* **165A** (1933) 433.
[20] L. Janssen and E. Barendrecht, *Electrochim. Acta* **24** (1979) 693.
[21] A. Coehn, *Z. Elektrochem.* **29** (1923) 1.
[22] A. Coehn and H. Neumann, *Z. Phys.* **20** (1923) 54.
[23] E. Dussan and R. Chow, *J. Fluid Mech.* **137** (1983) 1.
[24] E. Dussan, *J. Fluid Mech.* **151** (1985) 1.

[25] J. Venczel, *Electrochim. Acta* **15** (1970) 1909.
[26] N. Ibl and J. Venczel, *Metalloberflache* **24** (1970) 365.
[27] L. Janssen and J. Hoogland, *Electrochim. Acta* **18** (1973) 543.
[28] N. Ibl, *Chim. Ing. Tech.* **43** (1971) 202.
[29] D. Landolt, R. Acosta, R. Muller, and C. W. Tobias, *J. Electrochem. Soc.* **117** (1970) 839.
[30] E. L. Littauer, United States Patent 3,880,721 (1975).
[31] H. Vogt, Gas evolving electrodes, in *Comprehensive Treatise of Electrochemistry*, Vol. 6, Plenum Press, New York, 1983.
[32] R. E. Meredith and C. W. Tobias, *Advances in Electrochemistry and Electrochemical Engineering* 2, Interscience, New York, 1962, p. 15.
[33] J. C. Maxwell, *A Treatise on Electricity and Magnetism*, Vol. 1, 2nd edn., Clarendon Press, Oxford, 1881, p. 435.
[34] Lord Rayleigh, *Phil. Mag.* **34** (1892) 481.
[35] L. Sigrist, O. Dossenbach, and N. Ibl, *J. Appl. Electrochem.* **10** (1980) 223.
[36] R. E. DeLaRue and C. W. Tobias, *J. Electrochem. Soc.* **106** (1959) 827.
[37] Z. Hashin, *J. Composite Mater.* **2** (1963) 284.
[38] G. H. Neale and W. K. Nader, *Am. Inst. Chem. Engrs. J.* **19** (1973) 112.
[39] Yu. A. Buyevich, *Chem. Eng. Sci.* **29** (1974) 37.
[40] J. C. R. Turner, *Chem. Eng. Sci.* **31** (1976) 487.
[41] D. J. Jeffery, *Proc. R. Soc. London. Ser. A* **335** (1973) 355.
[42] R. W. O'Brien, *J. Fluid Mech.* **91** (1979) 17.
[43] W. I. Higuchi, *J. Phys. Chem.* **62** (1958) 649.
[44] C. A. R. Pearce, *Brit. J. Appl. Phys.* **6** (1955) 113.
[45] S. Prager, *Physica* **29** (1963) 129.
[46] Y. Chiew and E. Glandt, *J. Coll. Interface Sci.* **94** (1983) 90.
[47] R. E. DeLaRue, Electric conductivity of dispersed systems, Master's thesis, University of California, Berkeley, 1955.
[48] R. E. Meredith, Studies on the conductivity of dispersions, Doctoral Thesis, University of California, Berkeley, 1959.
[49] R. E. Meredith and C. W. Tobias, *J. Electrochem. Soc.* **108** (1961) 286.
[50] D. Bruggeman, *Ann. Phys.* (*Leipzig*) **24** (1935) 636.
[51] N. Clark, *Trans. Faraday Soc.* **44** (1948) 13.
[52] A. Slawinski, *J. de Chemie Physique* **23** (1926) 710.
[53] I. Runge, *Z. Tech. Phys.* **6** (1925) 61
[54] D. De Vries, *Bull. Inst. Int. Froid Annexe 1952-1* **32** (1952) 115.
[55] S. Churchill, in *Advances in Transport Processes IV*, Ed. by A. S. Mujamdar and R. A. Mashelkar, John Wiley, New York, 1986.
[56] D. McKenzie and R. McPhedran, *Nature* **265** (1977) 128.
[57] R. E. Meredith and C. W. Tobias, *J. Appl. Phys.* **31** (1960) 1270.
[58] M. Zuzovsky and H. Brenner, *J. Appl. Math. Phys.* **28** (1977) 979.
[59] R. McPhedran and D. McKenzie, *Proc. R. Soc. London Ser. A* **359** (1978) 45.
[60] D. McKenzie, R. McPhedran, and G. H. Derrick, *Proc. R. Soc. London Ser. A* **362** (1978) 211.
[61] A. Sangani and A. Acrivos, *Proc. R. Soc. London Ser. A* **386** (1983) 263.
[62] F. Hine, S. Yoshizawa, and S. Okada, *Denki Kagaku* **24** (1956) 370.
[63] S. Yoshizawa, F. Hine, and Z. Takehara, *J. Electrochem. Soc. Jpn.* **28** (1960) 88.
[64] F. Hine, M. Yasuda, R. Nakamura, and T. Noda, *J. Electrochem. Soc. Jpn.* **122** (1975) 1185.
[65] F. Hine and K. Murakami, *J. Electrochem. Soc. Jpn.* **127** (1980) 293.
[66] F. Hine and K. Murakami, *J. Electrochem. Soc.* **128** (1981) 65.
[67] F. Hine, M. Yasuda, Y. Ogata, and K. Hara, *J. Electrochem. Soc.* **131** (1984) 83.

[68] L. J. J. Janssen, J. J. M. Geraets, E. Barendrecht, and S. D. J. van Stralen, *Electrochim. Acta* **9** (1982) 1207.
[69] P. J. Sides and C. W. Tobias, *J. Electrochem. Soc.* **127** (1980) 288.
[70] P. J. Sides and C. W. Tobias, *J. Electrochem. Soc.* **129** (1982) 2715.
[71] O. Lanzi and R. F. Savinell, *J. Electrochem. Soc.* **130** (1983) 799.
[72] J. O'M. Bockris and M. Azzam, *Trans. Faraday Soc.* **48** (1952) 145.
[73] M. Breiter and Th. Guggenberger, *Z. Elektrochem.* **60** (1956) 594.
[74] K. Takata and H. Morishita, *Denki Kagaku* **32** (1964) 378.
[75] V. L. Kubasov and G. I. Volkov, *Sov. Electrochem.* **2** (1966) 665.
[76] W. Haupin, *J. Metals* **46** (1971) October.
[77] A. Kuhn and M. Stevenson, *Electrochim. Acta* **27** (1982) 329.
[78] L. J. J. Janssen and E. Barendrecht, *Electrochim. Acta* **28** (1983) 341.
[79] Ngoya, Gas evolution at horizontal electrodes, Ph.D. thesis, University of Trondheim, 1983.
[80] J. Newman, *J. Electrochem. Soc.* **117** (1970) 507.
[81] J. Newman, *Electrochemical Systems*, Prentice Hall, Englewood Cliffs, New Jersey, 1973.
[82] M. Breiter, H. Kammermaier, and C. Knorr, *Z. Electrochem.* **60** (1956) 37,119,455.
[83] C. Bon, Supersaturation at gas-evolving electrodes, Master's Thesis, University of California, Berkeley, 1970.
[84] S. Shibata, *Bull. Chem. Soc. Jpn.* **33** (1960) 1635.
[85] S. Shibata, *Electrochim. Acta* **23** (1978) 619.
[86] H. Vogt, *Electrochim. Acta* **25** (1980) 527.
[87] L. Janssen and E. Barendrecht, *Electrochim. Acta* **29** (1984) 1207.
[88] J. Thonstad, Abstracts of the 35th ISE Meeting, August 5-10, Berkeley, California, 1984.
[89] B. Roald and W. Beck, *J. Electrochem. Soc.* **98** (1951) 277.
[90] J. Venczel, Dissertation, ETH Zurich Nr. 3019, 1961.
[91] N. Ibl, R. Kind, and E. Adam, *Ann. Quim.* **71** (1975) 1008.
[92] R. Akire and P. Lu, *J. Electrochem. Soc.* **126** (1979) 2118.
[93] L. J. J. Janssen, *Electrochim. Acta* **23** (1978) 81.
[94] K. Stephan and H. Vogt, *Electrochim. Acta* **24** (1979) 11.
[95] D. Dees, Mass transfer at gas evolving surfaces in electrolysis, Ph.D. thesis with C. W. Tobias, University of California, Berkeley, 1983.

Index

Absolute potential, and Hansen and Kolb, 12
Absolute potential difference, 8
Acid-based properties of oxides, work of Kang and Shay, 174
Acidification, of layer adjacent to membrane, 125
Activity, versus concentration of sodium, in amalgams, 272
Additives, and bubble growth, 315
Ahlgren, and the Butler–Volmer reaction rate, 88
Albery
  approach to space charge recombination, 53
  extension of Gartner's model, 89
Alwitt
  account of aluminum oxide systems, 172
  films on aluminum, 175
Amalgam decomposition and materials balance, 277
Amalgam–solution interface, 274
Andrieux and Saveant, and potential sweeps, 184
Angerstein-Kozlowska, work on gold oxide, 206
Anodic charging curves, on platinum, 196
Aoki, and potential sweep, 184
Appleby
  contribution to membrane cells, 299
  and platinum oxide formation, 195

Bagotzky and Tarasevich
  only two broad peaks in an anodic

Bagotzky and Tarasevich (*cont.*)
  sweep on platinum, 195
  oxidation mechanisms, 195
  work on palladium oxide, 206
Balej and Spalek, and induction period for oxide growth, 199
Band–band recombination, in semiconductors, 74
Band theory, in photoelectrochemical kinetics, 73
Bard, and Fermi level pinning, 17
Battery oxides, 181
Beer
  development of ruthenium oxide layers, 171
  ruthenium oxide coatings, 266
Belanger and Vijh, work on platinum oxides, 197
Bilayer membranes, containing chlorophyll, 137
Bilayers, and proton transport, 151
Biological membranes, and electron motion, 145
Birss
  hydrous oxide growth, 229
  work on oxide films, 220
BLMs
  accumulation of quinones, 133
  affected by iodine, 149
Blumenfeld, and interactions at membranes, 162
Boardman and Robertson, co-enzyme Q traps, 152
Bockris
  model of electrode processes, 32
  and modification of Gurney model, 28

Bockris (cont.)
  views criticizing those of Gerischer, 35
Bockris and Argade, and the absolute metal–solution potential distribution, 13
Bockris and Khan, and Fermi level in solution, 9
Bockris and Uosaki
  model for electrode processes, 40
  model for electron transfer at illuminated surfaces, 38
Bolzan, and activity in dissolution of palladium, 205
Boundary conditions, in semiconductor-solution problems, 82
Breiter, and hydrogen concentration at cathodes, 338
Briggs, review on nickel and manganese, 234
Brine
  effects of impurities, 254
  purification, 257
Bruggeman, and test of Maxwell's theory, 324
Bubble break-off diameters, as a function of current density, 317
Bubble layer, and conductivity, 330
Bubble layer effects, summary, 340
Bubbles
  detachment, 312, 314
  electrical effects, 330
  growth, 306
  microscopic investigation of, 348
  and nucleation, 305
  photographs of, 309, 346
  and return from electrode, 313
  work of Sides, 303
  work of Tobias, 325
Bubbling, work of Sides and Tobias, 333
Buckley and Burke, and iridium films, 217
Bulk recombination centers, and Butler and Ginely, 52
Burke and Healey, reversible peaks involving ruthenium oxides, 230

Burke and McRann, hydrous layers and potentiostatic conditions, 211
Burke and O'Sullivan
  work on electrochromic transitions, 225
  work on rhodium–platinum alloys, 226
Burke and Roche, discussion of anionic products, 196
Burke and Twomey, and nickel metal base electrolyte, 234
Burke and Whelan
  growth on ruthenium, 230
  iridium films, 219
Butler
  model for photoelectrochemistry, 45
  model of the Schottky barrier, 42
  and photoelectrochemical kinetics, 43
Butler and Ginley, and bulk recombination centers, 52

Carrier length, and a simple model for determination, 45
Case Western Reserve University, and Eltech, $2.5MM grant, 270
Cathodes
  air-depolarized, 269
  water molecules at, 251
Cell design, its influence, 91
Cell technology and ion exchange membranes, 292
Cells, photoelectrochemical
  design, 84
  operation, 66
Characterization of oxide films on iridium, 216
Charge transfer
  and pathway, 51
  and surface recombination, 45
Chazalviel, and pathway for charge transfer, 51
Cheh and Tobias, and supersaturation of the solution in bubble growth, 306
Chemiosmotic model, 156
Chemiosmotic theory, and the potential across a membrane, 158
Chialvo, work on hydrous oxide growth, 204

Chiew and Glandt, and Maxwell's void
  fractions, 30
Chlor-alkali
  cell operation, complications, 281
  industry, 249
  operation, engineering aspects, 270
  plan, diagrammatic, 257
  production, 250, 287
Chloride, is there a positive ion?, 267
Chlorine
  electrode reactions, mechanisms of,
    265
  evolution, and Tafel lines, 264
  liquefaction, diagrammatic, 258
Chloronium ion, 267
Chlorophyll, and membranes, 137
Christov, and model for electrode
  processes, 32
Coalescence, scavenging, as shown by
  Sides and Tobias, 311
Co-enzyme Q, and proton exchange, 134
Conductance mechanisms, in bilayer
  membranes, 152
Conductivity
  of bubble layer, 331
  of bulk dispersions, 318
  of ordered arrangements, 325
  of random dispersions, 322
Contact, semiconductor–solution, 3
Continuum theory, in electrode
  processes, deduction of expression
  for, 34
Conway
  "charge enhancement factor," 218
  and hysteresis in oxide formation, 195
  and monolayer oxide growth on gold,
    210
  and platinum oxide formation, 195
Conway and Mozota, work on iridium
  oxide films, 223
Corrosion, of semiconductors, chemical
  engineering considerations, 86
Coupling mechanisms, for potentials in
  membranes, 155
Coupling routes, Mitchell's, a reserve
  route, 163
Current density distribution, in cells, 293

Current distribution
  around the dielectric, 332
  and height of polarization of elec-
    trodes, 336
  macroscopic, and Vogt, 319
Current efficiency, for chlorine evolution, 252
Cyclic voltammograms
  for hydrous-oxide-coated manganese
    electrode, 236
  for iron electrodes, 231
  for rhodium electrodes, 225

Damjanovic, and platinum oxide analysis, 197
Davis, and a computer program for surface solution equilibria, 90
Debye length, applied to semiconductors, 77
DeLaRue, and uniform dispersion of
  glass spheres, 321
Density of states
  as a function of electron energy, 31
  at semiconductor–solution interface, 28
Detachment, of bubbles, 312–314
De Vries, and corrections of diffusion
  coefficients, 325
Diaphragm, and mass transfer through,
  282
Dielectric, and current distribution
  around, 330
Diffusion-controlled growth of bubbles,
  308
Digman, and gel models for interfaces,
  190
Dispersions, random, and conductivity,
  321
DNA, and aging, 120
Dogonadze, and model for electrode
  processes, 32

Efficiencies, of photoelectrochemical
  cells, 95–97
Ehrenberg, and the possible chloronium
  ion, 267

Electrical effects
  and bubbles, 330
  gas evolution, 318
Electrocatalytic effects, and ruthenium oxide coatings, 266
Electrochemistry, of manganese systems, 235
Electrochromosome, in manganese films, 234
Electrode material and electrode processes, 263
Electrode potential, absolute, 8
Electrodes, semiconducting, 62
Electrolyte, interface with the metal, 234
Electrolyte-semiconductor interface in equilibrium, 7
Electrolytic resistance, in membranes, 290
Electronegativity, and the Schottky barrier, 15
Electron transfer, in respiratory chain, 114
Electron transfer effects, and membrane potentials, 135
Electron transport, nonenzymatic use of, 121
Eltech, and Case Western Reserve University, $2.5MM grant, 270
El Wakkad and El Din, work on palladium oxide, 206
Energy, of metal-semiconductor junction, 5
Energy and consumption of energy, according to Mitchell's theory, 156
Energy level, in redox couples, 23
Engineering aspects for chlor-alkali operation, 270
Equations
  bubble sizes, 320
  for semiconductors, some solutions, 87
  equilibrium, at semiconductor-electrolyte interface, 7
ESCA
  studies of iridium oxide, 218
  and technique on palladium oxide, 207

Fermi distribution, 81
Fermi level, 8, 9
  pinning, 17
  in solution, 8, 13
Films, hydrous, 169
Fine's theory, and lattice vibrations, 161
Formation factor in chlor-alkali diagram, 282
Forster and Zuber, and convection in bubble growth, 306
Frank-Condon principle, 26

Gärtner
  application of his equation to semiconductors, 88
  treatment for semiconductor-metal interface, 63
Gas bubbles
  and effect on concentration overpotential, 339
  and evolution of the diffusion layer, 343
  and hydrodynamic model, 344
Gas evolution
  electrical effects on, 318
  systems, dynamic, 334
Gaussian distribution
  in electrochemistry, invalid, 35, 36
  model for, 35
Generation, thermal, and recombination in trap sites, 75
Gerischer
  and extension of Gurney's equations, 23–25
  and Fermi level in solution, 9
  and inversion layer, 20
  model for electrode processes, 32
  criticized by Bockris, 35
  modification of Gurney's theory, 29
  similarity to Bockris's model, 35
  photoelectrochemical equations, 76
Glas and Westwater, and rapid-fire emission, 313
Gold, and monolayer behavior, 208
Gold, monolayer oxide growth, and Conway, 210

Index

Gold oxide, and Kozlowska, 206
Gomer and Tryson, and the absolute potential, 12
Gottesfeld, and AC impedance measurements, 221
Gottesfeld and Srinivasan, and potential sites in iridium films, 217
Grahame theory, a classical model for double layers, 189
Graphite anodes, 263
Green, and surface states, effect on semiconductor–solution interface, 23
Greenbaum, and recombination, 53
Grotthus-type transfer, in hydrous materials, 220
Growth
　of bubbles, 306
　on platinum films, and the potential cycling conditions, 200
Guibaly, and charge transfer model, 50
Gurney
　and foundation concepts in electrode kinetics, 23
　and model for electrode processes, 24, 32

Hansen and Kolb, and the absolute potential, 12
Healy
　and oxide structures, 189
　and parameters in Grahame model applied to interfaces, 189
　and polyelectrolyte theory for interfaces, 190
Heide and Westwater, pits and scratches, 305
Heller and Miller, and high efficiencies for redox cells, 94
Helmholtz layer
　and the potential difference at the interface, 16
　and the semiconductor–solution interface, 21
Heterogeneity, along the membrane, 159
Heterogeneous conductivities, predictions of, 327

Hole lifetimes, and grain size, 54
Hoogland, and bubbles, 346
Horowitz, and Reichman's treatment, 51
Hydration, and oxide battery systems, 240
Hydrodynamic model, for gas bubbles, 344
Hydrogen overvoltage, on graphite, 276
Hydrous films, generated on platinum, 203
Hydrous oxide films
　on iridium, 215
　and transport processes, 182
Hydrous oxide growth
　mechanisms 241
　not in strong bases, 223
　on platinum, 198
　pictorial, 175
　on ruthenium, studied by Birss, 229
Hydrous oxides, 170
　on platinum, 202
　structural aspects, 179

Ibl, and gas bubbles, 343
Ilkani and Berns, and bilayer lipid membrane, 147
Illuminated surfaces, model of Bockris and Uosaki, 38
Illumination, of photoelectrochemical cell, 67
Interface, semiconductor–electrolyte, 78
Interfacial reaction schemes, theories for, 79
Inversion layer, 20
Iodide concentration, and conductance effects, 150
Ion exchange membranes, 287
　and cell technology, 292
Iridium, and characterization of oxide films, 216
　work of Rand and Woods, 213
Iridium oxides, hydrous, and Conway, 218
Iron and cobalt, oxide growths on, 230
Iron, and slow step in corrosion reactions, 232

James
  and beneficial effects of grinding, in hydrous oxide growth, 198
  and difficulty in forming thick films on platinum, 199
  and reduction processes on platinum, 202
Janssen and Barendrecht, and hydrogen bubbles, 312
Janssen and Van Stralen, and transport of bubbles from nickel electrodes, 310
Japanese contributions, to membrane cell technology, 296

Kelly and Memming, and surface recombination, 50
Khan, and criticism of Gerischer's model, 30
Khan and Bockris, and models of photoelectrochemical kinetics, 52
Kinetics, for transport on hydrous oxide films, 183
Kozawa, and no cathodic peak in hydrous oxides, 199
Kuhn, and gas-evolving electrodes, 335

Landl, and bubble size, 317
Laplace equation, applied to semiconductor-solution interface, 83
Laplace transform methods, in potential sweep theory, 184
Laser and Bard
  computer program for semiconductor-solution interfaces, 90
  digital simulation technique, 43
Layers, unstirred near membranes, 162
Le Blanc, and transmembrane potentials, 122
Lemasson, and simple model for carrier length, 45
Levels, energy
  of electrolyte (and redox potential), 4
  in semiconductors, 2
Levich, and model for electrode processes, 32
Levine and Smith, and non-Nernstian behavior of surface potentials, 189

Liberman, and experiments on liposome membranes, 146
Lifetime of carriers, 53
Localized proton processing in membranes, 161
Lohman, and the absolute potential, 12

McCann, and recombinations, 89
McCann and Haneman
  and photoelectrochemistry, 49
  and recombination, 59
  and recombination kinetics, 53
McMullin *et al.*, and porosity of packed beds, 281
Manganese electrodes, and cyclic voltammograms, 234
Manganese systems and electrochemistry, 234
Marcus
  derivation of the equation for his model by Kita and Uosaki, 33
  and model for electrode processes, 32
Mass transfer
  through diaphragm, 282
  at gas-evolving electrodes, 342
Masters and Mauzerall, and bilayer lipid membrane of egg lecithin, 151
Material balance
  in amalgam decomposer, 277
  at semiconductor-solution interface, 80
Materials, choice of, 85
Maxwell
  equations on conductivity of bubbles and predictions, 329
  and high void fractions, 320
  and spheres of different sizes, 319
Mechanisms
  absent from Mitchell's theory, 161
  of chlorine electrode reactions, 265
  for motion of electrons and protons in membranes, 144
Membrane cell, 259
Membrane cell technology and Japanese contributions, 296
Membrane-electrolyte interface, and chlorophyll, 13
Membrane lipids
  and their oxidation, 127

Index

Membrane lipids (*cont.*)
  in respiratory chain, 119
Membranes
  and electrolytic resistance, 290
  and electron transfer effects, 113
  at mitochondria, and Mitchell's theory, 162
  natural, and Mitchell's theory, 160
  their performance characteristic, 88
  potentials for, 129
  and surface gas bubbles, 291
Membrane technology, its advantage, 295
Memming, and chart for redox reactions, 68
Menezes *et al.*, discussion of absorbative losses, 87
Mercury cell, 260
Mercury content, and slow dissolution, 262
Mercury-sodium, phase diagram, 270
Meredith and Tobias
  attack on Bruggeman, 323
  and bubbles, 319
Metal oxides, as anodes, 266
Micro convection model, for bubbles, 347
Migration, through diaphragms, 284
Mitchell's hypothesis, and ATP synthesis, 157
Mitchell's theory, 156
  and the consumption of energy, 156
  disagreement with experiment, 161
  as a phenomenological theory, 160
  simplicity, 157
Model, physical, for semiconductor-electrolyte interface, 65
Monolayer oxide growth, on platinum, 194, 197
Mott-Schottky plot, 22
Mozota and Conway
  and the lower oxide peak for iridium, 213
  work on iridium growth, 213

NADH, and oxygen at electrodes, 124
Nakabayashi, criticism of the Gaussian distribution, 36

Nernst-Einstein relationship, applied to semiconductors, 72
Newman
  numerical program for solving coupled differential equations, 90
  and numerical solutions for semiconductors, 90
  work on current distribution, 337
Nicholson and Shayne, and potential sweep for redox reactions, 185
Nickel and manganese, oxide films thereon, 233
Nicotinomide, and ubiquinones, 126
NMR, its use in interfacial studies, 126
Normal hydrogen electrode, potential, absolute, 11
Notation for membrane cells, 297
Nozik
  and hot electron theory, 39
  and an inversion level, 20
Nucleation, and bubbles, 305

O'Brien, and conductivity functions relating to bubble sizes, 320
Open circuit photovoltage, and Fermi level pinning, 18
Orazem and Newman, equations for cell design, 20
Ord and Ho, and platinum oxides, 197
Oxidation, of lipids, 120
Oxide electrodes, their thermodynamics, 177
Oxide films, on nickel and manganese, 234
Oxide-solution interfaces, theoretical model, 188
Oxide structures, investigated by Burns and Burns, 180
Oxide systems, reversible potentials, 177
Oxygen bubble growth and the cyclic mechanism of Sides and Tobias, 310
Oxygen evolution, near membranes, 295

Packed beds, and work of McMullin *et al.*, 281

Palladium
  dissolution, according to Rand and Woods, 206–207
  and oxide film growth, 205
Palladium oxide, work of El Wakkad and El Din, 206
Path of electron transport chain, 136
Peak potentials, and pH, for manganese films, 237
Peerce and Bard, potential sweeping, 184
Permeability, of bilayer membranes, 148
Perram, and the gel layer for the oxide–solution interface, 189
Peter
  and model for photoelectrochemistry, 51
  and surface recombination, 49
  pH dependence of oxide electrodes, 178
Phase diagram for mercury–sodium, 271
Phospholipid vessicles, 140
Photoelectrochemical cells
  chemical engineering description, 69
  economics, 198
  model of Khan and Bockris, 52
Phototransfer of protons and electrons, in bilayer membranes, 152
Pits and scratches, and Westwater, 305
Plastoquinones, 151
Platinum
  and hydrous films, 203
  and hydrous oxide growth, 195
  and monolayer oxidation, 190
  monolayers on, 194
  oxide formation, and Conway and co-workers, 195
  and potentiodynamic sweep, 193
Plesset and Zewick, and initial stage of bubble growth, 306
Polarization curves
  for amalgam electrodes, 273
  in current efficiency, 264
  on ruthenium oxide, 267
Polarization parameters, and current distribution, 330
Porous diaphragms, chemical engineering aspects of, 277

Porphyrin rings, and membranes, 138
Potential
  absolute, 9
  of the normal hydrogen electrode, 11
  cycling, and oxide layers, 213
  distribution, at the semiconductor-solution interface, 13
  of membrane, 114
  sweep methods, and Laviron, 184
Potential sweep techniques, applied to ruthenium, 228
Potentiodynamic sweep, for platinum oxides, 192
Prager, and entropy principles applied to the suspension of solid particles, 321
Protein matrix, and transport of protons through membranes, 154
Proton transfer, in biological membranes, hypothesis for mechanism of, 153
Proton transport, in bilayers, 157
Protons, in membranes, 142

Q traps, work of Robertson and Boardman, 152

Rajeshwar
  and charge transfer pathway, 57
  and surface recombination, 49
Raleigh
  equation, 325
  and heterogeneous conductivity, 320
  original expression for bubbles, 306
  theoretical work on conductivity ratios of cubic arrangements, 325
Rand and Woods
  and oxide coverages on palladium, 207
  and palladium dissolution, 205
  work on irridium, 213
Rate expression, for electron transfer, at illuminated surfaces, 37
Recombination
  kinetics, McCann and Haneman, 53
  and semiconductors, McCann, 90
  and space charge region, 51

Index

Redox components, including chlorophyll, 138
Redox couples, energy level discussion, 23
Redox potential
  and chlorophyll, 138
  and energy level in solution, 4
  measurement of, 6
  in mitochondrial respiratory chain, 117
  and potential drop at the semiconductor-solution interface, 16
Redox properties, of metal oxides, 176
Reichman, and Mott-Schottky treatment, 52
Reichman and Bard, and proton diffusion coefficients for hydrated tungstic oxide layers, 187
Reiss
  and absolute potential, 12
  and surface recombination, 88
Respiratory chain
  elements of, 116
  and membrane lipids, 119
Rhodium
  films, 224
  hydrous oxide growth on, 224
Rhodium-platinum alloys, work of Burke and O'Sullivan, 226
Roald and Beck, and magnesium dissolution, control by transport, 343
Rohrengel and Schulze, work on growth of oxide layers, 211
Ross, and fine structure in the initial stage of surface oxidation, 195
Ruthenium
  oxide
    coatings and electrocatalytic effects, 267
    and polarization curves, 267
    work of Burke and Mulcahy, 227

Salvador, and model determination of diffusion length, 45
Saveant, and linear potential sweep methods, 186
Scherson and Kolb, and peaks on noble metals, 241
Schlotter and Pickelmann, experiments on active oxides, 238

Schottky barrier, 14
  application to irradiated surfaces, 42
Schwartz-Christoffel transformations, 93
Scriven, and bubble growth, 306
Semiconductors
  and the Gartner equation, 88
  theory of, 1
  transport base description, 70
Semiconductor-electrolyte interface, 2, 78
Semiconductor-solution interface
  and Helmholtz layer, 21
  and original contributions of Green, 23
  potential difference at, 13
Shibata, and period rate interruption, 199
Shockley and Read, and space charge recombination, 53
Sides, work on bubbles, 304
Sides and Tobias, on bubbles, 333, 346
  and cyclic mechanism of oxygen bubble growth, 310
  observations of coalescence in bubbles, 344
  on scavenging, 311
Silly, review on energetics, 137
Sokolov
  and lipid hydropyroxides, 131
  and protons in membranes, 127
Sokolov and Shipunov, and transmembrane potentials, 123
Solid particles, as treated by minimum-entropy principles, 320
Space charge recombination, Albery's approach, 53
Space charge region, and recombination, 51
Spheres of equal size
  and Maxwell, 319
  random arrangement, 323
Sunlight, absorption at semiconductors, 64
Surface charge, and semiconductors, in respect to surface states, 19
Surface gas bubbles and membranes, 292

Surface recombination
  and charge transfer, 45
  and Memming, 50
  Peter's work, 50
  Rajeshwar's work, 50
Surface states
  effect on Mott–Schottky plot, 23
  foundational discussion by Green, 23

Tafel slopes, 239
Thermodynamics, and oxide electrodes, 177
Thin film photovoltaics, and Mitchell, 92
Tobias, and bubbles at electrodes, 325
Transmembrane potential
  and BLM conductance, 132
  in the chain NADH–coenzyme $Q-O_2$, 122
  and concentration ratios of redox couple, 141
  model for, 128
  and NADP, 123
  and proton concentration gradient, 143
Transport processes in hydrous oxide films, 182
Transport theory, application to semiconductors, 71
Trasatti
  and the absolute potential, 9, 12
  and gel models and interfaces, 189
Tungsten electrodes, oxide films thereon, 238
Turner, indication of a physical absurdity, 320

Ubiquinone
  and bilayer membranes, 152
  in respiratory chain, 118
Uosaki and Kita, work on photoelectrochemistry, 20

Venczel, and physics of gas evolution, 316

Vetter and Berndt
  oxides on gold, 208
  platinum oxide formation, 195
Vetter and Schultze, and platinum oxides, 197
Vibration–rotation levels, for electrode reactions, 27
Visscher and Blijlevens, and limiting coverage for platinum oxides, 198
Visscher and Devanathan, and diffusion of metal into films, 197
Vogt, and macroscopic current distribution, 318
Void fraction, in porous diaphragms, 280
Voltammograms
  for cycling ruthenium–platinum electrodes, 228
  for gold, 208
  for hydrous oxides on irridium

Water transport across membranes, 294
Weaver, and photoelectrochemical cells, 97
White, a general method for treating boundary conditions, 90
Wide gap semiconductors, and electrochemical theory, 44
Williams
  and the localized proton hypothesis, 162
  theory of localized proton transfer, 161
Wilson's model for photoelectrochemistry, 46, 48
Winograd, and XPS for ionidization of platinum, 201
Wright and Hunter, and Grahame model applied to colloids, 189

Yaguzhinsky, and transmembrane potentials, 122

RAYMOND H. FOGLER LIBRARY

DATE DUE

BOOKS ARE SUBJECT TO
    R TWO WEEKS